AF352948

Immunophenotyping in Autoimmune Diseases and Cancer

Immunophenotyping in Autoimmune Diseases and Cancer

Editors

Gábor J. Szebeni
László G. Puskás

MDPI • Basel • Beijing • Wuhan • Barcelona • Belgrade • Manchester • Tokyo • Cluj • Tianjin

Editors
Gábor J. Szebeni
Biological Research Centre
Hungary

László G. Puskás
Biological Research Centre
Hungary

Editorial Office
MDPI
St. Alban-Anlage 66
4052 Basel, Switzerland

This is a reprint of articles from the Special Issue published online in the open access journal *International Journal of Molecular Sciences* (ISSN 1422-0067) (available at: https://www.mdpi.com/journal/ijms/special_issues/Immunophenotyping).

For citation purposes, cite each article independently as indicated on the article page online and as indicated below:

LastName, A.A.; LastName, B.B.; LastName, C.C. Article Title. *Journal Name* **Year**, *Article Number, Page Range.*

ISBN 978-3-03943-466-4 (Hbk)
ISBN 978-3-03943-467-1 (PDF)

Contents

About the Editors

Gábor J. Szebeni, Ph.D. Dr. Szebeni studied the tumor microenvironment at the University of Szeged and Biological Research Centre in Szeged and received his Ph.D. in 2012. Briefly, he contributed to the dissection of T-cell apoptosis caused by tumor-derived galectin-1, and he showed the tumor-promoting effect of mesenchymal stem cell derived galectin-1. In order to work on cancer-related inflammation, he joined Prof. Antonio Sica's group in Milan as a postdoctoral fellow. They revealed the role of RorC1 in the cancer-driven emergency granulo-monocytopoiesis. Afterward, he worked on the investigation and development of anti-cancer therapeutics as a project leader at Avidin Ltd. He gave university lessons about 'Milestones in Cancer'. He successfully implemented the first mass cytometer in Hungary and currently leads the cytometry unit at the Biological Research Centre in Szeged, focusing on the dysregulation of immune homeostasis in human pathologies such as autoimmunity (RA, SLE, SSC), cancer (adenocarcinoma, breast cancer, AML), and inflammation (obesity, colitis, COPD).

László G. Puskás, Ph.D., D.Sc. Dr. Puskás received a Degree in Biology (1994) and a Ph.D. in Chemistry (1997) at the Medical Faculty of the University of Szeged, Hungary. He received his D.Sc. in Biology from the Hungarian Academy of Sciences (2006). He was a postdoctoral fellow in the field of biotechnology at Research Institute of Innovative Technology for the Earth (RITE), Kyoto, Japan, and was a guest researcher at the Microarray Facility, Leuven, Belgium. He has been the head of the Laboratory for Functional Genomics in the Biological Research Centre in Szeged, Hungary, since 1999. He became the scientific advisor and later the chief executive officer of Avidin Ltd., the chief executive officer of Avicor Ltd, and the chief executive officer of Aperus Pharma Co. Ltd. He is the author of more than 170 scientific papers and the main investigator of 13 patents. His main scientific contributions are the discovery and development of Q134R, a multitarget clinical candidate against Alzheimer's disease, and the development of single-cell digital RNA profiling and chemical microarray technologies.

Preface to "Immunophenotyping in Autoimmune Diseases and Cancer"

The mammalian immune system is a Janus-faced network of well-coordinated, highly specialized cells and biomolecules. The sensitive balance of reactive and suppressive signals maintains homeostasis in the physiological state. Under pathological conditions, however, responsiveness can escalate in autoimmune diseases or remain suppressed in cancer. In severe autoimmunity, the overwhelming immune response leads to devastating diseases such as rheumatoid arthritis, systemic lupus erythematosus, systemic sclerosis, and sclerosis multiplex. Conversely, immune cells, e.g., professional antigen-presenting cells and cytotoxic T cells, are not able to execute their physiological function in cancer. Recent progress in fluorescence flow cytometry and mass cytometry has contributed to a high-dimensional resolution of the complex immunophenotype both in autoimmune diseases and in cancer. We called authors to publish their latest achievements associated with the perturbance of the regulation of immune activation and the discovery of rare subpopulations of both innate and adaptive immune players in autoimmune diseases and cancer. The online version of the Special Issue can be found at https://www.mdpi.com/journal/ijms/special_issues/Immunophenotyping.

Gábor J. Szebeni, László G. Puskás
Editors

Review

Endoplasmic Reticulum Stress Pathway, the Unfolded Protein Response, Modulates Immune Function in the Tumor Microenvironment to Impact Tumor Progression and Therapeutic Response

Manuel U. Ramirez [1], Salvador R. Hernandez [2], David R. Soto-Pantoja [3,4] and Katherine L. Cook [3,4,*]

[1] Department of Physiology and Pharmacology, Wake Forest University Health Sciences, Winston-Salem, NC 27157, USA; muramire@wakehelath.edu
[2] Digital Illustrator, Winston Salem, NC 27103, USA; SalvadorHernandez87@gmail.com
[3] Department of Surgery, Wake Forest School of Medicine, Winston-Salem, NC 27157, USA; dsotopan@wakehealth.edu
[4] Department of Cancer Biology, Wake Forest University Health Sciences, Winston Salem, NC 27157, USA
[*] Correspondence: klcook@wakehealth.edu; Tel.: +01-336-716-2234

Received: 1 October 2019; Accepted: 9 December 2019; Published: 25 December 2019

Abstract: Despite advances in cancer therapy, several persistent issues remain. These include cancer recurrence, effective targeting of aggressive or therapy-resistant cancers, and selective treatments for transformed cells. This review evaluates the current findings and highlights the potential of targeting the unfolded protein response to treat cancer. The unfolded protein response, an evolutionarily conserved pathway in all eukaryotes, is initiated in response to misfolded proteins accumulating within the lumen of the endoplasmic reticulum. This pathway is initially cytoprotective, allowing cells to survive stressful events; however, prolonged activation of the unfolded protein response also activates apoptotic responses. This balance is key in successful mammalian immune response and inducing cell death in malignant cells. We discuss how the unfolded protein response affects cancer progression, survival, and immune response to cancer cells. The literature shows that targeting the unfolded protein response as a monotherapy or in combination with chemotherapy or immunotherapies increases the efficacy of these drugs; however, systemic unfolded protein response targeting may yield deleterious effects on immune cell function and should be taken into consideration. The material in this review shows the promise of both approaches, each of which merits further research.

Keywords: unfolded protein response; Inositol-requiring enzyme 1 (IRE1); PKR-like endoplasmic reticulum kinase (PERK); Glucose-regulated protein 78 (GRP78); Activating transcription factor 6 (ATF6); immune cells; T cell; macrophage; tumor microenvironment

---

1. Introduction

For a cell to become cancerous, it must overcome several evolutionary obstacles [1]. Among these, a cancerous cell must proliferate readily and avoid immune destruction [2,3]. Achieving this state is complex. While mutations to tumor suppressors and proto-oncogenes contribute to these changes, over the past few decades it has become obvious that cancer cells also repurpose several endogenous survival systems to assist in their formation and progression.

We must consider both the tumor cells and their microenvironment to understand how tumor–host interactions drive transformation and carcinogenesis, and subvert these survival systems. The tumor microenvironment has been extensively investigated (reviewed in [1,4,5]). Of note are the tumor immune infiltrates, metabolite availability, and stress effects of the tumor microenvironment. Tumors have been referred to as "wounds that never heal" [6]. Healing mechanisms associated with normal tissue injury promote tumor formation and metastasis. Immune infiltrating cells induce wound-like inflammation in the tumor microenvironment, further assisting in the development of malignant cancers. Poor perfusion of solid tumors leads to high levels of hypoxia and low metabolite availability, and the healing response responds to these via angiogenesis [7,8]. Leaky vasculature results in high osmolarity in the tumor microenvironment. Recent studies have even demonstrated that microbial flora in cancers differ from normal tissues, and these enhance tumor progression [9].

The combination of these factors creates a 'perfect storm' to subvert evolutionary pathways and repurpose them to be procancer. One such system of interest is the unfolded protein response (UPR). The UPR is an evolutionarily conserved mechanism discovered somewhat serendipitously [10]. In the mid-1970s studies found that virally transformed cells increased the expression of protein p78. Unknowingly, Hass and Wabl identified the same protein in 1983 [11]. This protein was localized to the endoplasmic reticulum (ER) and bound unsecreted Ig heavy chains, thus named binding immunoglobulin protein (BiP). In 1987, Lee et al. reported that p78 expression in highly proliferative transformed cells was due to media glucose depletion [12]. The protein was named glucose-regulated protein 78 (GRP78). This protein was also identified as heat shock 70 kDa protein 5 (HSPA5). Additional research led to the discovery that BiP and GRP78 were the same protein, and bound to unfolded or incomplete Ig intermediates, identifying GRP78 as the first ER chaperone protein. GRP78 has since become recognized as the primary regulator of the UPR. The UPR is triggered by accumulation of unfolded proteins in the ER, conditions that are often found in highly proliferative, secretory, or pathogen-infected cells [13].

Recent evidence indicates that the UPR is critical in multiple systems, such as cell differentiation, proliferation, immune response, and cell maintenance. This review focuses on the role of the UPR in tumor microenvironment stress, its effect on cancer cell progression, and immune response to cancer cells.

2. Body

2.1. ER Stress and UPR Signaling

When stressed, the ER is overwhelmed with an accumulation of proteins due to improper folding, insufficient glycosylation, and/or inhibited transport [14]. These often arise due to a sudden increase in protein expression leading to insufficient chaperone proteins, saturation of the ER lumen space, or insufficient nutrients for post-translational modification. These situations are often found in both immune and cancer cells. This accumulation and subsequent ER stress results in canonical UPR signaling through three different proteins: IRE1, PERK, and ATF6.

Each of the three proteins initiate an 'arm' of UPR signaling and are thought to be regulated, in part, by association with ER membrane-bound GRP78. Current understanding of this system suggests that, under non-stressed conditions, IRE1, PERK, and ATF6 are primarily bound to GRP78, maintaining an inactive state and preventing UPR signaling. While it does not directly assist in folding, GRP78 binds unfolded proteins to maintain them in a foldable state. When unfolded proteins accumulate, i.e., induction of 'ER stress', GRP78 releases IRE1, PERK, and ATF6, preferentially binding unfolded polypeptide chains. This release allows each of these three proteins to initiate their portion of the UPR. The pathway and function of these arms and their effects are briefly described below.

IRE1: Inositol-requiring enzyme 1, also known as endoplasmic reticulum to nucleus signaling 1 (ERN1). IRE1 oligomerizes and autophosphorylates upon release from GRP78. Phosphorylation activates an endonuclease domain that cleaves an intron from the X-box binding protein 1 (*XBP-1*) mRNA. This cleaved mRNA is then translated to the transcription factor XBP-1s. It is unclear whether cleavage occurs in the ER or the nucleus, as IRE1 has been found in the inner nuclear envelope [15]. XBP-1s induces the expression of ER chaperone proteins and ER-associated protein degradation (ERAD) proteins, and induces differentiation of metabolic regulators. XBP-1s also induces *XBP-1* transcription in the form of self-regulation. In addition to effects mediated by XBP-1s, IRE1 continues nuclease function in the ER, degrading ribosomal-associated mRNA through regulated IRE1-dependent decay (RIDD). This degradation prevents the translation and further accumulation of unfolded proteins. IRE1 also contains a kinase function, which phosphorylates c-Jun N-terminal Kinase (JNK), contributing to apoptosis under prolonged UPR signaling [16]. Although GRP78 association is the primary inhibitor of IRE1 activation, there is evidence for alternate methods of IRE1 activation, including direct binding by unfolded proteins [17].

PERK: Protein kinase R (PKR)-like endoplasmic reticulum kinase, or eukaryotic translation initiation factor 2-alpha kinase 3 (EIF2AK3). Release from GRP78 suppression induces PERK oligomerization and transphosphorylation similar to IRE1. PERK then phosphorylates eukaryotic translation initiating factor 2A (eIF2α), preventing the formation of ribosomal pre-initiation complexes and reducing cap-dependent protein translation. An open reading frame in the 5′-untranslated region of activating transcription factor 4 (*ATF4*) mRNA allows cap-independent translation during eIF2α phosphorylation. PERK activity thus increases *ATF4* function to propagate UPR signaling. ATF4 expression results in products enhancing metabolic changes and ERAD, in concert with transcription products from XBP-1s activity. Prolonged UPR leads to cell cycle arrest and, under certain conditions, apoptosis via CCAAT-enhancer-binding protein homologous protein (CHOP) expression downstream of PERK activation. Independent of PERK-mediated ATF4 expression, PERK activation results in an antioxidant response via nuclear factor erythroid 2-related factor 2 (NRF2)-induced expression of genes containing antioxidant response elements (AREs) in their promoters [18].

ATF6: Activating transcription factor 6. ATF6 translocates to the Golgi complex upon GRP78 release. Golgi-localized site-1 and site-2 proteases (S1P and S2P) then cleave ATF6, releasing a cytosolic basic leucine zipper (bZIP) domain. This bZIP domain translocates to the nucleus and induces the transcription of ER chaperones, lipid biosynthesis, and ERAD proteins. These allow expansion of the ER, reducing the density of unfolded proteins and increasing chaperone protein availability, further assisting with reducing the unfolded protein burden and ER stress. Additionally, like XBP-1s, ATF6 induces XBP-1 expression for UPR autoregulation. Prolonged ATF6 activation also leads to a form of CHOP-independent apoptosis.

For this review, we will be focusing on UPR signaling through IRE1, PERK, and ATF6 (Figure 1). This is a simplified model of UPR signaling, omitting numerous additional proteins involved in glycosylation, folding, and quality control. IRE1, PERK, and ATF6 signaling pathways work together to reduce ER burden. While this traditional role of UPR is widely agreed upon, recent research suggests that this model requires further refinement and may not be applicable in all cell types, particularly in immune and cancer cells, both of which have atypical expression needs.

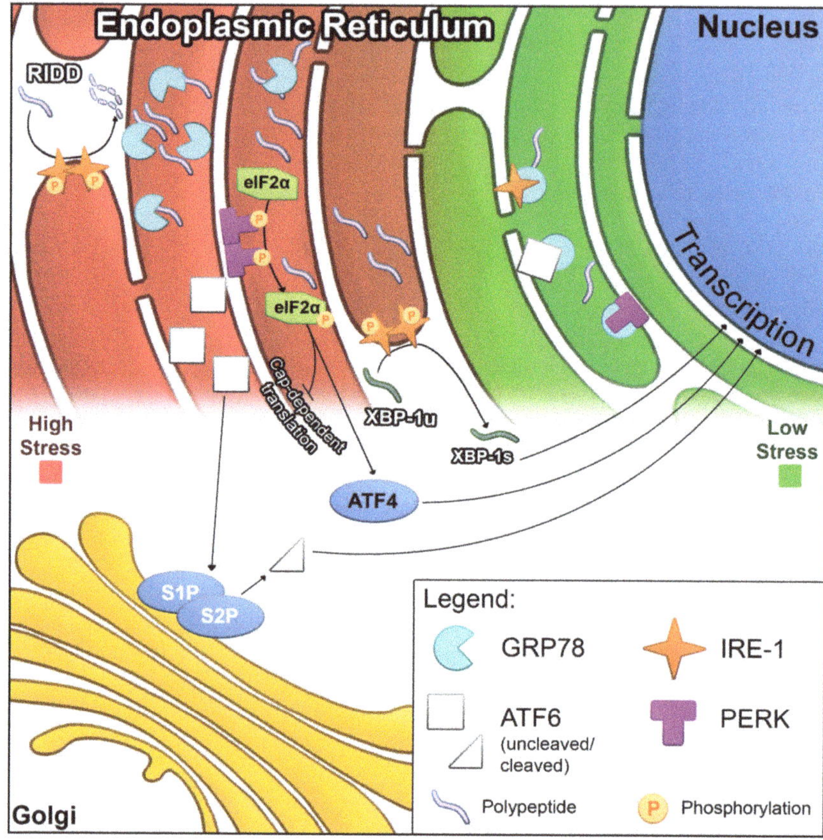

Figure 1. ER-stress induced UPR signaling. Summary mapping of the UPR signaling pathways and locations in which they occur. Each of the three 'arms' of UPR signaling are bound by inhibition due to GRP78 sequestration (right, green ER). Under ER stress, GRP78 binds unfolded proteins, releasing IRE1, ATF6, and PERK (left, red ER).

2.2. ER Stress and the UPR in the Tumor Microenvironment

UPR signaling is frequently upregulated in the tumor microenvironment due to inflammatory factors, the high metabolic rate of cancer cells, elevated hypoxia, and poor nutrient availability. In prostate cancer, tumor cells induce an UPR in the local microenvironment, termed Transmissible ER Stress (TERS), leading to an UPR in neighboring cells [19]. What secreted factors are responsible for TERS are unclear, though TERS appears to be dependent upon Toll-like Receptor 4 (TLR4) activation [20]. It is likely that transmissible ER stress will be found in other cancers as well.

Much like inflammation, UPR in the tumor microenvironment increases tumorgenicity and is associated with a stem-like phenotype, proliferation, angiogenesis, and survival during starvation or hypoxic conditions [21–27] (reviewed in Figure 2). Increased UPR in neighboring tissues supports tumor development via Wnt signaling. Wnt signaling reduces pro-apoptotic UPR signaling in prostate cancer cells [19]. The UPR may further assist in metastasis of circulating cancer cells to hypoxic regions. UPR signaling is increased in bone metastases of breast, lung, and prostate cancers [28–30]. However, the role of the UPR is not clear; there are also reports that UPR activation, through increased ER stress, can induce immunogenic or apoptotic cancer cell death [22,31–33]. Simultaneously, reducing UPR signaling can induce immunogenicity and clearance of cancer cells [34–36]. There is a balance to UPR signaling that allows cancer progression, without activating cell death pathways.

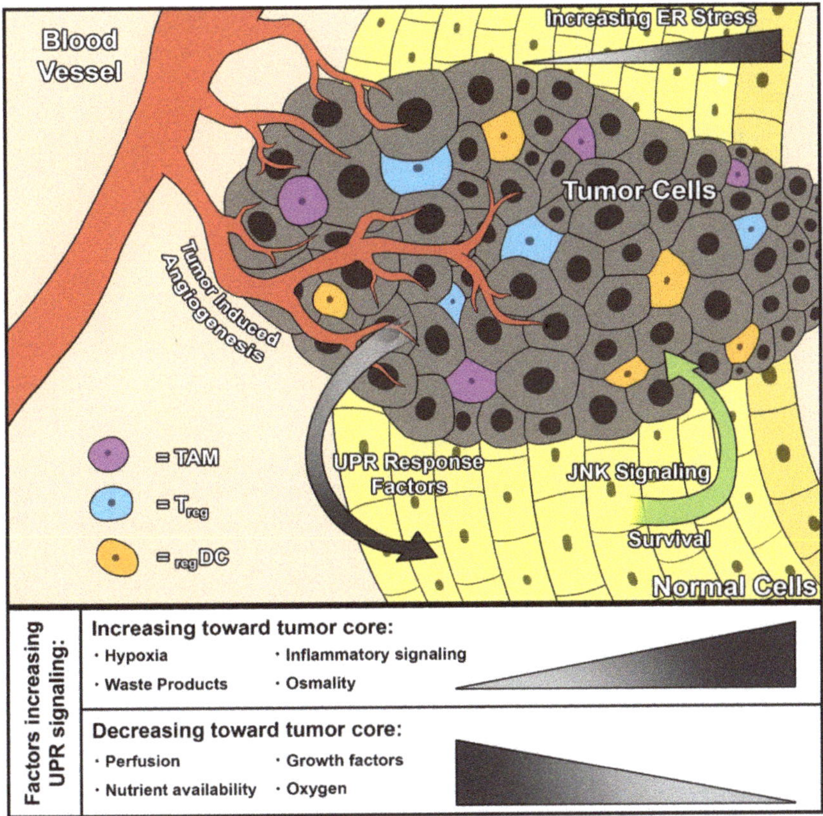

Figure 2. UPR signaling in a tumor and the surrounding microenvironment. The composition of the tumor and its microenvironment promote UPR signaling in cancer cells via various mechanisms. Poor perfusion, low oxygen, and reduced growth factor availability contribute to increased UPR signaling. The ability of cancer cells to exert 'transmissible UPR signaling' to the surrounding normal tissue has been associated with increased survival via JNK signaling. TAM, tumor associated macrophage; Treg, regulatory T-cell; regDC, regulatory dendritic cell; JNK, c-Jun N-terminal kinase.

The role of UPR signaling in various cancers has been a topic of interest for many years, though parsing the exact role of UPR signaling has been difficult. For example, activation of the UPR is reported to prevent apoptosis in prostate cancer cells [37,38], but in another report, UPR was downregulated in murine models of prostate cancer [39]. Interestingly, the arms of UPR signaling had divergent effects in androgen-dependent prostate cancer; IRE1α activity and *XBP-1* expression were increased, but PERK activation was reduced [40]. UPR activation in the microenvironment also induces resistance to bortezomib and paclitaxel in prostate cancer cells [19]. Although UPR activity has a clear role in prostate cancer progression, the subtleties of its activation and how they affect survival are not well understood. Investigating the characteristics of the UPR in prostate and colorectal cancers may still yield effective therapeutic targets [41].

Our group and others have shown breast cancers, that overexpress GRP78 in response to chemotherapies, exhibit resistance to said chemotherapies [35,42–44], resistance to anti-estrogen therapies [13,45], and increased tumor anti-immunity [35,36]. This overexpression response is associated with hypoxic [44] and triple-negative [46] breast cancers. UPR-induced resistance may be downstream of MYC (cellular Myelocytomatosis; or *c-Myc*) [26], though *MYC* activation is unlikely to be the only contributor. Expression of ATF6α is correlated with resistance to chemotherapy and reduced time to breast cancer reoccurrence [47]. UPR stress aids age-related breast tumor development, and overcoming estrogen receptor-positive status is associated with UPR induction [48]. Recently, Ypt-interacting protein 1A (Yip1A) regulation of IRE1 and PERK signaling facilitated survival in cervical cancer cells [49].

Although the intricacies of UPR signaling in cancer microenvironment have not yet been deciphered, there is a clear trend that optimal levels of UPR signaling confer immune protection and may assist in the progression of cancer. Due to these findings, investigation of UPR targeting in cancer therapies is of particular interest.

2.3. Tumor Immunology

Among the many challenges of cancer formation is immune escape. Mutations in cancer cells lead to increased antigen presentation which stimulates an immune response. Indeed, tumors arise with greater frequency and grow more rapidly in immunodeficient models [50,51]. To overcome this, cancer cells must maintain presentation as 'self' or have sufficient immunosuppressive capability in their microenvironment; often, a combination of both. A model used to describe how cancer cells can overcome immune system destruction is 'immunoediting'. In this model, the immune system effectively prevents tumor development, most often by destruction of tumor cells. In some cases, the immune system can only delay tumor cell growth. This stasis, rather than destruction, allows time for further mutations and resistance to occur, thus selecting for tumor cells resistant to the immune system [50,52]. This model gives rise to an interesting parallel between cancer cells developing immunoresistance and microbes developing antibiotic resistance.

To overwhelm immune response to novel and mutated antigens, cancer cells can simply overexpress 'self' markers, inactivating cytotoxic cells [53–55]. Beyond identifying as 'self', cancer cells can subvert immune system effects to promote growth. While the immune system can induce cancer cell death and quiescence [56–60], it can also promote tumor development and establish a favorable tumor microenvironment [57,61–68]. Which of these effects occur is highly dependent on the microenvironment, what immune infiltrates are present, and the cancer type [69–74]. Infiltrates to consider are antigen-presenting cells (APCs), including tumor-invading macrophages (TAMs), neutrophils (TANs), and dendritic cells (DCs). In specific contexts, these cells are associated with survival in multiple cancer types [58–60,68,75,76]; however, they can also promote cancer progression in different contexts.

A characteristic of the tumor microenvironment strongly associated with tumor progression and inhibition of therapeutic efficacy is inflammation [63,64,77]. Inflammatory cell infiltrates, including macrophage subtypes, regulatory T-cells (Tregs), neutrophils, and myeloid-derived suppressor cells (MDSCs), create a microenvironment that suppresses cytotoxic cell activity [72,78–86]. Counterintuitively, many of these inflammatory infiltrates are recruited by B and T cells in efforts to increase immune response, but instead promote tumor development [83,84,87].

Beyond immunosuppression, the inflammatory environment generated by these immune cells assists in establishing a metastatic niche, inducing cancer cell "stemness" [88]. These cells can induce stemness by multiple means, including adjusting metabolism and further enhancing immunosuppression and tolerance in tumor microenvironment [61,79,82,84–86,89,90]. This state includes many features associated with undifferentiated cells, inducing proliferative potential, survival and durability advantages, migratory potential via epithelial-to-mesenchymal transition (EMT), and inducing angiogenesis of solid tumors [74,81]. Increased 'stemness' is additionally associated with increased metastatic potential, more malignant cancers, and poorer patient outcomes [91].

Another aspect that must be considered is the function of UPR signaling in tumor immunology. In the context of a highly stressful tumor microenvironment, the function of UPR signaling in individual immune cell types is worth consideration (summarized in Table 1).

2.4. UPR Signaling in Dendritic (Myeloid-Derived Suppressor) Cells

Myeloid-derived suppressor cells (DCs) are key regulators in immune cell response [92]. Once mature, DCs are tissue-imbedded and function as APCs. DCs additionally release cytokines to recruit and/or activate other immune cells. DCs regulate T-cell activation through a combination of presenting foreign and dead cell antigens. These steps are important in immune cells recognizing which pathogens and mutated cells must be cleared, and preventing an autoimmune response to healthy self-cells [93].

The ability of DCs to induce immune response requires maturation, presentation of antigens, and cytokine expression/release. Successful maturation of DCs is dependent upon IRE1α activation and *XBP-1* splicing. Inhibition of IRE1 function or abrogation of XBP-1S signaling reduces successful maturation and increases apoptotic death in DCs [94]. High mobility group box 1 protein (HMGB1), known to instigate DC maturation, induces GRP78 expression and XBP-1 signaling [95]. DCs that do mature under these conditions fail to stimulate T-cell proliferation due to decreased CD80, CD86, and major histocompatibility complex (MHC)-II expression and cytokine secretion [94,95].

Inhibition of UPR signaling in matured DCs also reduces their ability to activate T cells. CD80, CD86, MHC-II, and cytokine secretion are all reduced in XBP-1-inhibited mature DCs [95,96]. Inhibition of XBP-1 signaling may prevent cytokine expression and antigen presentation due to compensatory RIDD hyperactivity [97]. IRE1α/XBP-1 signaling is prioritized in DC function and it is generally believed that the PERK/ATF6 arms of the UPR do not significantly contribute to DC function [97]. The inhibition of CHOP, however, specifically prevents the expression of the inflammatory cytokine, interleukin-23 (IL-23) [96]. These findings suggest that the role of UPR signaling in antigen-presenting DCs may be more complex than currently appreciated.

Two additional classes of DCs are functionally defective DCs, which do not successfully migrate or present antigens, and regulatory DCs (regDCs), which are immunosuppressive (reviewed in [98]). Unlike APC DCs, regDCs are associated with progression in cancers and poorer prognosis [99,100]. Tumor-supporting regDCs were reported as early as 1997 [101]. In that study, Enk et al. reported that tumor-suppressive APC DCs converted to immunosuppressive regDCs. RegDCs exist in normal tissues to prevent a hyperactive immune response, along with other forms of regulatory or alternatively activated immune cells (discussed further below). Much like normal tissues, tumor-infiltrating DCs can be 'reprogrammed' into regDCs by the tumor microenvironment [102].

DCs in the tumor microenvironment can affect responsiveness or tolerance to therapies [101]. In one study, DCs were isolated from patients exhibiting metastatic melanomas, of which some metastases were responsive to chemotherapy and some metastases remained progressive. In vitro, the DCs isolated from responsive metastases activated T lymphocytes five-fold higher than DCs from progressive metastases, suggesting that the latter exhibited a regulatory DC phenotype and induced chemotherapeutic resistance. Another group showed that regulatory DCs promote metastatic expansion of pancreatic ductal adenocarcinoma [103]. Metastatic sites were enriched in regulatory DCs (CD11b$^+$CD11c$^+$MHC-II$^+$CD24$^+$CD64^{low}F4/80^{low}) that produced Treg cells via secretion of programmed death ligand 2 (PD-L2), a T-cell checkpoint inhibitor ligand. This study also showed the depletion of regulatory DCs, blockade of PD-L2-reduced expansion, and metastatic pancreatic ductal adenocarcinoma in vivo. The role of the UPR in regulatory DCs has yet to be determined; it may play a key role in their formation. Given the role of regulatory DCs in tumor progression and immunotherapy resistance, this area requires further investigation.

Table 1. UPR signaling in immune cells. A summary of how UPR signaling and each arm function in immune cell differentiation, activation, and function. Areas yet to be investigated are marked with "?".

Immune Cell Type	Cell Sub-Type	General UPR Signaling	UPR Signaling Component			
			GRP78	PERK	ATF6	IRE-1
DCs	DCs	Required for development and functional antigen presentation and cytokine secretion. Inhibition leads to cell death.	Increased expression during maturation downstream of HMGB1 signaling.	PERK considered to have no association; however, CHOP function required for successful IL-23 secretion.	No associations found upon testing.	Increased during maturation downstream HMGB1 signaling; Required for mature function, CD80, CD86, MHC-1, and cytokine secretion.
	Reg DCs	Unknown.	?	?	?	?
Macrophages	Macrophages	Required for trafficking, function, and M1/M2 polarization.	Increased expression in maturation; decreased expression in target cells increases macrophage efficacy.	Required for mature function; Inhibiting PERK increases M1 polarization; ATF4 function associated with M1/M2 macrophage balance; maintains function during stress signaling.	?	XBP-1 splicing increased; function required for inflammatory response; signaling associated with survival.
	Microglia	Required for function.	?	Required for mature function.	?	?
	Foam cell	Induces CD36 expression, positive-feedback cycle in formation and eventual cell death.	?	Increased function leads to GSK3a/b signaling; CHOP function induces cell death.	?	?
T cell	T cell	Required for various stages of differentiation, maturation, activation, and cytotoxic functions; also required for trafficking and homing. Excessive function associated with T-cell exhaustion.	Increased expression during differentiation.	?	Signaling increased during differentiation, function, and immune response.	XBP-1 splicing increased during differentiation.
	T helper	Required for differentiation but inhibited upon maturation.	Increased expression during differentiation.	?	Signaling increased during differentiation.	XBP-1 splicing increased during differentiation.
	Treg	Unknown.	?	?	?	?

2.5. UPR Signaling in Macrophages

Macrophages are another class of APC immune cells that regulate the innate immune response. Macrophages rival DCs in antigen presentation to and activation of T cells. This is characteristic of M1 ('killer' or 'classically activated') macrophages, which induce inflammation and anti-antigen responses. M1 macrophages form from circulating monocytes that infiltrate tissues in response to chemoattractants. M1 macrophages then release inflammatory factors; phagocytize pathogens, cell debris, and unhealthy cells; and present antigens to activate T cells. In contrast, M2 ('repair' or 'alternatively activated') macrophages suppress inflammatory responses and inhibit T-cell activation. In these contexts, macrophages need to induce sufficient immune response without deleterious effects to host tissues. Poor regulation of this process, or loss of M1 and M2 macrophage function, is associated with progression or initiation of several diseases, including cancers [103].

Macrophages were found to initiate UPR signaling upon differentiation [104]. This study examined both peripheral blood and atherosclerotic-infiltrating macrophages, the latter are comparable to tumor-infiltrating macrophages. Although initiation timing was unclear, GRP78 expression was increased and *XBP-1* transcripts were primarily found in their spliced forms upon monocyte infiltration and macrophage differentiation [104]. GRP78 heterozygous macrophages can differentiate and mature, but expend more energy and have reduced capacity for inflammation [105]. Spliced *XBP-1* expression is also associated with survival in macrophages, potentially through UPR regulating macrophage metabolism via induced autophagy [106]. Importantly, UPR signaling is required to maintain the ratio of inflammatory M1 to suppressive M2 cells via ATF4 expression [51].

Macrophage-induced inflammatory response is associated with IRE-1 in models of arthritis, while data in microglia (neurological tissue equivalent of macrophages) show that PERK activation is required [107,108]. The UPR is further necessary for successful macrophage response under induced ER stress. Chemically induced ER stress prevents macrophage function, but upregulation of PERK/ATF4 signaling can compensate for this effect [109].

Loss of UPR function in macrophages is strongly correlated with disease, including fibrosis [110], obesity-induced inflammatory disease [111,112], tuberculosis [113], and fatty liver disease [114]. These studies associate macrophage-induced inflammation via UPR signaling in these diseases, such that inhibition of UPR signaling abrogated or ablated disease.

Atherosclerosis, narrowing of arteries by plaque lesions, is a macrophage-associated disease in which the role of UPR signaling has been extensively studied (reviewed in [115]). In atherosclerosis, M1 macrophages are recruited to arterial plaques, whereupon they accumulate lipids, becoming foam cells. Foam cells accumulate free cholesterol due to the presence of low-density lipoprotein (LDL), which then induces UPR signaling [116–118]. UPR signaling induces CD36 expression, increasing the uptake of oxidized LDL and leading to further upregulation of all UPR signaling arms [119]. PERK activation leads to Glycogen Synthase Kinase 3 Alpha/Beta (GSK3a/b) signaling, further inducing lipid accumulation [120]. These pathways lead to a macrophage–foam cell–UPR self-potentiating cycle. Perpetuated UPR signaling leads to foam cell death via the CHOP pathway [113].

The changes found in atherosclerosis plaques are similar to those in the tumor microenvironment. Fittingly, macrophages have been associated with regulation and immune response in the cancer microenvironment. Increased M1 macrophages are associated with clearance and good prognosis, while M2 macrophages are immunosuppressive and procarcinogenic. Our lab has shown that UPR signaling mediates lipid metabolism in breast cancer, and as a result, macrophages infiltrate into breast cancers [35]. Reducing whole body GRP78 levels by antisense morpholino injection increased macrophage infiltration of breast tumors and reduced the expression of CD47 ("do not eat me"/"self") signaling in tumor samples. These results were replicated with the administration of linoleic acid, the polyunsaturated fatty lipid cleared downstream of GRP78 activation. Our lab has also demonstrated that inhibition of PERK, but not GRP78 or IRE1 inhibition, is responsible for the increased proliferation of M1 macrophages and cancer cell clearance in melanoma [36]. This study showed that cancer cell UPR activity regulates macrophage response. The inhibition of GRP78 or IRE1 in cancer cells increased macrophage-mediated clearance. The effects of UPR signaling in cancer cells, and their ability to induce UPR signaling in the microenvironment, doubly inhibit macrophage-mediated immune response to cancers.

2.6. UPR Signaling in T Cells

T cells are lymphocytes that participate in the adaptive immune response. The T-cell family includes various lymphocytes that mature in the thymus. Each of these cell types contribute to an immune/apoptotic (CD4$^+$ helper, CD8$^+$ killer, memory, or natural killer) or immunosuppressive (regulatory) role. In proper orchestration, T cells and the APCs that activate them induce an effective response against invading pathogens and malignant cells. Dysregulation of T cells can lead to a compromised immune system or autoimmune diseases.

The UPR is required for successful T-cell formation and activation. Interestingly, different lymphocyte classes exhibit distinct patterns of UPR signaling during differentiation. UPR signaling is activated upon differentiation of both B and T cells. In the presence of a differentiation stimulus, both B and T cells increase GRP78 protein levels, initiate *XBP-1* transcript cleavage, and induce ATF6 signaling [121–124]. The inhibition of GRP78, ATF6, or XBP-1 signaling pathways greatly reduces plasma cell differentiation and efficacy upon maturation [121,125]. Cell fate determines whether UPR signaling is maintained. For example, early B-cells exhibit UPR signaling, but it is absent in mature B-cells. Similarly, CD4$^-$/CD8$^-$ progenitor T-cells do not exhibit an UPR, but greatly increase UPR during maturation as CD4$^+$/CD8$^+$ T-cells. Upon differentiation to CD4$^+$ T-cells, the UPR is once again repressed [122].

Unlike B cells and CD4$^+$ T-cells, mature CD8$^+$ T-cells maintain UPR signaling [122]. XBP-1 signaling downstream of IRE1 is increased during acute infection, and inhibition of XBP-1 signaling prevents terminal differentiation and immune response in CD8$^+$ T-cells [123]. T-cell trafficking and homing under oxidative stress also requires UPR signaling [126]. Inhibited signaling, specifically via the GRP78, ATF6, and XBP-1 pathways, greatly reduces plasma cell differentiation and efficacy upon maturation [121,125].

Another area in which the UPR plays a role is T-cell exhaustion [127]. This is a state in which sufficient stimulation does not induce T-cell activation, and thus, the T cell will not proliferate and/or generate the cytolytic compounds required for inducing targeted cell death [128]. The causes for this abnormality may be varied. We do know that a lack of appropriate metabolites and inhibitory signals contribute to this exhausted phenotype. T-cell exhaustion is a concern in numerous diseases, including cancers [127,129–132]. As stated previously, the microenvironment is hostile and frequently features hypoxia, low metabolite availability, inflammation, and transmissible ER stress responses. All these factors induce UPR signaling, which is directly associated with T-cell exhaustion in models of infectious disease [123,128,133]. T-cell exhaustion in the tumor microenvironment has become an area of interest and potential immunotherapeutic target [127,129,130,132,134].

The specific roles of UPR signaling in T-cell differentiation and activity are incompletely understood. Similarly, the role of UPR signaling in the microenvironment and during activation of helper T-cells has yet to be investigated.

2.7. UPR Signaling and Cancer-Associated Fibroblasts

Cancer-associated fibroblasts (CAFs) play a significant role in the development, protection, and metastasis of cancers [135]. CAFs are known to regulate tumor-associated immune cells and warrant mention [136]. Recent findings suggest that UPR signaling plays a large role in the generation of the tumor environment, including the differentiation of CAFs [137]. In turn, CAFs have been shown to stimulate non-small-cell lung cancer invasion by upregulating GRP78 expression [138]. As a regulator of the tumor microenvironment and immune cells, the role of the UPR in CAFs and their function should be further investigated.

2.8. Implications for UPR-Targeting Drugs in Cancer Therapy

Interestingly, the efficacy of some chemotherapeutic agents may be due to previously unknown effects on UPR signaling. Triptolide activates UPR signaling (IRE1 and PERK) in breast cancer, inducing cell death, and simultaneously reduces the expression of GRP78. This may be characteristic of several chemotherapeutic agents, for example, nemorosone and ONC212 in pancreatic cancer and nelfinavir in ovarian cancer [139–141].

The associations among the UPR, development of clinically diagnosed cancer, and chemotherapeutic resistance have increased interest in targeting the UPR as a strategy for cancer therapy [142]. There is a delicate balance between surviving ER stress and UPR-initiated apoptosis in cancer cells [143]. Disrupting this balance via UPR inhibition [47,48,142,144–148] induces cell death via apoptotic means or immunogenic clearance. Conversely, the overstimulation of the IRE1 and PERK/CHOP pathways [37,149–155] effectively induces cancer cell apoptosis, likely through pro-apoptotic effects of CHOP. An indirectly activation of UPR signaling by inducing the generation of reactive oxygen species via small molecule therapy leads to cancer cell death in xenografts [153].

Altering UPR signaling may resensitize cancers to chemotherapeutic agents and may increase the efficacy of as yet unknown chemotherapeutic agents. To date, several studies have shown that disrupting UPR signaling increases drug sensitivity. These include reports of abrogating UPR signaling with concurrent drug treatment in murine xenografts [156] and in vivo colorectal cancer models [157,158], and resensitizing breast cancer cells to chemotherapy and immunotherapy [35,36,43,45]. Alternatively, inducing UPR signaling sensitizes non-small-cell lung cancer to doxorubicin [159], instigates ovarian cancer cell death when paired with mifepristone [160], and increases the efficacy of viral antineoplastic therapies [161]. Inducing the UPR was shown to sensitize ovarian cancer cells to chemotherapy via increased JNK signaling [162], which may implicate the IRE1 signaling arm.

There is a growing interest in immunotherapies for cancer treatment. The immune system regularly clears mutated and senesced cells from the body. The goal of immunotherapy is to re-enable the immune system to recognize and clear cancer cells. In general, accomplishing this goal means that immunotherapy must alter cancer cells to present antigens resulting in clearance, or prevent immunosuppressive effects exhibited by cancers. The latter generally focuses on "self" markers expressed by cancers, including cytotoxic T-lymphocyte-associated protein 4 (CTLA-4), programmed cell death protein 1 (PD-1), and lymphocyte activation gene 3 (LAG3). These are checkpoint inhibitor proteins. When immune cell–cancer cell interactions engage these receptors, immune cells cannot begin to activate or proliferate. Our lab has demonstrated that targeting the UPR could induce immune response both through increasing antigen presentation and preventing immune cell inhibition. Targeting GRP78 induces the accumulation of immunogenic polyunsaturated lipids in breast cancer models, inducing macrophage infiltration and clearance [35]. Resistance to (CTLA-4) immunotherapy was also associated with increased UPR signaling [36]. Peripheral blood mononuclear cells (PBMCs) of melanoma patients were collected prior to and post-development of resistance to CTLA-4 immunotherapy with ipilimumab. Arginase 1 (Arg-1) was increased in resistant PBMCs, indicating a shift from M1 cancer-clearing macrophages to M2 immune-inhibitory macrophages. These PBMCs exhibited increased PERK and IRE-1 expression, suggesting that UPR signaling induced the shift to M2 macrophages and subsequent resistance to immunotherapy.

3. Conclusions

While the mechanism is still incompletely understood, our knowledge in the functions and activity of the unfolded protein response allows us to examine its role in complex contexts [14,163]. Each of the three arms of the UPR—IRE1, PERK, and ATF6—exhibit unique effects dependent upon the cellular context. Immune cells are particularly reliant upon UPR to handle the stress of rapid division and expression of critical proteins. UPR regulates immune function both in induction of pathogen response and inhibition of autoimmunity [164]. Another context in which UPR is of particular interest is tumor development and tumor microenvironment [21,23,29,31,32,39,48,165–168]. The role of UPR has been investigated in many types of cancer, suggesting that targeting UPR will be a viable strategy regardless of cancer origin and mutations. Indeed, there are numerous studies targeting UPR as a cancer therapy [27,30,37,48,139–141,154,156,162,167,169], to increase chemotherapy efficacy [33,34,44, 48,141,142,153,159–161,165,170–172], and to enhance immunotherapy [34,36,53,132,173–176].

A better understanding of the roles of each UPR arm in cell and organism homeostasis has the potential to increase the understanding of numerous diseases and this requires further investigation. Given the potential of cancer immunotherapy, understanding the function of UPR in each immune cell type and how this affects their response to cancer cells is of particular interest. In addition to immunotherapy, targeting the UPR shows great promise for increasing the selectivity and efficacy of cancer therapy, and may be a key target in overcoming cancer resistance to chemotherapies. By enhancing efficacy as an adjuvant treatment, targeting the UPR may decrease required concentrations of chemotherapies and therefore off-target effects. Similarly, increasing immunogenicity and clearance of cancer cells by targeting the UPR is likely to be effective with minimal side effects. For these reasons, the mechanisms and role of the UPR in cancer cells, immune response, and how to best target these pathways are high priority targets in furthering cancer treatment.

Author Contributions: Conceptualization, K.L.C. and D.R.S.-P.; writing—original draft preparation, M.U.R.; writing—review and editing, M.U.R., K.L.C., S.H. and D.R.S.-P.; visualization, S.R.H.; supervision, K.L.C.; project administration, K.L.C. All authors have read and agreed to the published version of the manuscript.

Funding: This work was funded by Susan G. Komen Career Catalyst Research Grant, CCR18547795 to K.L.C. and an IRACDA PRIME K12 fellowship, 1K12-GM102773 (PI-A. Howlett) to M.R.

Conflicts of Interest: The authors declare no conflicts of interest.

References

1. Hanahan, D.; Weinberg, R.A. Hallmarks of cancer: The next generation. *Cell* **2011**, *144*, 646–674. [CrossRef] [PubMed]

2. Feitelson, M.A.; Arzumanyan, A.; Kulathinal, R.J.; Blain, S.W.; Holcombe, R.F.; Mahajna, J.; Marino, M.; Martinez-Chantar, M.L.; Nawroth, R.; Sanchez-Garcia, I.; et al. Sustained proliferation in cancer: Mechanisms and novel therapeutic targets. *Semin. Cancer Biol.* **2015**, *35*, S25–S54. [CrossRef] [PubMed]

3. Messerschmidt, J.L.; Prendergast, G.C.; Messerschmidt, G.L. How Cancers Escape Immune Destruction and Mechanisms of Action for the New Significantly Active Immune Therapies: Helping Nonimmunologists Decipher Recent Advances. *Oncologist* **2016**, *21*, 233–243. [CrossRef] [PubMed]

4. Quail, D.F.; Joyce, J.A. Microenvironmental regulation of tumor progression and metastasis. *Nat. Med.* **2013**, *19*, 1423–1437. [CrossRef] [PubMed]

5. McGee, H.M.; Jiang, D.; Soto-Pantoja, D.R.; Nevler, A.; Giaccia, A.J.; Woodward, W.A. Targeting the Tumor Microenvironment in Radiation Oncology: Proceedings from the 2018 ASTRO-AACR Research Workshop. *Clin. Cancer Res.* **2019**, *25*, 2969–2974. [CrossRef]

6. Dvorak, H.F. Tumors: Wounds that do not heal-redux. *Cancer Immunol. Res.* **2015**, *3*, 1–11. [CrossRef]

7. Muz, B.; de la Puente, P.; Azab, F.; Azab, A.K. The role of hypoxia in cancer progression, angiogenesis, metastasis, and resistance to therapy. *Hypoxia (Auckl)* **2015**, *3*, 83–92. [CrossRef]

8. McCann, J.V.; Xiao, L.; Kim, D.J.; Khan, O.F.; Kowalski, P.S.; Anderson, D.G.; Pecot, C.V.; Azam, S.H.; Parker, J.S.; Tsai, Y.S.; et al. Endothelial miR-30c suppresses tumor growth via inhibition of TGF-beta-induced Serpine1. *J. Clin. Investig.* **2019**, *130*, 1654–1670. [CrossRef]

9. Fulbright, L.E.; Ellermann, M.; Arthur, J.C. The microbiome and the hallmarks of cancer. *PLoS Pathog.* **2017**, *13*, e1006480. [CrossRef]

10. Ma, Y.; Hendershot, L.M. The unfolding tale of the unfolded protein response. *Cell* **2001**, *107*, 827–830. [CrossRef]

11. Haas, I.G.; Wabl, M. Immunoglobulin heavy chain binding protein. *Nature* **1983**, *306*, 387–389. [CrossRef] [PubMed]

12. Lee, A.S. Coordinated regulation of a set of genes by glucose and calcium ionophores in mammalian cells. *Trends Biochem. Sci.* **1987**, *12*, 20–23. [CrossRef]

13. Clarke, R.; Cook, K.L. Unfolding the Role of Stress Response Signaling in Endocrine Resistant Breast Cancers. *Front. Oncol.* **2015**, *5*, 140. [CrossRef] [PubMed]

14. Clarke, R.; Cook, K.L.; Hu, R.; Facey, C.O.; Tavassoly, I.; Schwartz, J.L.; Baumann, W.T.; Tyson, J.J.; Xuan, J.; Wang, Y.; et al. Endoplasmic reticulum stress, the unfolded protein response, autophagy, and the integrated regulation of breast cancer cell fate. *Cancer Res.* **2012**, *72*, 1321–1331. [CrossRef] [PubMed]

15. Lee, K.; Tirasophon, W.; Shen, X.; Michalak, M.; Prywes, R.; Okada, T.; Yoshida, H.; Mori, K.; Kaufman, R.J. IRE1-mediated unconventional mRNA splicing and S2P-mediated ATF6 cleavage merge to regulate XBP1 in signaling the unfolded protein response. *Genes Dev.* **2002**, *16*, 452–466. [CrossRef] [PubMed]

16. Urano, F.; Wang, X.; Bertolotti, A.; Zhang, Y.; Chung, P.; Harding, H.P.; Ron, D. Coupling of stress in the ER to activation of JNK protein kinases by transmembrane protein kinase IRE1. *Science* **2000**, *287*, 664–666. [CrossRef] [PubMed]

17. Gardner, B.M.; Walter, P. Unfolded proteins are Ire1-activating ligands that directly induce the unfolded protein response. *Science* **2011**, *333*, 1891–1894. [CrossRef]

18. Cullinan, S.B.; Diehl, J.A. Coordination of ER and oxidative stress signaling: The PERK/Nrf2 signaling pathway. *Int. J. Biochem. Cell Biol.* **2006**, *38*, 317–332. [CrossRef]

19. Rodvold, J.J.; Chiu, K.T.; Hiramatsu, N.; Nussbacher, J.K.; Galimberti, V.; Mahadevan, N.R.; Willert, K.; Lin, J.H.; Zanetti, M. Intercellular transmission of the unfolded protein response promotes survival and drug resistance in cancer cells. *Sci. Signal.* **2017**, *10*. [CrossRef]

20. Mahadevan, N.R.; Rodvold, J.; Sepulveda, H.; Rossi, S.; Drew, A.F.; Zanetti, M. Transmission of endoplasmic reticulum stress and pro-inflammation from tumor cells to myeloid cells. *Proc. Natl. Acad. Sci. USA* **2011**, *108*, 6561–6566. [CrossRef]

21. Jain, B.P. An Overview of Unfolded Protein Response Signaling and Its Role in Cancer. *Cancer Biother. Radiopharm.* **2017**, *32*, 275–281. [CrossRef] [PubMed]

22. Rufo, N.; Garg, A.D.; Agostinis, P. The Unfolded Protein Response in Immunogenic Cell Death and Cancer Immunotherapy. *Trends Cancer* **2017**, *3*, 643–658. [CrossRef] [PubMed]

23. Wouters, B.G.; Koritzinsky, M. Hypoxia signalling through mTOR and the unfolded protein response in cancer. *Nat. Rev. Cancer* **2008**, *8*, 851–864. [CrossRef] [PubMed]

24. Tsai, Y.C.; Weissman, A.M. The Unfolded Protein Response, Degradation from Endoplasmic Reticulum and Cancer. *Genes Cancer* **2010**, *1*, 764–778. [CrossRef] [PubMed]

25. Tsachaki, M.; Mladenovic, N.; Stambergova, H.; Birk, J.; Odermatt, A. Hexose-6-phosphate dehydrogenase controls cancer cell proliferation and migration through pleiotropic effects on the unfolded-protein response, calcium homeostasis, and redox balance. *FASEB J.* **2018**, *32*, 2690–2705. [CrossRef] [PubMed]

26. Shajahan-Haq, A.N.; Cook, K.L.; Schwartz-Roberts, J.L.; Eltayeb, A.E.; Demas, D.M.; Warri, A.M.; Facey, C.O.; Hilakivi-Clarke, L.A.; Clarke, R. MYC regulates the unfolded protein response and glucose and glutamine uptake in endocrine resistant breast cancer. *Mol. Cancer* **2014**, *13*, 239. [CrossRef] [PubMed]

27. Saito, S.; Furuno, A.; Sakurai, J.; Sakamoto, A.; Park, H.R.; Shin-Ya, K.; Tsuruo, T.; Tomida, A. Chemical genomics identifies the unfolded protein response as a target for selective cancer cell killing during glucose deprivation. *Cancer Res.* **2009**, *69*, 4225–4234. [CrossRef]

28. Nagelkerke, A.; Bussink, J.; Mujcic, H.; Wouters, B.G.; Lehmann, S.; Sweep, F.C.; Span, P.N. Hypoxia stimulates migration of breast cancer cells via the PERK/ATF4/LAMP3-arm of the unfolded protein response. *Breast Cancer Res.* **2013**, *15*. [CrossRef]

29. Bartkowiak, K.; Effenberger, K.E.; Harder, S.; Andreas, A.; Buck, F.; Peter-Katalinic, J.; Pantel, K.; Brandt, B.H. Discovery of a novel unfolded protein response phenotype of cancer stem/progenitor cells from the bone marrow of breast cancer patients. *J. Proteome Res.* **2010**, *9*, 3158–3168. [CrossRef]

30. Bartkowiak, K.; Kwiatkowski, M.; Buck, F.; Gorges, T.M.; Nilse, L.; Assmann, V.; Andreas, A.; Muller, V.; Wikman, H.; Riethdorf, S.; et al. Disseminated Tumor Cells Persist in the Bone Marrow of Breast Cancer Patients through Sustained Activation of the Unfolded Protein Response. *Cancer Res.* **2015**, *75*, 5367–5377. [CrossRef]

31. Corazzari, M.; Gagliardi, M.; Fimia, G.M.; Piacentini, M. Endoplasmic Reticulum Stress, Unfolded Protein Response, and Cancer Cell Fate. *Front. Oncol.* **2017**, *7*, 78. [CrossRef] [PubMed]

32. Shen, X.; Xue, Y.; Si, Y.; Wang, Q.; Wang, Z.; Yuan, J.; Zhang, X. The unfolded protein response potentiates epithelial-to-mesenchymal transition (EMT) of gastric cancer cells under severe hypoxic conditions. *Med. Oncol.* **2015**, *32*, 447. [CrossRef] [PubMed]

33. Mujumdar, N.; Banerjee, S.; Chen, Z.; Sangwan, V.; Chugh, R.; Dudeja, V.; Yamamoto, M.; Vickers, S.M.; Saluja, A.K. Triptolide activates unfolded protein response leading to chronic ER stress in pancreatic cancer cells. *Am. J. Physiol. Gastrointest. Liver Physiol.* **2014**, *306*, G1011–G1020. [CrossRef] [PubMed]

34. Cook, K.L.; Soto-Pantoja, D.R. "UPRegulation" of CD47 by the endoplasmic reticulum stress pathway controls anti-tumor immune responses. *Biomark. Res.* **2017**, *5*, 26. [CrossRef] [PubMed]

35. Cook, K.L.; Soto-Pantoja, D.R.; Clarke, P.A.; Cruz, M.I.; Zwart, A.; Warri, A.; Hilakivi-Clarke, L.; Roberts, D.D.; Clarke, R. Endoplasmic Reticulum Stress Protein GRP78 Modulates Lipid Metabolism to Control Drug Sensitivity and Antitumor Immunity in Breast Cancer. *Cancer Res.* **2016**, *76*, 5657–5670. [CrossRef] [PubMed]

36. Soto-Pantoja, D.R.; Wilson, A.S.; Clear, K.Y.; Westwood, B.; Triozzi, P.L.; Cook, K.L. Unfolded protein response signaling impacts macrophage polarity to modulate breast cancer cell clearance and melanoma immune checkpoint therapy responsiveness. *Oncotarget* **2017**, *8*, 80545–80559. [CrossRef]

37. Misra, U.K.; Pizzo, S.V. Modulation of the unfolded protein response in prostate cancer cells by antibody-directed against the carboxyl-terminal domain of GRP78. *Apoptosis* **2010**, *15*, 173–182. [CrossRef]

38. Thornton, M.; Aslam, M.A.; Tweedle, E.M.; Ang, C.; Campbell, F.; Jackson, R.; Costello, E.; Rooney, P.S.; Vlatkovic, N.; Boyd, M.T. The unfolded protein response regulator GRP78 is a novel predictive biomarker in colorectal cancer. *Int. J. Cancer* **2013**, *133*, 1408–1418. [CrossRef]

39. So, A.Y.; de la Fuente, E.; Walter, P.; Shuman, M.; Bernales, S. The unfolded protein response during prostate cancer development. *Cancer Metastasis Rev.* **2009**, *28*, 219–223. [CrossRef]

40. Sheng, X.; Arnoldussen, Y.J.; Storm, M.; Tesikova, M.; Nenseth, H.Z.; Zhao, S.; Fazli, L.; Rennie, P.; Risberg, B.; Waehre, H.; et al. Divergent androgen regulation of unfolded protein response pathways drives prostate cancer. *EMBO Mol. Med.* **2015**, *7*, 788–801. [CrossRef]

41. Storm, M.; Sheng, X.; Arnoldussen, Y.J.; Saatcioglu, F. Prostate cancer and the unfolded protein response. *Oncotarget* **2016**, *7*, 54051–54066. [CrossRef] [PubMed]

42. Scriven, P.; Coulson, S.; Haines, R.; Balasubramanian, S.; Cross, S.; Wyld, L. Activation and clinical significance of the unfolded protein response in breast cancer. *Br. J. Cancer* **2009**, *101*, 1692–1698. [CrossRef] [PubMed]

43. Wang, J.; Yin, Y.; Hua, H.; Li, M.; Luo, T.; Xu, L.; Wang, R.; Liu, D.; Zhang, Y.; Jiang, Y. Blockade of GRP78 sensitizes breast cancer cells to microtubules-interfering agents that induce the unfolded protein response. *J. Cell Mol. Med.* **2009**, *13*, 3888–3897. [CrossRef] [PubMed]

44. Notte, A.; Rebucci, M.; Fransolet, M.; Roegiers, E.; Genin, M.; Tellier, C.; Watillon, K.; Fattaccioli, A.; Arnould, T.; Michiels, C. Taxol-induced unfolded protein response activation in breast cancer cells exposed to hypoxia: ATF4 activation regulates autophagy and inhibits apoptosis. *Int. J. Biochem. Cell Biol.* **2015**, *62*, 1–14. [CrossRef]

45. Cook, K.L.; Clarke, P.A.; Clarke, R. Targeting GRP78 and antiestrogen resistance in breast cancer. *Future Med. Chem.* **2013**, *5*, 1047–1057. [CrossRef]

46. Rajapaksa, G.; Thomas, C.; Gustafsson, J.A. Estrogen signaling and unfolded protein response in breast cancer. *J. Steroid Biochem. Mol. Biol.* **2016**, *163*, 45–50. [CrossRef]

47. Andruska, N.; Zheng, X.; Yang, X.; Helferich, W.G.; Shapiro, D.J. Anticipatory estrogen activation of the unfolded protein response is linked to cell proliferation and poor survival in estrogen receptor alpha-positive breast cancer. *Oncogene* **2015**, *34*, 3760–3769. [CrossRef]

48. Cook, K.L.; Clarke, P.A.; Parmar, J.; Hu, R.; Schwartz-Roberts, J.L.; Abu-Asab, M.; Warri, A.; Baumann, W.T.; Clarke, R. Knockdown of estrogen receptor-alpha induces autophagy and inhibits antiestrogen-mediated unfolded protein response activation, promoting ROS-induced breast cancer cell death. *FASEB J.* **2014**, *28*, 3891–3905. [CrossRef]

49. Taguchi, Y.; Horiuchi, Y.; Kano, F.; Murata, M. Novel prosurvival function of Yip1A in human cervical cancer cells: Constitutive activation of the IRE1 and PERK pathways of the unfolded protein response. *Cell Death Dis.* **2017**, *8*, e2718. [CrossRef]

50. Teng, M.W.; Swann, J.B.; Koebel, C.M.; Schreiber, R.D.; Smyth, M.J. Immune-mediated dormancy: An equilibrium with cancer. *J. Leukoc. Biol.* **2008**, *84*, 988–993. [CrossRef]

51. Kim, J.H.; Lee, E.; Friedline, R.H.; Suk, S.; Jung, D.Y.; Dagdeviren, S.; Hu, X.; Inashima, K.; Noh, H.L.; Kwon, J.Y.; et al. Endoplasmic reticulum chaperone GRP78 regulates macrophage function and insulin resistance in diet-induced obesity. *FASEB J.* **2018**, *32*, 2292–2304. [CrossRef] [PubMed]

52. Schreiber, R.D.; Old, L.J.; Smyth, M.J. Cancer immunoediting: Integrating immunity's roles in cancer suppression and promotion. *Science* **2011**, *331*, 1565–1570. [CrossRef] [PubMed]

53. Park, H.J.; Kusnadi, A.; Lee, E.J.; Kim, W.W.; Cho, B.C.; Lee, I.J.; Seong, J.; Ha, S.J. Tumor-infiltrating regulatory T cells delineated by upregulation of PD-1 and inhibitory receptors. *Cell Immunol.* **2012**, *278*, 76–83. [CrossRef] [PubMed]

54. Hu, F.; Yu, X.; Wang, H.; Zuo, D.; Guo, C.; Yi, H.; Tirosh, B.; Subjeck, J.R.; Qiu, X.; Wang, X.Y. ER stress and its regulator X-box-binding protein-1 enhance polyIC-induced innate immune response in dendritic cells. *Eur. J. Immunol.* **2011**, *41*, 1086–1097. [CrossRef] [PubMed]

55. Shields, J.D.; Kourtis, I.C.; Tomei, A.A.; Roberts, J.M.; Swartz, M.A. Induction of lymphoidlike stroma and immune escape by tumors that express the chemokine CCL21. *Science* **2010**, *328*, 749–752. [CrossRef] [PubMed]

56. Romero, I.; Garrido, F.; Garcia-Lora, A.M. Metastases in immune-mediated dormancy: A new opportunity for targeting cancer. *Cancer Res.* **2014**, *74*, 6750–6757. [CrossRef]

57. Senovilla, L.; Vacchelli, E.; Galon, J.; Adjemian, S.; Eggermont, A.; Fridman, W.H.; Sautes-Fridman, C.; Ma, Y.; Tartour, E.; Zitvogel, L.; et al. Trial watch: Prognostic and predictive value of the immune infiltrate in cancer. *Oncoimmunology* **2012**, *1*, 1323–1343. [CrossRef]

58. Koelzer, V.H.; Canonica, K.; Dawson, H.; Sokol, L.; Karamitopoulou-Diamantis, E.; Lugli, A.; Zlobec, I. Phenotyping of tumor-associated macrophages in colorectal cancer: Impact on single cell invasion (tumor budding) and clinicopathological outcome. *Oncoimmunology* **2016**, *5*, e1106677. [CrossRef]

59. Zhang, M.; He, Y.; Sun, X.; Li, Q.; Wang, W.; Zhao, A.; Di, W. A high M1/M2 ratio of tumor-associated macrophages is associated with extended survival in ovarian cancer patients. *J. Ovarian Res.* **2014**, *7*, 19. [CrossRef]

60. Zhang, Q.W.; Liu, L.; Gong, C.Y.; Shi, H.S.; Zeng, Y.H.; Wang, X.Z.; Zhao, Y.W.; Wei, Y.Q. Prognostic significance of tumor-associated macrophages in solid tumor: A meta-analysis of the literature. *PLoS ONE* **2012**, *7*, e50946. [CrossRef]

61. Mantovani, A. Molecular pathways linking inflammation and cancer. *Curr. Mol. Med.* **2010**, *10*, 369–373. [CrossRef] [PubMed]

62. Grivennikov, S.I.; Greten, F.R.; Karin, M. Immunity, inflammation, and cancer. *Cell* **2010**, *140*, 883–899. [CrossRef] [PubMed]

63. Dvorak, H.F. Tumors: Wounds that do not heal. Similarities between tumor stroma generation and wound healing. *N. Engl. J. Med.* **1986**, *315*, 1650–1659. [PubMed]

64. Schafer, M.; Werner, S. Cancer as an overhealing wound: An old hypothesis revisited. *Nat. Rev. Mol. Cell Biol* **2008**, *9*, 628–638. [CrossRef]

65. Porta, C.; Larghi, P.; Rimoldi, M.; Totaro, M.G.; Allavena, P.; Mantovani, A.; Sica, A. Cellular and molecular pathways linking inflammation and cancer. *Immunobiology* **2009**, *214*, 761–777. [CrossRef]

66. Yu, H.; Huang, X.; Liu, X.; Jin, H.; Zhang, G.; Zhang, Q.; Yu, J. Regulatory T cells and plasmacytoid dendritic cells contribute to the immune escape of papillary thyroid cancer coexisting with multinodular non-toxic goiter. *Endocrine* **2013**, *44*, 172–181. [CrossRef]

67. Liu, J.; Zhang, N.; Li, Q.; Zhang, W.; Ke, F.; Leng, Q.; Wang, H.; Chen, J.; Wang, H. Tumor-associated macrophages recruit CCR6+ regulatory T cells and promote the development of colorectal cancer via enhancing CCL20 production in mice. *PLoS ONE* **2011**, *6*, e19495. [CrossRef]

68. Edin, S.; Wikberg, M.L.; Oldenborg, P.A.; Palmqvist, R. Macrophages: Good guys in colorectal cancer. *Oncoimmunology* **2013**, *2*, e23038. [CrossRef]

69. Bogolyubova, A.V.; Belousov, P.V. Inflammatory Immune Infiltration in Human Tumors: Role in Pathogenesis and Prognostic and Diagnostic Value. *Biochemistry* **2016**, *81*, 1261–1273. [CrossRef]

70. Pages, F.; Galon, J.; Dieu-Nosjean, M.C.; Tartour, E.; Sautes-Fridman, C.; Fridman, W.H. Immune infiltration in human tumors: A prognostic factor that should not be ignored. *Oncogene* **2010**, *29*, 1093–1102. [CrossRef]

71. Ferrone, C.; Dranoff, G. Dual roles for immunity in gastrointestinal cancers. *J. Clin. Oncol.* **2010**, *28*, 4045–4051. [CrossRef] [PubMed]

72. Johansson, M.; Denardo, D.G.; Coussens, L.M. Polarized immune responses differentially regulate cancer development. *Immunol. Rev.* **2008**, *222*, 145–154. [CrossRef] [PubMed]

73. Schiavoni, G.; Gabriele, L.; Mattei, F. The tumor microenvironment: A pitch for multiple players. *Front. Oncol.* **2013**, *3*, 90. [CrossRef] [PubMed]

74. Marvel, D.; Gabrilovich, D.I. Myeloid-derived suppressor cells in the tumor microenvironment: Expect the unexpected. *J. Clin. Investig.* **2015**, *125*, 3356–3364. [CrossRef]

75. Preynat-Seauve, O.; Contassot, E.; Schuler, P.; French, L.E.; Huard, B. Melanoma-infiltrating dendritic cells induce protective antitumor responses mediated by T cells. *Melanoma Res.* **2007**, *17*, 169–176. [CrossRef]

76. Eruslanov, E.B.; Bhojnagarwala, P.S.; Quatromoni, J.G.; Stephen, T.L.; Ranganathan, A.; Deshpande, C.; Akimova, T.; Vachani, A.; Litzky, L.; Hancock, W.W.; et al. Tumor-associated neutrophils stimulate T cell responses in early-stage human lung cancer. *J. Clin. Investig.* **2014**, *124*, 5466–5480. [CrossRef]

77. De Visser, K.E.; Eichten, A.; Coussens, L.M. Paradoxical roles of the immune system during cancer development. *Nat. Rev. Cancer* **2006**, *6*, 24–37. [CrossRef]

78. Solinas, G.; Germano, G.; Mantovani, A.; Allavena, P. Tumor-associated macrophages (TAM) as major players of the cancer-related inflammation. *J. Leukoc. Biol.* **2009**, *86*, 1065–1073. [CrossRef]

79. Qian, B.Z.; Pollard, J.W. Macrophage diversity enhances tumor progression and metastasis. *Cell* **2010**, *141*, 39–51. [CrossRef]

80. Mougiakakos, D.; Johansson, C.C.; Trocme, E.; All-Ericsson, C.; Economou, M.A.; Larsson, O.; Seregard, S.; Kiessling, R. Intratumoral forkhead box P3-positive regulatory T cells predict poor survival in cyclooxygenase-2-positive uveal melanoma. *Cancer* **2010**, *116*, 2224–2233. [CrossRef]

81. Ostrand-Rosenberg, S.; Sinha, P. Myeloid-derived suppressor cells: Linking inflammation and cancer. *J. Immunol.* **2009**, *182*, 4499–4506. [CrossRef] [PubMed]

82. Coffelt, S.B.; Lewis, C.E.; Naldini, L.; Brown, J.M.; Ferrara, N.; De Palma, M. Elusive identities and overlapping phenotypes of proangiogenic myeloid cells in tumors. *Am. J. Pathol.* **2010**, *176*, 1564–1576. [CrossRef] [PubMed]

83. DeNardo, D.G.; Andreu, P.; Coussens, L.M. Interactions between lymphocytes and myeloid cells regulate pro- versus anti-tumor immunity. *Cancer Metastasis Rev.* **2010**, *29*, 309–316. [CrossRef] [PubMed]

84. Egeblad, M.; Nakasone, E.S.; Werb, Z. Tumors as organs: Complex tissues that interface with the entire organism. *Dev. Cell* **2010**, *18*, 884–901. [CrossRef] [PubMed]

85. Murdoch, C.; Muthana, M.; Coffelt, S.B.; Lewis, C.E. The role of myeloid cells in the promotion of tumour angiogenesis. *Nat. Rev. Cancer* **2008**, *8*, 618–631. [CrossRef] [PubMed]

86. De Palma, M.; Murdoch, C.; Venneri, M.A.; Naldini, L.; Lewis, C.E. Tie2-expressing monocytes: Regulation of tumor angiogenesis and therapeutic implications. *Trends Immunol.* **2007**, *28*, 519–524. [CrossRef] [PubMed]

87. Biswas, S.K.; Mantovani, A. Macrophage plasticity and interaction with lymphocyte subsets: Cancer as a paradigm. *Nat. Immunol.* **2010**, *11*, 889–896. [CrossRef]

88. Plaks, V.; Kong, N.; Werb, Z. The cancer stem cell niche: How essential is the niche in regulating stemness of tumor cells? *Cell Stem Cell* **2015**, *16*, 225–238. [CrossRef]

89. Daurkin, I.; Eruslanov, E.; Stoffs, T.; Perrin, G.Q.; Algood, C.; Gilbert, S.M.; Rosser, C.J.; Su, L.M.; Vieweg, J.; Kusmartsev, S. Tumor-associated macrophages mediate immunosuppression in the renal cancer microenvironment by activating the 15-lipoxygenase-2 pathway. *Cancer Res.* **2011**, *71*, 6400–6409. [CrossRef]

90. Joyce, J.A.; Pollard, J.W. Microenvironmental regulation of metastasis. *Nat. Rev. Cancer* **2009**, *9*, 239–252. [CrossRef]

91. Li, Y.; Rogoff, H.A.; Keates, S.; Gao, Y.; Murikipudi, S.; Mikule, K.; Leggett, D.; Li, W.; Pardee, A.B.; Li, C.J. Suppression of cancer relapse and metastasis by inhibiting cancer stemness. *Proc. Natl. Acad. Sci. USA* **2015**, *112*, 1839–1844. [CrossRef] [PubMed]

92. Segura, E. Review of Mouse and Human Dendritic Cell Subsets. *Methods Mol. Biol.* **2016**, *1423*, 3–15. [PubMed]

93. Worbs, T.; Hammerschmidt, S.I.; Forster, R. Dendritic cell migration in health and disease. *Nat. Rev. Immunol.* **2017**, *17*, 30–48. [CrossRef] [PubMed]

94. Iwakoshi, N.N.; Pypaert, M.; Glimcher, L.H. The transcription factor XBP-1 is essential for the development and survival of dendritic cells. *J. Exp. Med.* **2007**, *204*, 2267–2275. [CrossRef] [PubMed]

95. Zhu, X.M.; Yao, F.H.; Yao, Y.M.; Dong, N.; Yu, Y.; Sheng, Z.Y. Endoplasmic reticulum stress and its regulator XBP-1 contributes to dendritic cell maturation and activation induced by high mobility group box-1 protein. *Int J. Biochem Cell Biol.* **2012**, *44*, 1097–1105. [CrossRef]

96. Goodall, J.C.; Wu, C.; Zhang, Y.; McNeill, L.; Ellis, L.; Saudek, V.; Gaston, J.S. Endoplasmic reticulum stress-induced transcription factor, CHOP, is crucial for dendritic cell IL-23 expression. *Proc. Natl. Acad. Sci. USA* **2010**, *107*, 17698–17703. [CrossRef]

97. Osorio, F.; Tavernier, S.J.; Hoffmann, E.; Saeys, Y.; Martens, L.; Vetters, J.; Delrue, I.; De Rycke, R.; Parthoens, E.; Pouliot, P.; et al. The unfolded-protein-response sensor IRE-1alpha regulates the function of CD8alpha+ dendritic cells. *Nat. Immunol.* **2014**, *15*, 248–257. [CrossRef]

98. Shurin, G.V.; Ma, Y.; Shurin, M.R. Immunosuppressive mechanisms of regulatory dendritic cells in cancer. *Cancer Microenviron.* **2013**, *6*, 159–167. [CrossRef]

99. Ma, Y.; Shurin, G.V.; Gutkin, D.W.; Shurin, M.R. Tumor associated regulatory dendritic cells. *Semin. Cancer Biol.* **2012**, *22*, 298–306. [CrossRef]

100. Shurin, M.R.; Naiditch, H.; Zhong, H.; Shurin, G.V. Regulatory dendritic cells: New targets for cancer immunotherapy. *Cancer Biol. Ther.* **2011**, *11*, 988–992. [CrossRef]

101. Enk, A.H.; Jonuleit, H.; Saloga, J.; Knop, J. Dendritic cells as mediators of tumor-induced tolerance in metastatic melanoma. *Int. J. Cancer* **1997**, *73*, 309–316. [CrossRef]

102. Liu, Q.; Zhang, C.; Sun, A.; Zheng, Y.; Wang, L.; Cao, X. Tumor-educated CD11bhighIalow regulatory dendritic cells suppress T cell response through arginase I. *J. Immunol.* **2009**, *182*, 6207–6216. [CrossRef] [PubMed]

103. Shapouri-Moghaddam, A.; Mohammadian, S.; Vazini, H.; Taghadosi, M.; Esmaeili, S.A.; Mardani, F.; Seifi, B.; Mohammadi, A.; Afshari, J.T.; Sahebkar, A. Macrophage plasticity, polarization, and function in health and disease. *J. Cell Physiol.* **2018**, *233*, 6425–6440. [CrossRef] [PubMed]

104. Dickhout, J.G.; Lhotak, S.; Hilditch, B.A.; Basseri, S.; Colgan, S.M.; Lynn, E.G.; Carlisle, R.E.; Zhou, J.; Sood, S.K.; Ingram, A.J.; et al. Induction of the unfolded protein response after monocyte to macrophage differentiation augments cell survival in early atherosclerotic lesions. *FASEB J.* **2011**, *25*, 576–589. [CrossRef] [PubMed]

105. Ye, R.; Jung, D.Y.; Jun, J.Y.; Li, J.; Luo, S.; Ko, H.J.; Kim, J.K.; Lee, A.S. Grp78 heterozygosity promotes adaptive unfolded protein response and attenuates diet-induced obesity and insulin resistance. *Diabetes* **2010**, *59*, 6–16. [CrossRef]

106. Tian, P.G.; Jiang, Z.X.; Li, J.H.; Zhou, Z.; Zhang, Q.H. Spliced XBP1 promotes macrophage survival and autophagy by interacting with Beclin-1. *Biochem. Biophys. Res. Commun.* **2015**, *463*, 518–523. [CrossRef]

107. Qiu, Q.; Zheng, Z.; Chang, L.; Zhao, Y.S.; Tan, C.; Dandekar, A.; Zhang, Z.; Lin, Z.; Gui, M.; Li, X.; et al. Toll-like receptor-mediated IRE1alpha activation as a therapeutic target for inflammatory arthritis. *EMBO J.* **2013**, *32*, 2477–2490. [CrossRef]

108. Guthrie, L.N.; Abiraman, K.; Plyler, E.S.; Sprenkle, N.T.; Gibson, S.A.; McFarland, B.C.; Rajbhandari, R.; Rowse, A.L.; Benveniste, E.N.; Meares, G.P. Attenuation of PKR-like ER Kinase (PERK) Signaling Selectively Controls Endoplasmic Reticulum Stress-induced Inflammation Without Compromising Immunological Responses. *J. Biol. Chem.* **2016**, *291*, 15830–15840. [CrossRef]

109. Srivastava, R.K.; Li, C.; Chaudhary, S.C.; Ballestas, M.E.; Elmets, C.A.; Robbins, D.J.; Matalon, S.; Deshane, J.S.; Afaq, F.; Bickers, D.R.; et al. Unfolded protein response (UPR) signaling regulates arsenic trioxide-mediated macrophage innate immune function disruption. *Toxicol. Appl. Pharmacol.* **2013**, *272*, 879–887. [CrossRef]

110. Ayaub, E.A.; Kolb, P.S.; Mohammed-Ali, Z.; Tat, V.; Murphy, J.; Bellaye, P.S.; Shimbori, C.; Boivin, F.J.; Lai, R.; Lynn, E.G.; et al. GRP78 and CHOP modulate macrophage apoptosis and the development of bleomycin-induced pulmonary fibrosis. *J. Pathol.* **2016**, *239*, 411–425. [CrossRef]

111. Isa, S.A.; Ruffino, J.S.; Ahluwalia, M.; Thomas, A.W.; Morris, K.; Webb, R. M2 macrophages exhibit higher sensitivity to oxLDL-induced lipotoxicity than other monocyte/macrophage subtypes. *Lipids Health Dis.* **2011**, *10*, 229. [CrossRef] [PubMed]

112. Shan, B.; Wang, X.; Wu, Y.; Xu, C.; Xia, Z.; Dai, J.; Shao, M.; Zhao, F.; He, S.; Yang, L.; et al. The metabolic ER stress sensor IRE1alpha suppresses alternative activation of macrophages and impairs energy expenditure in obesity. *Nat. Immunol.* **2017**, *18*, 519–529. [CrossRef] [PubMed]

113. Tabas, I.; Seimon, T.; Timmins, J.; Li, G.; Lim, W. Macrophage apoptosis in advanced atherosclerosis. *Ann. NY Acad. Sci.* **2009**, *1173*, E40–E45. [CrossRef] [PubMed]

114. Malhi, H.; Kropp, E.M.; Clavo, V.F.; Kobrossi, C.R.; Han, J.; Mauer, A.S.; Yong, J.; Kaufman, R.J. C/EBP homologous protein-induced macrophage apoptosis protects mice from steatohepatitis. *J. Biol. Chem.* **2013**, *288*, 18624–18642. [CrossRef] [PubMed]

115. Zhou, A.X.; Tabas, I. The UPR in atherosclerosis. *Semin. Immunopathol.* **2013**, *35*, 321–332. [CrossRef]

116. Zhou, J.; Lhotak, S.; Hilditch, B.A.; Austin, R.C. Activation of the unfolded protein response occurs at all stages of atherosclerotic lesion development in apolipoprotein E-deficient mice. *Circulation* **2005**, *111*, 1814–1821. [CrossRef]

117. Myoishi, M.; Hao, H.; Minamino, T.; Watanabe, K.; Nishihira, K.; Hatakeyama, K.; Asada, Y.; Okada, K.; Ishibashi-Ueda, H.; Gabbiani, G.; et al. Increased endoplasmic reticulum stress in atherosclerotic plaques associated with acute coronary syndrome. *Circulation* **2007**, *116*, 1226–1233. [CrossRef]

118. Yao, S.; Yang, N.; Song, G.; Sang, H.; Tian, H.; Miao, C.; Zhang, Y.; Qin, S. Minimally modified low-density lipoprotein induces macrophage endoplasmic reticulum stress via toll-like receptor 4. *Biochim. Biophys. Acta* **2012**, *1821*, 954–963. [CrossRef]

119. Yao, S.; Miao, C.; Tian, H.; Sang, H.; Yang, N.; Jiao, P.; Han, J.; Zong, C.; Qin, S. Endoplasmic reticulum stress promotes macrophage-derived foam cell formation by up-regulating cluster of differentiation 36 (CD36) expression. *J. Biol. Chem.* **2014**, *289*, 4032–4042. [CrossRef]

120. McAlpine, C.S.; Werstuck, G.H. Protein kinase R-like endoplasmic reticulum kinase and glycogen synthase kinase-3alpha/beta regulate foam cell formation. *J. Lipid Res.* **2014**, *55*, 2320–2333. [CrossRef]

121. Gass, J.N.; Gifford, N.M.; Brewer, J.W. Activation of an unfolded protein response during differentiation of antibody-secreting B cells. *J. Biol. Chem.* **2002**, *277*, 49047–49054. [CrossRef] [PubMed]

122. Brunsing, R.; Omori, S.A.; Weber, F.; Bicknell, A.; Friend, L.; Rickert, R.; Niwa, M. B- and T-cell development both involve activity of the unfolded protein response pathway. *J. Biol. Chem.* **2008**, *283*, 17954–17961. [CrossRef] [PubMed]

123. Kamimura, D.; Bevan, M.J. Endoplasmic Reticulum Stress Regulator XBP-1 Contributes to Effector CD8+ T Cell Differentiation during Acute Infection. *J. Immunol.* **2008**, *181*, 5433–5441. [CrossRef] [PubMed]

124. Iwakoshi, N.N.; Lee, A.H.; Glimcher, L.H. The X-box binding protein-1 transcription factor is required for plasma cell differentiation and the unfolded protein response. *Immunol. Rev.* **2003**, *194*, 29–38. [CrossRef]

125. Reimold, A.M.; Iwakoshi, N.N.; Manis, J.; Vallabhajosyula, P.; Szomolanyi-Tsuda, E.; Gravallese, E.M.; Friend, D.; Grusby, M.J.; Alt, F.; Glimcher, L.H. Plasma cell differentiation requires the transcription factor XBP-1. *Nature* **2001**, *412*, 300–307. [CrossRef]

126. Li, S.; Zhu, G.; Yang, Y.; Jian, Z.; Guo, S.; Dai, W.; Shi, Q.; Ge, R.; Ma, J.; Liu, L.; et al. Oxidative stress drives CD8(+) T-cell skin trafficking in patients with vitiligo through CXCL16 upregulation by activating the unfolded protein response in keratinocytes. *J. Allergy Clin. Immunol.* **2017**, *140*, 177–189. [CrossRef]

127. Catakovic, K.; Klieser, E.; Neureiter, D.; Geisberger, R. T cell exhaustion: From pathophysiological basics to tumor immunotherapy. *Cell Commun. Signal.* **2017**, *15*, 1. [CrossRef]

128. Durward-Diioia, M.; Harms, J.; Khan, M.; Hall, C.; Smith, J.A.; Splitter, G.A. CD8+ T cell exhaustion, suppressed gamma interferon production, and delayed memory response induced by chronic Brucella melitensis infection. *Infect. Immun.* **2015**, *83*, 4759–4771. [CrossRef]

129. Jiang, Y.; Li, Y.; Zhu, B. T-cell exhaustion in the tumor microenvironment. *Cell Death Dis.* **2015**, *6*, e1792. [CrossRef]

130. Pauken, K.E.; Wherry, E.J. Overcoming T cell exhaustion in infection and cancer. *Trends Immunol.* **2015**, *36*, 265–276. [CrossRef]

131. Wang, J.C.; Xu, Y.; Huang, Z.M.; Lu, X.J. T cell exhaustion in cancer: Mechanisms and clinical implications. *J. Cell Biochem.* **2018**, *119*, 4279–4286. [CrossRef] [PubMed]

132. Zarour, H.M. Reversing T-cell Dysfunction and Exhaustion in Cancer. *Clin. Cancer Res.* **2016**, *22*, 1856–1864. [CrossRef] [PubMed]

133. Lin, J.H.; Walter, P.; Yen, T.S. Endoplasmic reticulum stress in disease pathogenesis. *Annu. Rev. Pathol.* **2008**, *3*, 399–425. [CrossRef] [PubMed]

134. Im, S.J.; Hashimoto, M.; Gerner, M.Y.; Lee, J.; Kissick, H.T.; Burger, M.C.; Shan, Q.; Hale, J.S.; Lee, J.; Nasti, T.H.; et al. Defining CD8+ T cells that provide the proliferative burst after PD-1 therapy. *Nature* **2016**, *537*, 417–421. [CrossRef] [PubMed]

135. Shiga, K.; Hara, M.; Nagasaki, T.; Sato, T.; Takahashi, H.; Takeyama, H. Cancer-Associated Fibroblasts: Their Characteristics and Their Roles in Tumor Growth. *Cancers* **2015**, *7*, 2443–2458. [CrossRef]

136. Liu, T.; Han, C.; Wang, S.; Fang, P.; Ma, Z.; Xu, L.; Yin, R. Cancer-associated fibroblasts: An emerging target of anti-cancer immunotherapy. *J. Hematol. Oncol.* **2019**, *12*, 86. [CrossRef]

137. Peng, Y.; Li, Z.; Li, Z. GRP78 secreted by tumor cells stimulates differentiation of bone marrow mesenchymal stem cells to cancer-associated fibroblasts. *Biochem. Biophys. Res. Commun.* **2013**, *440*, 558–563. [CrossRef]

138. Yu, T.; Guo, Z.; Fan, H.; Song, J.; Liu, Y.; Gao, Z.; Wang, Q. Cancer-associated fibroblasts promote non-small cell lung cancer cell invasion by upregulation of glucose-regulated protein 78 (GRP78) expression in an integrated bionic microfluidic device. *Oncotarget* **2016**, *7*, 25593–25603. [CrossRef]

139. Holtrup, F.; Bauer, A.; Fellenberg, K.; Hilger, R.A.; Wink, M.; Hoheisel, J.D. Microarray analysis of nemorosone-induced cytotoxic effects on pancreatic cancer cells reveals activation of the unfolded protein response (UPR). *Br. J. Pharmacol.* **2011**, *162*, 1045–1059. [CrossRef]

140. Bruning, A.; Burger, P.; Vogel, M.; Rahmeh, M.; Gingelmaiers, A.; Friese, K.; Lenhard, M.; Burges, A. Nelfinavir induces the unfolded protein response in ovarian cancer cells, resulting in ER vacuolization, cell cycle retardation and apoptosis. *Cancer Biol. Ther.* **2009**, *8*, 226–232. [CrossRef]

141. Lev, A.; Lulla, A.R.; Wagner, J.; Ralff, M.D.; Kiehl, J.B.; Zhou, Y.; Benes, C.H.; Prabhu, V.V.; Oster, W.; Astsaturov, I.; et al. Anti-pancreatic cancer activity of ONC212 involves the unfolded protein response (UPR) and is reduced by IGF1-R and GRP78/BIP. *Oncotarget* **2017**, *8*, 81776–81793. [CrossRef] [PubMed]

142. Lee, D.H.; Jung Jung, Y.; Koh, D.; Lim, Y.; Lee, Y.H.; Shin, S.Y. A synthetic chalcone, 2′-hydroxy-2,3,5′-trimethoxychalcone triggers unfolded protein response-mediated apoptosis in breast cancer cells. *Cancer Lett.* **2016**, *372*, 1–9. [CrossRef] [PubMed]

143. Maurel, M.; McGrath, E.P.; Mnich, K.; Healy, S.; Chevet, E.; Samali, A. Controlling the unfolded protein response-mediated life and death decisions in cancer. *Semin. Cancer Biol.* **2015**, *33*, 57–66. [CrossRef] [PubMed]

144. Ojha, R.; Amaravadi, R.K. Targeting the unfolded protein response in cancer. *Pharmacol. Res.* **2017**, *120*, 258–266. [CrossRef] [PubMed]

145. Nagelkerke, A.; Bussink, J.; Sweep, F.C.; Span, P.N. The unfolded protein response as a target for cancer therapy. *Biochim. Biophys. Acta* **2014**, *1846*, 277–284. [CrossRef] [PubMed]

146. Matsumura, K.; Sakai, C.; Kawakami, S.; Yamashita, F.; Hashida, M. Inhibition of cancer cell growth by GRP78 siRNA lipoplex via activation of unfolded protein response. *Biol. Pharm. Bull.* **2014**, *37*, 648–653. [CrossRef]

147. Lee, D.H.; Kim, C.G.; Lim, Y.; Shin, S.Y. Aurora kinase A inhibitor TCS7010 demonstrates pro-apoptotic effect through the unfolded protein response pathway in HCT116 colon cancer cells. *Oncol. Lett.* **2017**, *14*, 6571–6577. [CrossRef]

148. Simard, J.C.; Durocher, I.; Girard, D. Silver nanoparticles induce irremediable endoplasmic reticulum stress leading to unfolded protein response dependent apoptosis in breast cancer cells. *Apoptosis* **2016**, *21*, 1279–1290. [CrossRef]

149. Wang, M.; Shim, J.S.; Li, R.J.; Dang, Y.; He, Q.; Das, M.; Liu, J.O. Identification of an old antibiotic clofoctol as a novel activator of unfolded protein response pathways and an inhibitor of prostate cancer. *Br. J. Pharmacol.* **2014**, *171*, 4478–4489. [CrossRef]

150. Fribley, A.M.; Miller, J.R.; Brownell, A.L.; Garshott, D.M.; Zeng, Q.; Reist, T.E.; Narula, N.; Cai, P.; Xi, Y.; Callaghan, M.U.; et al. Celastrol induces unfolded protein response-dependent cell death in head and neck cancer. *Exp. Cell Res.* **2015**, *330*, 412–422. [CrossRef]

151. Chien, W.; Ding, L.W.; Sun, Q.Y.; Torres-Fernandez, L.A.; Tan, S.Z.; Xiao, J.; Lim, S.L.; Garg, M.; Lee, K.L.; Kitajima, S.; et al. Selective inhibition of unfolded protein response induces apoptosis in pancreatic cancer cells. *Oncotarget* **2014**, *5*, 4881–4894. [CrossRef] [PubMed]

152. Maddalo, D.; Neeb, A.; Jehle, K.; Schmitz, K.; Muhle-Goll, C.; Shatkina, L.; Walther, T.V.; Bruchmann, A.; Gopal, S.M.; Wenzel, W.; et al. A peptidic unconjugated GRP78/BiP ligand modulates the unfolded protein response and induces prostate cancer cell death. *PLoS ONE* **2012**, *7*, e45690. [CrossRef] [PubMed]

153. Shin, S.Y.; Lee, J.M.; Lee, M.S.; Koh, D.; Jung, H.; Lim, Y.; Lee, Y.H. Targeting cancer cells via the reactive oxygen species-mediated unfolded protein response with a novel synthetic polyphenol conjugate. *Clin. Cancer Res.* **2014**, *20*, 4302–4313. [CrossRef] [PubMed]

154. Burton, L.J.; Rivera, M.; Hawsawi, O.; Zou, J.; Hudson, T.; Wang, G.; Zhang, Q.; Cubano, L.; Boukli, N.; Odero-Marah, V. Muscadine Grape Skin Extract Induces an Unfolded Protein Response-Mediated Autophagy in Prostate Cancer Cells: A TMT-Based Quantitative Proteomic Analysis. *PLoS ONE* **2016**, *11*, e0164115. [CrossRef] [PubMed]

155. Sidhu, A.; Miller, J.R.; Tripathi, A.; Garshott, D.M.; Brownell, A.L.; Chiego, D.J.; Arevang, C.; Zeng, Q.; Jackson, L.C.; Bechler, S.A.; et al. Borrelidin Induces the Unfolded Protein Response in Oral Cancer Cells and Chop-Dependent Apoptosis. *ACS Med. Chem. Lett.* **2015**, *6*, 1122–1127. [CrossRef] [PubMed]

156. Huang, H.; Liu, H.; Liu, C.; Fan, L.; Zhang, X.; Gao, A.; Hu, X.; Zhang, K.; Cao, X.; Jiang, K.; et al. Disruption of the unfolded protein response (UPR) by lead compound selectively suppresses cancer cell growth. *Cancer Lett.* **2015**, *360*, 257–268. [CrossRef] [PubMed]

157. Mokarram, P.; Albokashy, M.; Zarghooni, M.; Moosavi, M.A.; Sepehri, Z.; Chen, Q.M.; Hudecki, A.; Sargazi, A.; Alizadeh, J.; Moghadam, A.R.; et al. New frontiers in the treatment of colorectal cancer: Autophagy and the unfolded protein response as promising targets. *Autophagy* **2017**, *13*, 781–819. [CrossRef]

158. Liang, G.; Fang, X.; Yang, Y.; Song, Y. Knockdown of CEMIP suppresses proliferation and induces apoptosis in colorectal cancer cells: Downregulation of GRP78 and attenuation of unfolded protein response. *Biochem. Cell Biol.* **2018**, *96*, 332–341. [CrossRef]

159. Zhao, X.; Yang, Y.; Yao, F.; Xiao, B.; Cheng, Y.; Feng, C.; Duan, C.; Zhang, C.; Liu, Y.; Li, H.; et al. Unfolded Protein Response Promotes Doxorubicin-Induced Nonsmall Cell Lung Cancer Cells Apoptosis via the mTOR Pathway Inhibition. *Cancer Biother. Radiopharm.* **2016**, *31*, 347–351. [CrossRef]

160. Zhang, L.; Hapon, M.B.; Goyeneche, A.A.; Srinivasan, R.; Gamarra-Luques, C.D.; Callegari, E.A.; Drappeau, D.D.; Terpstra, E.J.; Pan, B.; Knapp, J.R.; et al. Mifepristone increases mRNA translation rate, triggers the unfolded protein response, increases autophagic flux, and kills ovarian cancer cells in combination with proteasome or lysosome inhibitors. *Mol. Oncol.* **2016**, *10*, 1099–1117. [CrossRef]

161. Prasad, V.; Suomalainen, M.; Pennauer, M.; Yakimovich, A.; Andriasyan, V.; Hemmi, S.; Greber, U.F. Chemical induction of unfolded protein response enhances cancer cell killing through lytic virus infection. *J. Virol.* **2014**, *88*, 13086–13098. [CrossRef] [PubMed]

162. Zheng, G.F.; Cai, Z.; Meng, X.K.; Zhang, Y.; Zhu, W.; Pang, X.Y.; Dou, L. Unfolded protein response mediated JNK/AP-1 signal transduction, a target for ovarian cancer treatment. *Int. J. Clin. Exp. Pathol.* **2015**, *8*, 6505–6511.

163. Schroder, M.; Kaufman, R.J. The mammalian unfolded protein response. *Annu. Rev. Biochem.* **2005**, *74*, 739–789. [CrossRef]

164. Todd, D.J.; Lee, A.H.; Glimcher, L.H. The endoplasmic reticulum stress response in immunity and autoimmunity. *Nat. Rev. Immunol.* **2008**, *8*, 663–674. [CrossRef] [PubMed]

165. Davies, M.P.; Barraclough, D.L.; Stewart, C.; Joyce, K.A.; Eccles, R.M.; Barraclough, R.; Rudland, P.S.; Sibson, D.R. Expression and splicing of the unfolded protein response gene XBP-1 are significantly associated with clinical outcome of endocrine-treated breast cancer. *Int. J. Cancer* **2008**, *123*, 85–88. [CrossRef]

166. Galmiche, A.; Sauzay, C.; Chevet, E.; Pluquet, O. Role of the unfolded protein response in tumor cell characteristics and cancer outcome. *Curr. Opin. Oncol.* **2017**, *29*, 41–47. [CrossRef] [PubMed]

167. Harrington, P.E.; Biswas, K.; Malwitz, D.; Tasker, A.S.; Mohr, C.; Andrews, K.L.; Dellamaggiore, K.; Kendall, R.; Beckmann, H.; Jaeckel, P.; et al. Unfolded Protein Response in Cancer: IRE1alpha Inhibition by Selective Kinase Ligands Does Not Impair Tumor Cell Viability. *ACS Med. Chem. Lett.* **2015**, *6*, 68–72. [CrossRef]

168. Manalo, R.V.M. Anastasis and the ER stress response: Solving the paradox of the unfolded protein response in cancer. *Med. Hypotheses* **2017**, *109*, 25–27. [CrossRef]

169. Tameire, F.; Verginadis, I.I.; Koumenis, C. Cell intrinsic and extrinsic activators of the unfolded protein response in cancer: Mechanisms and targets for therapy. *Semin. Cancer Biol.* **2015**, *33*, 3–15. [CrossRef]

170. Fan, L.X.; Liu, C.M.; Gao, A.H.; Zhou, Y.B.; Li, J. Berberine combined with 2-deoxy-d-glucose synergistically enhances cancer cell proliferation inhibition via energy depletion and unfolded protein response disruption. *Biochim. Biophys. Acta* **2013**, *1830*, 5175–5183. [CrossRef]

171. Ruan, Q.; Han, S.; Jiang, W.G.; Boulton, M.E.; Chen, Z.J.; Law, B.K.; Cai, J. AlphaB-crystallin, an effector of unfolded protein response, confers anti-VEGF resistance to breast cancer via maintenance of intracrine VEGF in endothelial cells. *Mol. Cancer Res.* **2011**, *9*, 1632–1643. [CrossRef] [PubMed]

172. Saha, S.; Bhanja, P.; Partanen, A.; Zhang, W.; Liu, L.; Tome, W.; Guha, C. Low intensity focused ultrasound (LOFU) modulates unfolded protein response and sensitizes prostate cancer to 17AAG. *Oncoscience* **2014**, *1*, 434–445. [CrossRef] [PubMed]

173. Hilakivi-Clarke, L.; Warri, A.; Bouker, K.B.; Zhang, X.; Cook, K.L.; Jin, L.; Zwart, A.; Nguyen, N.; Hu, R.; Cruz, M.I.; et al. Effects of In Utero Exposure to Ethinyl Estradiol on Tamoxifen Resistance and Breast Cancer Recurrence in a Preclinical Model. *J. Natl. Cancer Inst.* **2017**, *109*. [CrossRef]

174. Klionsky, D.J.; Abdelmohsen, K.; Abe, A.; Abedin, M.J.; Abeliovich, H.; Acevedo Arozena, A.; Adachi, H.; Adams, C.M.; Adams, P.D.; Adeli, K.; et al. Guidelines for the use and interpretation of assays for monitoring autophagy (3rd edition). *Autophagy* **2016**, *12*, 1–222. [CrossRef] [PubMed]

175. Luo, B.; Lee, A.S. The critical roles of endoplasmic reticulum chaperones and unfolded protein response in tumorigenesis and anticancer therapies. *Oncogene* **2013**, *32*, 805–818. [CrossRef] [PubMed]
176. Rodvold, J.J.; Mahadevan, N.R.; Zanetti, M. Immune modulation by ER stress and inflammation in the tumor microenvironment. *Cancer Lett.* **2016**, *380*, 227–236. [CrossRef]

International Journal of *Molecular Sciences*

Review

The Evolving Role of CD8$^+$CD28$^-$ Immunosenescent T Cells in Cancer Immunology

Wei X. Huff [1], Jae Hyun Kwon [1], Mario Henriquez [1], Kaleigh Fetcko [2] and Mahua Dey [1,*

[1]   Department of Neurosurgery, Indiana University School of Medicine, Indianapolis, IN 46202, USA; wxia@iupui.edu (W.X.H.); kwonjaeh@iu.edu (J.H.K.); mhenriqu@iu.edu (M.H.)

[2]   Department of Neurology, University of Illinois at Chicago School of Medicine, Chicago, IL 60612, USA; kaleighfetcko@gmail.com

*   Correspondence: mdey@iu.edu; Tel.: +1-317-274-2601

Received: 1 May 2019; Accepted: 6 June 2019; Published: 8 June 2019

Abstract: Functional, tumor-specific CD8$^+$ cytotoxic T lymphocytes drive the adaptive immune response to cancer. Thus, induction of their activity is the ultimate aim of all immunotherapies. Success of anti-tumor immunotherapy is precluded by marked immunosuppression in the tumor microenvironment (TME) leading to CD8$^+$ effector T cell dysfunction. Among the many facets of CD8$^+$ T cell dysfunction that have been recognized—tolerance, anergy, exhaustion, and senescence—CD8$^+$ T cell senescence is incompletely understood. Naïve CD8$^+$ T cells require three essential signals for activation, differentiation, and survival through T-cell receptor, costimulatory receptors, and cytokine receptors. Downregulation of costimulatory molecule CD28 is a hallmark of senescent T cells and increased CD8$^+$CD28$^-$ senescent populations with heterogeneous roles have been observed in multiple solid and hematogenous tumors. T cell senescence can be induced by several factors including aging, telomere damage, tumor-associated stress, and regulatory T (Treg) cells. Tumor-induced T cell senescence is yet another mechanism that enables tumor cell resistance to immunotherapy. In this paper, we provide a comprehensive overview of CD8$^+$CD28$^-$ senescent T cell population, their origin, their function in immunology and pathologic conditions, including TME and their implication for immunotherapy. Further characterization and investigation into this subset of CD8$^+$ T cells could improve the efficacy of future anti-tumor immunotherapy.

Keywords: CD8$^+$CD28$^-$ T cells; cancer immunology; glioblastoma; immunotherapy; malignant glioma; cancer

---

1. Introduction

The conflict between cancer and the immune system has long been established [1]. Immunotherapies are being investigated to augment the anti-tumor effects of the immune system and promote long-term cancer control [2]. CD8$^+$ cytotoxic T cells (CTLs) are the main players driving the adaptive immune response against cancer and execute tumor-specific immune responses, rendering them the primary endpoint to most immunotherapies [3,4]. Establishment of effective antigen-specific CD8$^+$ T cells enabled preliminary clinical success of cancer vaccines, oncolytic viruses, adoptive cellular therapy, and checkpoint inhibitors in several cancers including melanoma, lung cancer, renal cell cancer, Hodgkin's lymphoma, etc. [5–7]. Unfortunately, despite their promise, the efficacy of these treatments varies depending on the type and location of the tumor, and has been ineffective in poorly immunogenic cancers such as glioblastoma (GBM) [7–13].

A variety of T cell deficiencies have been identified in immunosuppressive tumors that contribute to the ultimate ineffectiveness of CD8$^+$ CTL-mediated tumor killing [14]. Immune tolerance, anergy, and exhaustion of CD8$^+$ T cells have been studied extensively in the past [14–17]. While the concept of immune senescence, defined by terminally differentiated cells in cell cycle arrest after extensive replication or in response to cellular damage or stress, has been well established with aging and chronic infections [18–21], our knowledge of its role in cancer is still in its early stages. CD28 is an indispensable costimulatory molecule needed for the activation of T cells and its role is critical to the proper activation of CD8[+] CTLs [22]. Current evidence shows that the downregulation of CD28 is a hallmark of senescent CD8[+] T cells and CD28[−] senescent T cells display immunosuppressive functions in cancer [23–26].

In this review, we will focus on the recent advances in our understanding of CD8[+]CD28[−] T cells. We first will discuss cellular senescence and the evolution of the CD28[−] T cells. Next, we will review the significance of CD8[+]CD28[−] T cells in multiple disease processes, including transplant, autoimmune disease, chronic viral infection, and cancer, including CNS tumors. Finally, we will discuss the functional implications of CD28[−] T cells in onco-immunology and the important areas of future investigation on novel immunotherapeutic strategies.

2. Role of CD8[+] T Cells in Cancer Immunology

CD8[+] T cells are a subset of lymphocytes committed to detecting peptide antigens presented by major histocompatibility complex (MHC) class I molecules (Figure 1) [27,28]. CD8[+] T cells arise from common lymphoid progenitors that migrate from the bone marrow to the thymus where they pass through a series of distinct phases of maturation [29,30]. The naïve CD8[+] T cell pool is comprised of polyclonal T cells that express CD28, CCR7, and CD62L, the latter two allowing them to recirculate between blood and secondary lymphoid organs [31,32]. Initial priming of CD8[+] T cells involves T cell receptor (TCR) recognition of peptide/MHC complexes presented by professional antigen presenting cells (APCs), such as dendritic cells (DCs). DCs also express surface markers CD70 and CD80/CD86 for binding to CD27 and CD28 receptors expressed on CD8[+]T cells. This provides a critical secondary signal for T cell activation. Host cells, including cancer cells, can serve as targets for previously activated CD8[+] T cells by processing and presenting antigenic tumor peptides by MHC class I.

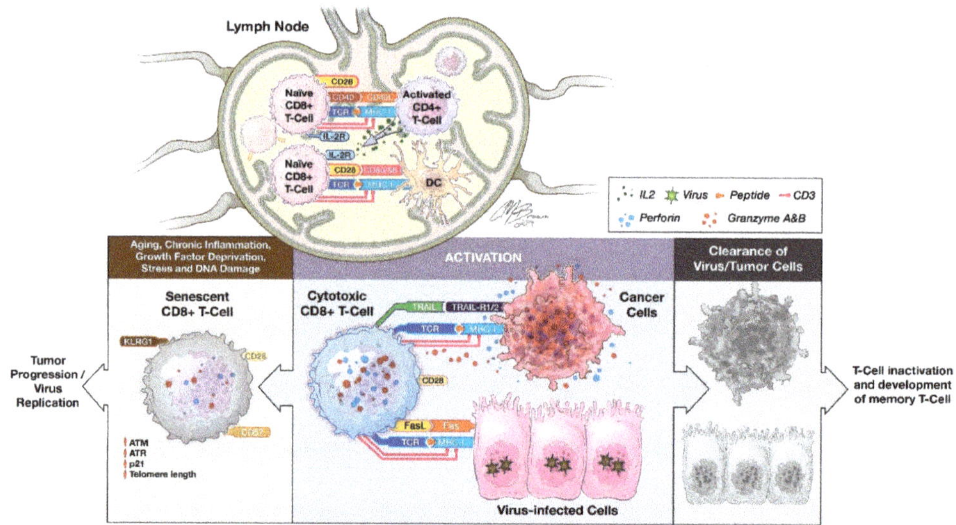

Figure 1. The priming and inactivation of CD8$^+$ T cells. The interaction between TCRs and the peptide-MHC complex is the first step toward antigen-induced CD8$^+$ T cell activation. This creates a site of extensive contact between T cells and APC, also called immunological synapses, where binding of CD28 on T cells with CD80/CD86 on APCs transduces a pivotal secondary costimulatory signal to complete the priming of naïve CD8$^+$ T cells. In addition, CD4$^+$ T helper (Th) cells when activated by DCs acquire not only the synapse-composed MHC class II and costimulatory molecules (CD54 and CD80), but also the bystander peptide-MHC-I complex from DC and become CD4$^+$ Th-APCs, resulting in direct CD4$^+$ T–CD8$^+$ T cell interactions and subsequently delivery of CD40L signaling to CD40-expressing CD8$^+$ T cells [21]. Furthermore, CD4 Th cells also secrete cytokines, such as IL-2, which promotes the differentiation of naïve CD8$^+$ T cells into effector CTLs and memory CD8$^+$ T cells. CTLs destroy antigen-specific target cells (such as cancer cells or viral infected host cells) via pathways including granule exocytosis, Fas ligand, and TRAIL-mediated apoptosis leading to tumor control or virus clearance and subsequent physiological T-cell inactivation as well as memory CD8$^+$ T-cell formation [33]. Whereas pathological T-cell inactivation or conversion of CTL to CD8$^+$ senescent T cells leads to tumor progression and virus replication.

In addition, CD4$^+$ T helper cells interact with CD8$^+$ T cells and modulate CD8$^+$ T cell activation [34–36]. Activated CD4$^+$ T helper cells can secrete a variety of cytokines, such as interferon-gamma (IFN-γ) and IL-2, and facilitate CD8$^+$ T cells' optimal proliferation and activation [28,37]. CD4$^+$ T cells could also help with DC maturation for expression of costimulatory molecules and secretion of cytokines that contribute to CD8$^+$ T cell priming [38]. A similar mechanism is also carried out by natural killer (NK) cells, showing that there is collaboration between CD4$^+$ T cells with NK and DCs for induction of CD8$^+$ T cell priming [28,38].

Upon activation, effector CD8$^+$ CTLs destroy antigen-expressing cancer cells primarily utilizing two main pathways: granule exocytosis (such as perforin and granzyme) and death ligand/receptor-mediated apoptosis (such as Fas ligand and TRAIL) [39]. Additionally, activated CD8$^+$ T cells release IFN-γ and tumor necrosis factor alpha (TNF-α) to induce cytotoxicity in the target cells and stimulate M1 macrophage-mediated anti-tumor response [28]. In multiple solid tumors, tumor-infiltrating CD8$^+$ CTLs can be used as a prognostic factor [40–48]. For example, in breast cancer, significantly increased CD8$^+$ T cells at tumor sites have been shown to have an inverse correlation with advanced tumor stages and a positive correlation with clinical outcomes [41,49,50]. Similar findings of a favorable prognosis associated with the accumulation of tumor-infiltrating CD8$^+$ T cells were reported in colorectal, oral squamous cell, pancreatic, and ovarian carcinomas [43–45,47,48,51].

3. CD28 Costimulatory Receptor

The CD28 costimulatory receptor, a 44-kDa membrane glycoprotein, is expressed on nearly all human T lymphocytes at birth [52]. Binding of the CD28 receptor on T cells provides an essential second signal alongside TCR ligation for naïve T cell activation. CD28 signaling has diverse effects on T cell function, including orchestrating membrane raft trapping at the immunological synapse, transcriptional changes, downstream post-translational modifications, and actin cytoskeletal remodeling [52–54]. This leads to intracellular biochemical events such as survival and proliferation signals, induction of IL-2, activation of telomerase, stabilization of mRNA for several cytokines, increased glucose metabolism, and enhanced T cell migration and homing [52,55,56].

CD28 family of costimulatory molecules also includes ICOS, CTLA-4, PD-1, PD1H, TIGIT, and BTLA [41]. This family of receptors and ligands has considerable complexity in both binding pattern and biological effects. For instance, CD28 (activating) and CTLA-4 (inhibitory) are highly homologous and compete for the same ligands (CD80 and CD86) and regulate immune response by providing opposing effects [51,52].

The critical role of CD28 in induction of immune response was demonstrated in mice treated with CD28 antagonist, which induced antigen specific tolerance and prevented the progression of autoimmune diseases and organ graft rejection [57]. This has led to the development of abatacept

(Orencia® Bristol-Myers Squibb, New York, NY, USA) [58] and belatacept (Nulojix® Bristol-Myers Squibb New York, NY, USA) [59], a modified antibody composed of Fc region of IgG1 fused to the extracellular domain of CTLA-4, which bind to CD80/86 on APCs and block the costimulatory signaling by CD28. Abatacept and belatacept are used clinically to treat rheumatoid arthritis and organ transplant rejection, respectively [56,60].

On the other hand, the use CD28 agonists to awaken T cells from the tolerant state could lead to new therapies to re-activate the immune system for the treatment of infectious disease [61] and cancer [62]. Although, in a phase I trial of systemic administration of CD28 superagonist monoclonal antibodies (mAb) (TGN1412), uncontrolled CD28 signaling led to a potent induction of downstream immune activation independent of TCR-CD3 complex resulting in catastrophic systemic inflammatory syndromes in six volunteer subjects [63]. Investigation of these unexpected serious adverse events have led to better design of clinical trials and appreciation of differences in CD28 expression and regulation between species (critical for transitioning preclinical testing to clinical investigations) [64,65]. An updated CD28 superagonist TAB08 is under clinical testing for rheumatoid arthritis [64]. In addition, localized or targeted use of anti-CD28 mAbs has much potential such as incorporating the intracellular costimulatory domain of CD28 into CAR (chimeric antigen receptor) T cells for adoptive transfer immunotherapy and the use of CD28 agonist aptamer with tumor vaccine [66–68].

Importantly, the use of these therapeutics are in clinical trials for a variety of disease states including solid neoplasms and inflammatory diseases (Table 1). Although previous experience with CD28 agonist mAbs has been disappointing, headway is being made in their use in solid tumors and rheumatoid arthritis. Perhaps out of an abundance of caution, current clinical trials for the use of CD28 agonists are testing their safety, efficacy, and tolerability in patients and are undergoing dose escalation studies. Fortunately, significant progress has been made into CAR-T cell therapy incorporating costimulatory domains and has led to the FDA-approval of the CAR-T cell therapy tisagenlecleucel (KYMRIAH® Novartis Pharmaceuticals, Basel, Switzerland) for relapse or refractory acute lymphoblastic leukemia patients [69,70].

Table 1. Clinical trials related to the therapeutic use of CD28 manipulation, such as CAR-T cell therapy and monoclonal antibodies.

Malignancy	Phase	N	Trial Name	Clinical Trial Identifier	Therapeutics	References
Relapsed or Refractory Acute Lymphoblastic Leukemia	1	5	Chimeric Antigen Receptor (CAR)-Modified T Cell Therapy in Treating Patients with Acute Lymphoblastic Leukemia	NCT02186860	Third Gen CAR-T cells containing CD28+CD137	[71]
Glioblastoma	1	17	CMV-Specific Cytotoxic T Lymphocytes Expressing CAR Targeting HER2 in Patients with GBM (HERT-GBM)	NCT01109095	Second Gen CMV-selected CAR-T cells against HER2 containing CD28.zeta signaling domain	[72]
Rheumatoid Arthritis	1/2	18	Safety, Tolerability, Pharmacodynamics and Efficacy Study of TAB08 in Patients with Rheumatoid Arthritis	NCT01990157	TAB08	
Solid Neoplasms	1	38	Dose Escalation Study of TAB08 in Patients with Advanced Solid Neoplasms (TAB08)	NCT03006029	TAB08	[73]
Systemic Lupus Erythematosus	2	730	Safety and Efficacy Study of a Biologic to Treat Systemic Lupus Erythematosus	NCT02265744	Lulizumab pegol (monoclonal antibody against CD28)	

Furthermore, recent progress in the manipulation of other costimulatory molecule such as ICOS, CD137 (4-1BB), OX40, and GITR has also demonstrated tremendous therapeutic potential [74].

Activation of ICOS, CD137, and OX40 via mAbs and aptamers improved T cell proliferation, function, and overall antitumor response [74–77]. Targeting of GITR exhibited effects on both effector and regulatory T cells. Ligation of constitutively expressed GITR on regulatory T cells caused depletion in their number, loss lineage stability and immunosuppressive function [78,79], while GITR agonists work synergistically with PD-1 blockage to promote CD4 and CD8 accumulation in murine ovarian cancer [80]. Blockade of inhibitory receptor CTLA-4 have been shown to be effective in enhancing CD28 signaling and augmenting ICOS stimulation [74,81]. Combination of checkpoint inhibitors and costimulatory agonists presents an exciting avenue of cancer treatment and several clinical studies are currently investigating their benefit [74].

4. Cellular Senescence in the Immune System

Cellular senescence is a state of cell cycle arrest in respond to cellular damage or stress to prevent neoplastic transformation [82]. Cellular senescence have been identified in areas of physiological homeostasis, such as development [82], wound healing [83], and placental natural killer lymphocytes [84]. However, cellular senescence also contributes to the loss of function associated with aging and age-related disease as well as chronic viral infection, neurodegenerative disease, and cancer [18,85,86]. Two categories of cellular senescence have been described in literature: the first is aging associated, telomere-dependent replicative senescence and the second is stress-induced premature senescence, also known as telomere-independent senescence [82,87]. Oncogene-induced senescence is one of the well-described mechanisms for premature senescence [87,88]. Regardless of the initiating mechanism, cells that undergo senescence survive by exhibiting a variety of phenotypical and molecular features (Figure 2), including morphological changes, cell division blockage, change of sensitivity against apoptosis, metabolic dysfunction, and a specialized secretory activity termed senescence associated secretory phenotype (SASP) [20]. Additional characteristics include nuclear p16 and p21 expression [89–91], DNA damage [92], senescence associated heterochromatin foci (SAHF) [93], and increased lysosomal senescence-associated β-galactosidase (SA-β-gal) activity [91]. Recently, lipofuscin accumulation was also established as a hallmark of senescent cells [94].

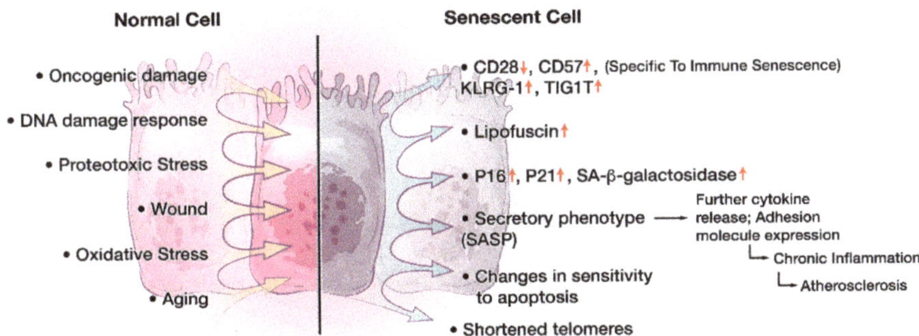

Figure 2. Phenotypical and molecular changes in cellular senescence. A variety of intracellular (DNA damage, oncogenes, etc.) and/or extracellular signals (oxidative stress, chronic inflammation, etc.) can induce cellular senescence. Senescent cells exhibit numerous characteristics including but not limited to cell cycle arrest, increased nuclear p16 and p21 expression, increased lysosomal SA-β-gal activity, shortened telomere, and increased lipofuscin. Senescent cells can also present as a specialized secretory phenotype termed senescence associated secretory phenotype (SASP). Of particular interest, senescent immune cells present with lowered expression of CD28 and CD27 but heightened expression of CD57, KLRG-1, TIGIT, and other NK-cell associated surface receptors.

Cellular senescence also occurs in the human immune system [18,95]. The effectiveness of the immune response declines with age particularly in the elderly population [96,97]. Immune deficiencies start to appear in DCs, NK cells, and monocytes/macrophages with aging, and it was proposed that myeloid-derived suppressor cells (MDSC) could also induce senescence in T and B cells compartment in diverse inflammatory conditions [96,97]. Importantly, lymphocytes, especially T cells, show the most considerable changes with aging [95,98]. Among the various complex features that contribute to the aging-associated changes in T cell immunity, the accumulation of CD28$^-$ T cells is one of the hallmark phenomena in T cell immunosenescence [26,99,100]. TME can also induce senescence in tumor-infiltrating T cells [14].

5. Role of CD8$^+$CD28$^-$ T Cells

Although CD28 is expressed on the majority of the CD8$^+$ T cells at birth [52], normal aging process and activation of CD8$^+$ T cells invariably leads to CD28 downregulation [101]. CD8$^+$CD28$^-$ cells represent a distinct population distinguishable from the general population of CD8$^+$CD28$^+$ T cells [99], which are known for their crucial role in the clearance of cancer and intracellularly infected cells, in terms of their phenotype and function [102]. *In vitro* studies showed purified CD28$^+$ T cells progressively lose CD28 during each successful stimulation, with the CD8$^+$ T cells losing their CD28 more rapidly than the CD4$^+$ T cells [26,103,104]. The differential rate of CD28 loss is associated with the rapid inactivation of telomerase and CD8$^+$ T cells reach replicative senescence faster than CD4$^+$ T cells, at which stage T cells are no longer able to enter mitosis but still remain viable [105]. Thus, these CD8$^+$CD28$^-$ T cells are defined as senescent T cells. Less than 50% of the CD8$^+$ T cell compartment of elderly or chronically infected individuals are CD28$^+$ while up to 80% of CD4$^+$ T cells maintain their CD28 expression even in the centenarians [26,103]. Interestingly, a large proportion of CD8$^+$CD28$^-$ T cells of elderly persons also have lower levels of CD8 expression [106,107]. Although the significance of this observation is unknown, downregulation of the expression of CD8 and CD4 molecules is characteristic for activated T cells, suggesting that those CD8^{low}CD28$^-$ T cells subset represent senescent lymphocytes that are chronically activated from either common persistent antigens (in the setting of aging) or persistent infection or inflammation (in the setting of cancer) [25,108].

6. Characteristics of CD8$^+$CD28$^-$ Senescent T cells

CD8$^+$CD28$^-$ T cells are highly oligoclonal and terminally differentiated effector lymphocytes that have lost their capacity to undergo cell division [23,108]. They are functionally heterogeneous and their characteristics vary depending on the context where they are found (Figure 3) [23,108]. They also express a variety of other NK cell-related receptors including KIR, NKG2D, CD56, CD57, CD94, and Fc-γ receptor IIIa and have features crossing the border between innate and adaptive immunity [109,110]. Alterations in the costimulatory receptor NKG2D signaling and expression levels in CD8$^+$ T cells can lead to autoimmune conditions that are either TCR dependent or TCR-independent [111–113]. Gained expression of CD57, also known as HNK-1 (human natural killer-1), is a common feature associated with circulating senescent T cells, and increased CD8$^+$CD28$^-$CD57$^+$ senescent T cells were identified in multiple pathological conditions, including HIV infection, multiple myeloma, lung cancer, and chronic inflammation conditions such as diabetes and obesity [99,114,115]. Although expression of CD57 is linked to antigen-induced apoptosis of CD8$^+$ T cells [116], the acquisition of CD94 has been reported to confer resistance to apoptosis in CD8$^+$CD28$^-$ T cells. [117] Similarly, CD8$^+$CD28$^-$ T cells are often associated with the lack of perforin, rendering them ineffective Ag-specific killers in chronic viral infections [21,118–120]. On the other hand, in certain disease processes such as chronic obstructive pulmonary disease (COPD) and rheumatoid arthritis, they have been reported to express increased levels of cytotoxic mediators, perforin and granzyme B, and pro-inflammatory cytokines, IFN-γ and TNFα, where CD8$^+$CD28$^-$ T cells can cause significant damages to normal surrounding tissue in an antigen-nonspecific manner [121].

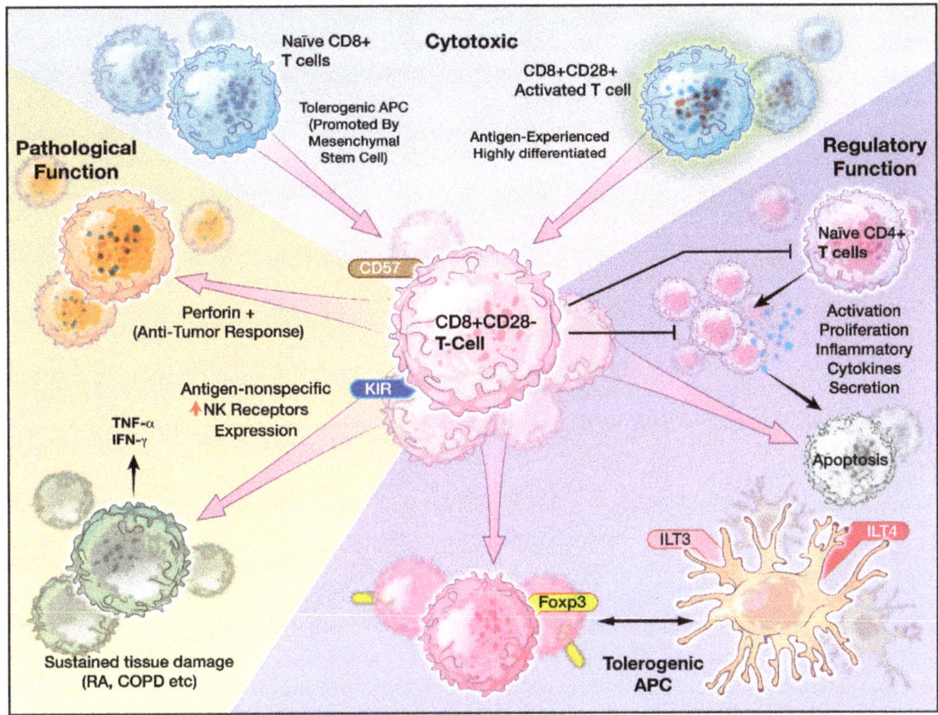

Figure 3. The heterogeneous functions of CD8$^+$CD28$^-$ T cells. CD8$^+$CD28$^-$ T cells originate from activated CD8$^+$CD28$^+$ T cells or from interaction with tolerogenic APCs. CD8$^+$CD28$^-$ T cells exhibit both cytotoxic and immunoregulatory phenotypes and vary in pathological states such as across different cancer types or inflammatory/autoimmune conditions.

CD8$^+$CD28$^-$ T cells are also shown to be immunosuppressive and function as regulatory T cells [122–125]. For example, CD8$^+$CD28$^-$ T cells directly inhibit Ag-presenting function of DCs by inducing inhibitory receptors, such as immunoglobulin like transcript 3 (ILT3) and ILT4, which leads DCs to be immune tolerant than immunogenic [122,126]. Such tolerogenic DCs anergize alloreactive CD4$^+$CD25$^+$ T cells and convert them into regulatory T cells, which in turn, continue the immunosuppressive cascade by tolerizing other DCs and amplify T cell immunosenescence [126,127]. *In vivo*, CD8$^+$CD28$^-$ T cells have been directly correlated with the suppression of antigen-specific T cell responses in patients with plasma cell dyscrasia [123]. Therefore, their characteristics and functions in immunity range from reduced antigen-specific killing to enhanced cytotoxic abilities and from crossing innate immunity function to promoting immune regulation.

7. CD8$^+$CD28$^-$ T cells in Pathologic Conditions

CD8$^+$CD28$^-$ T cells play a significant role in pathological conditions [18,99,100,121]. High populations of these cells have been associated with chronic viral infections including human immunodeficiency virus (HIV), hepatitis C virus, cytomegalovirus (CMV), and human parvovirus B19 [99]. Shortened telomeres, reduced IL-2 production, and increased IL-6 production were observed in these cells [19]. The loss of CD28 also serves as a prognostic indicator for viral infection. For instance, increased frequency of CD8$^+$CD28$^-$ T cells in the early stage of HIV infection correlates with faster progression to AIDS [99]. Additionally, higher levels of CD8$^+$CD28$^-$ T cells are associated with subclinical carotid artery disease in HIV-infected women [99]. Older CMV seropositive individuals, who had higher number of CD8$^+$CD28$^-$ cells, responded poorly to vaccines and had early mortality compared to aged-matched CMV seronegative counterparts [19].

A heterogeneous role was reported for CD8$^+$CD28$^-$ T cells in autoimmune diseases [99,108]. Senescent T cell population is increased in patients with Grave's disease and ankylosing spondylitis and these cells' cytotoxicity contributes to autoimmune response [128,129]. In rheumatoid arthritis patients, clinical response to abatacept, a CD80/86-CD28 T cell co-stimulation modulator, is associated with a concomitant decrease in CD8$^+$CD28$^-$ T cells, suggesting prognostic value for this phenotype [56]. In contrast, patients with systemic lupus erythematosus were found to have reduced CD8$^+$CD28$^-$ T cells [130]. Similarly, in a mice model for multiple sclerosis, adoptive transfer of CD8$^+$CD28$^-$ regulatory T cells have been shown to prevent autoimmune encephalomyelitis [131]. Though treated as a single phenotype, CD8$^+$CD28$^-$ T cells represent a heterogeneous group that has differential activities in different pathologic conditions. In solid organ transplant recipients, CD8$^+$CD28$^-$ T cells have been found to undergo oligoclonal expansion and play a suppressive role and promote allograft tolerance [108,132]. Elevated CD8$^+$CD28$^-$ T cell population in liver transplant patients are associated with better graft function and reduced rejection rates [133] and contribute to reducing immunosuppressant dosage [124]. In addition, the presence of CD8$^+$CD28$^-$ T cells is associated with decreased CD80/86 expression and increased inhibitory receptor (ILT3, ILT4) in circulating APCs, implying an immunosuppressive role of this subset [126,133].

8. CD8$^+$CD28$^-$ T cells and Cancer

The cycle of anti-tumor immunity starts with the presentation of cancer antigens released from cancer cell turnover. Resident tissue DCs or lymph nodes residing DCs capture cancer antigens and present the antigens in the form of peptide-MHC I complex to activate naïve CD8$^+$ T cells. Activated effector CD8$^+$ T cells travel through blood and lymphatic to reach tumor beds where they execute cancer-specific killing. This leads to further endogenous antigen release and DC activation, thereby closing the cycle for anti-tumor immune response [1,28].

The presence of lymphoid aggregates is linked with improved responses to cancer therapies such as standard cytotoxic therapies, vaccine-based treatments, or immune checkpoint blockades. [5,134] Immunologically 'hot' tumors, such as melanomas and lung cancers, are thus more amenable to control than 'cold' tumors, i.e., tumors with diminished T cell infiltrates, such as GBM. [135,136] This drives modern cancer therapy to investigate how to redirect the TME to attract the right types of immune infiltrates.

Effector arm insufficiency, especially CD8$^+$ T cell dysfunction, is a hallmark of inadequate anti-tumor immune response [16]. Four forms of T cell dysfunction—tolerance, anergy, exhaustion, and senescence—have all been reported in cancer microenvironment [17,35,137]. Immune tolerance is a physiological process where the body eliminates self-reacting T cells. Tumor cells, such as GBM, can mimic peripheral tolerance and facilities FasL-mediated deletion of T cells [17]. Tolerance can be enforced by TGF-β and IL-10 secreted by Tregs that are recruited in the TME as well as cancer cells. [17,138,139]. T cell anergy is a T cell hypo-responsive state with low IL-2 production and poor proliferative capacity [17]. It results from the lack of co-stimulation of TCRs through CD28. This is due to the competitive binding of CTLA-4 expressed on Tregs to CD80 and CD86 on the APCs [14,17,139]. Anergic T cells display very little to no effector function, but expression on inhibitory markers is unclear [15]. T cell exhaustion occurs after excessive and continuous stimulation of the TCR and inflammatory cytokines like IFNα/β. Exhausted T cells have diminished proliferative capacity and have poor cytokine production and effector function. However, they express high levels of inhibitory receptors, or immune checkpoints, such as PD-1, CTLA-4, TIM3, LAG3, etc. [15,140]. The degree of T cell exhaustion can vary with tumor types as well. It is more severe in GBM compared to other cancers such as breast, lung, and melanoma [137]. T cell senescence can be distinguished from anergy and exhaustion in their origins and their surface receptors. For example, senescent T cells express fewer CD28 but more NK receptors, whereas exhausted or anergic T cells express more inhibitory receptors such as PD-1 and CTLA-1 (Figure 4). While anergic and exhausted T cells are hyporesponsive and hypofunctional, senescent T cells were considered metabolically active in their physiological or pathological environment despite being in cell cycle arrest (Figure 3). Though both T cell anergy and T cell exhaustion in natural occurrence are considered reversible, T cell senescence was considered irreversible until recently [14].

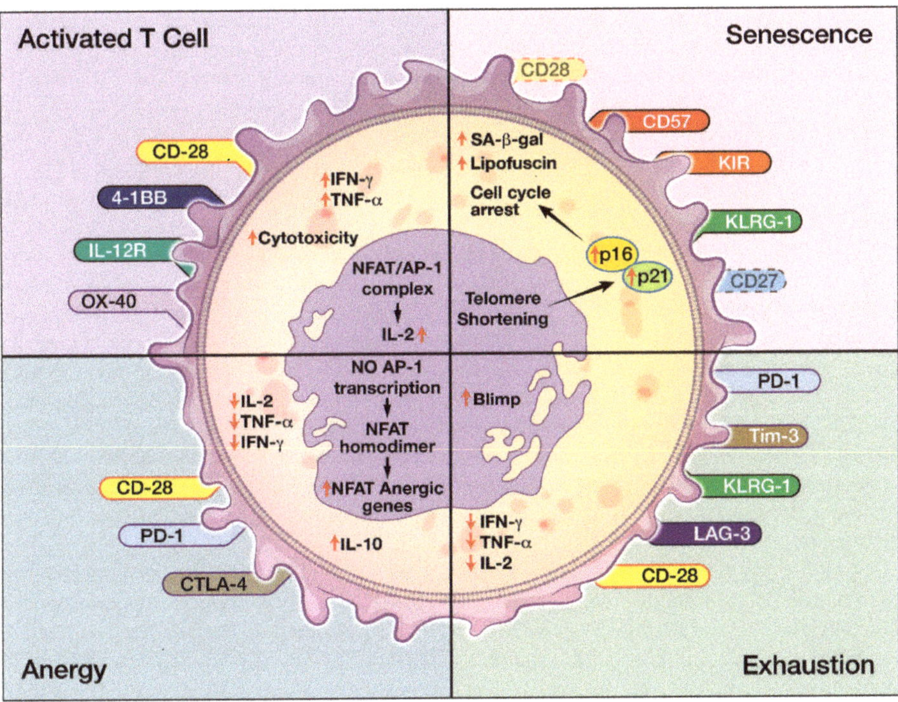

Figure 4. The many facets of C8+ T cell dysfunction. In comparison to activated effector T cells, T cell anergy, exhaustion, and senescence can be distinguished by their unique expression or lack of expression of surface receptors as well as different levels of intracellular cytokines, such as IL-2.

Targeting dysfunctional effector T cells has revolutionized the paradigm of tumor immunotherapy and immune checkpoint inhibitors set a paramount example [141]. Utilizing tumor dysregulation of immune checkpoint expression in exhausted dysfunctional T cells, immune checkpoint inhibitors were developed to promote protumor immune landscape [141]. Ipilimumab, CTLA-4 inhibitor, the first immune checkpoint inhibitor approved by FDA, was used to treat patients with advanced melanoma and has demonstrated improved survival when given with gp100 melanoma vaccine [142]. PD-1 inhibitors, pembrolizumab and nivolumab, and PD-L1 inhibitors, atezolizumab, durvalumab, and avelumab soon followed showing promising results. Pembrolizumab and nivolumab demonstrated 40–45% objective response rate in melanoma and non-small cell lung cancer [143]. In urothelial bladder cancer, use of PD-1/PD-L1 inhibitors showed overall response rate between 13% and 24% [144]. With metastatic brain disease, the combination of ipilimumab and nivolumab showed 56% intracranial response with 19% complete response from metastatic melanoma, and pembrolizumab have been shown to demonstrate intracranial activity against melanoma and NSCLC metastasis [145]. Continued investigation of the safety and efficacy of these novel immunomodulating drugs are ongoing in various malignancies [146]. However, it has been reported that the presence of functionally aberrant senescent T cells with loss of CD27 and CD28 and gained expression of CD57 cells was associated with resistance to checkpoint inhibitor blockade [8]. Therefore, senescent T cell phenotypes are possible predictive biomarkers for clinical response to checkpoint inhibitor therapy and potential targets to improve checkpoint inhibitor efficacy.

Increased CD8$^+$CD28$^-$ senescent populations displaying heterogeneous roles have been observed in multiple solid and hematogenous tumors [24]. This immunosuppressive phenotype was initially observed in patients with plasma cell dyscrasia, where increased number of CD8$^+$CD28$^-$ T cells was present in the bone marrow (i.e., TME) and the amount directly correlated with the suppression of antigen-specific T cell response [123]. Similarly, in patients with lung cancer, the CD8$^+$CD28$^-$ T cells express elevated Foxp3 and have been shown to play an immunoregulatory role [147]. High levels of CD8$^+$CD28$^-$ T cells were found in patients with advanced stages of non-small-cell lung cancer. Their numbers declined with resection of the tumor, and the decreased level of CD8$^+$CD28$^-$ T cells correlates with favorable prognosis in tumor management [148]. In contrast, the CD8$^+$CD28$^-$ T cell populations in melanoma patients express perforin, where they contribute to anti-tumor immune response [149].

CD8+CD28- T cell senescence is triggered by a variety of biological processes including telomere damage, Treg cells and tumor-associated stresses [150]. Naturally occurring CD4$^+$CD25^{hi}Foxp3$^+$ Treg (nTreg) and tumor-derived γδ Treg cells can induce responder T cell senescence as an immunosuppressive mechanism [127,151]. Senescent T cells induced by Treg cells have phenotypic changes, including expression of SA-β-Gal, downregulation of co-stimulatory molecules CD27 and CD28, and promotion of cell cycle and growth arrest in G0/G1 phase [127,151]. Importantly, CD8$^+$CD28$^-$ senescent T cells induced by Treg cells have potent regulatory activities [150] and deepen immunosuppression in TME [152]. Therefore, CD8$^+$CD28$^-$ senescent T cells are important mediators and amplifiers of immunosuppression mediated by Treg cells. The blockage of Treg-induced senescence in responder immune cells is critical in controlling tumor immunosuppression and restoring effector T cell function.

One of the mechanisms for Treg-induced CD8+ T cell immunosenescence is mediated by nuclear kinase ataxia-telangiectasia mutated protein (ATM)-associated DNA damage in responder T cells triggered by glucose competition [150]. MAP ERK1/2 and p38 signaling functionally cooperate with transcription factors STAT1/STAT3 to control Treg-induced senescence in responder T cells [150]. Utilizing these mechanisms, Treg-induced T cell senescence was successfully prevented by inhibiting the DNA damage response and/or STAT signaling in a recent *in vivo* mice study [150]. Another study has shown that senescent T cells are in fact able to regain function by inhibiting the p38 MAPK pathway [153]. Furthermore, human Toll-like receptor 8 (TLR8) signaling can directly target multiple types of tumors and prevent tumor-induced cell senescence through modulation of levels of endogenous secondary messenger cAMP in tumor cells [154].

9. CD8$^+$CD28$^-$ T cells and Glioblastoma

Despite being isolated in the intracranial compartment by the blood brain barrier, GBM, the most common and aggressive primary brain tumor in adults, demonstrates a remarkable level of immunosuppression [155]. Current standard of care for patients with GBM includes surgery, temozolomide chemotherapy, radiotherapy, and corticosteroids, all of which have potent immunosuppressive effects. Tumor cells express surface ligands such as PD-L1 and CD95 (Fas/apoptosis antigen 1) that can lead to T cell suppression via apoptosis and immunosuppressive cytokines like TGF-β, IL-10, and other tolerance factors [139]. Tumor-associate macrophages, modified neutrophils, and Foxp3$^+$ Tregs, are also recruited by the tumor to promote its progression [156–158].

T cell dysfunctions including tolerance, anergy, and exhaustion have also been well documented in GBM [14,17]. However, despite of promise of checkpoint inhibitors in the treatment of several solid tumors, their therapeutic efficacy in GBM remains to be validated. Phase III clinical trial Checkmate 143 reported that PD-1 monoclonal antibody (nivolumab) monotherapy failed to demonstrate survival benefits compared to bevacizumab in recurrent GBM patients who were previously treated with chemotherapy and radiotherapy [17,69,155]. Ongoing clinical trials are investigating the tolerability and drug toxicity in combination treatment and in patients with newly diagnosed GBM patients as well as recurrent GBM patients. The muted response to immune activators seen thus far highlights to the need for novel strategies to boost immunity to GBM.

The role of T cell senescence in GBM has been reported but is yet to be fully elucidated. The presence of circulating senescent $CD4^+CD28^-CD57^+$ T cells was correlated with poor prognosis in GBM patients [159]. $CD8^+CD28^-Foxp3^+$ T cells, which have been found in other cancers to cause APC dysfunction [160], were also identified in TME from patients with GBM [161]. The APCs isolated from these patients displayed dysfunctional phenotype associated with high levels of ILT2, ILT3, and ILT4 and low levels of CD40, CD80, and CD86 [162]. It is speculated that $CD8^+CD28^-$ T cells help sculpt an immunosuppressive environment in a similar fashion in GBM.

The potential pro-tumoral effect of $CD8^+CD28^-$ cells can also be inferred from worse prognosis observed in older GBM patients [163]. Since $CD8^+CD28^-$ cells are derived from the general population of immature $CD8^+CD28^+$ T cells originating from the thymus [99], the production of immature $CD8^+CD28^+$ decreases as thymic involution occurs through aging, but also with cancer [164]. This decrease in immature $CD8^+CD28^+$ cells due to thymic senescence has also been associated with poor outcome in GMB patients [17].

10. Implications of $CD8^+CD28^-$ T Cells for the Future of Immunotherapy

Since success of immunotherapy largely relies on addressing effector arm dysfunction, as evident from the success of checkpoint inhibitors, future investigations into new treatment methods should explore strategies to deplete or inhibit regulatory $CD8^+CD28^-$ T cells and reverse T cell senescence as an adjuvant for more effective immunotherapy. There are four main approaches to rejuvenate T cell pools (1) replacement of senescent cells, (2) reprogramming of the senescent cells to be functional, (3) adoptive transfer of proficient T cells, and (4) restoration of naïve T cell pool [165].

Replacement strategies include selectively removing senescent cells from the circulation and then subsequent expansion of memory and effector T cells. Removal of senescent cells is of particular importance, not only due to their own dysfunction but also due to their ability to and spreads senescence in bystander cells [166]. A possible approach for their removal is to promote selective apoptosis in senescent T cells. In a recent study, an engineered peptide that interferes with FOXO4/p53 interaction induced p53-mediated intrinsic apoptosis in senescent fibroblasts and neutralized doxorubicin-induced chemotoxicity *in vivo* [167]. Whether this also can be used in inducing apoptosis of senescent T cell remains to be investigated. Targeting commonly known senescent cell anti-apoptotic pathways such as Bcl-2 and ephrins in senescent T cells is also warranted [168]. Homeostatic expansion in the form of autologous stem cell transplantation has been used to reconstitute functional naïve, memory, and affect T cell pools in both autoimmune diseases and hematologic malignancies [169,170]. More recently, senolytic treatment of amyloid-beta (Aβ) peptide -associated senescent oligodendrocyte progenitor cell in mice with Alzheimer's disease showed successful selective removal of senescent cells from the plaque, reduced neuroinflammation, lessened Aβ load, and ameliorated cognitive deficits [171]. This is of particular interest as a successful application of senolytic therapy in the CNS pathology, such as GBM. $CD8^+CD28^-$ cells replacement strategies are still in early phases of development, however their successful implementation has the potential to complement the current paradigm of immunotherapies.

Re-programming involves differentiating T cells away from dysfunctional states by enhancing telomerase activity to extend cellular lifespan and preclude replicative senescence [172]. For example, restoring CD28 expression slows replicative senescence in human T cells through increased telomerase activity to increase proliferative potential [173]. Additionally, pharmacological inhibition of SRC homology 2 domain-containing phosphatase-1 (SHP-1), a key regulator of T cell signal transduction machinery, improved TCR/CD28 signaling and successfully improve T cell functions in elderly donors [174]. Aptamers, short, single-stranded DNA or RNA molecules, have been engineered to target immune costimulatory receptors (CD28, OX40 and 4-1BB) and have shown to improve T cell activation and induce antitumor response. Aptamers have the benefits of being chemically synthesized, versatility of targeting motifs, high penetration rate, and ease of neutralization [75]. Re-programming of T cells present the most promising avenue for anti-CD8$^+$CD28$^-$ therapy with wide selection of potential targets and treatment mechanisms.

Adoptive transfer is to bypass the co-stimulation requirement to re-differentiate pluripotent stem cells into naïve and cytotoxic T cells to fight malignancy [175]. The development of T cell adoptive immunotherapy using the third generation of CAR technology by incorporating the intracellular costimulatory domains to bypass the requirement for CD28 activation is underway. The third generation CARs T cells were investigated in hematological malignancies and xenograft model of solid tumors and have shown preclinical success [176–180]. While other types of immunotherapy can cause systemic side effects, antigen specificity of CAR therapy limits the adverse effect of immunotherapy to its targets, and they are reversible when the target cell is eliminated, or the engraftment of the CAR T cells is terminated [181]. However, its high specificity can be a weakness in heterogeneous tumors with high mutational profiles since CAR therapy can select for cells negative for the targeted antigen [155]. Such was the case for IL-13Rα2 CAR therapy for GBM where recurrence occurred 7.5 months after treatment despite shrinking all lesions by 77–100% [155].

Finally, restoring and maintaining the thymic environment reverse effects of thymic involution. Using bioengineered thymus organoids with the help of growth-promoting factors and cytokines such as IL-21, it has been shown that significant immune restorative function and rejuvenation of the peripheral T cell pools were achieved in murine models [182]. Unfortunately, current understanding of thymic restoration is not complete enough for clinical implementation, and there are still unanswered questions regarding its feasibility in establishing functional naïve T cell production [183,184]. The safe removal of senescent CD8$^+$ T cells and restoration or differential induction of functional CD8$^+$ cytotoxic T cells would add a promising mechanism to defend host against cancer invasion and fight immunotolerance of malignancy.

There is theoretical concern that reversing the growth arrest by selective blockage of senescent T cells carries a risk of malignancy, which is less of a concern for targeting functionally exhausted T cells [185]. Nevertheless, one may argue that increased theoretical lifetime risk of malignancy is outweighed by the potential immediate benefit of extending the life expectancy, even by a few months to years, in the battle against aggressive cancers, such as GBM with a median overall survival of only 20.6 months. Furthermore, the benefits of targeting both immunosenescence and exhaustion may be more evident with reduced dose requirement for each, thereby reducing risks of drug-associated adverse events. Potential synergistic efficacy to boost immunity against cancer may also be implemented as already seen with GITR stimulation/PD-1 blockade and CTLA-4/ICOS stimulation [74,80,81].

11. Conclusions

In summary, functional, tumor specific CD8$^+$ cytotoxic T cells drive the adaptive immune response to cancer and are the primary endpoint to most immunotherapies. However, the promise of current cancer immunotherapy has been limited by marked immunosuppression in the TME defined by CD8$^+$ T cell dysfunction, especially in immune 'cold' cancers, such as GBM. Among the many facets of CD8$^+$ T cell dysfunction, including tolerance, anergy, and exhaustion, CD8$^+$ T cell senescence, as represented by the CD8$^+$CD28$^-$ population, is an emerging field and their presence has been described in many cancers. CD8$^+$CD28$^-$ T cells contribute to tumor immunosuppression and resistance to immunotherapy. Further characterization and investigation into this subset of CD8$^+$ T cells will provide novel targets for effective immunotherapy and successful cancer control.

Author Contributions: Conceptualization, Supervision, and Funding Acquisition: M.D.; Writing-Original Draft Preparation and Writing-Review & Editing: W.X.H., J.H.W., M.H., K.F., and M.D.

Funding: This work was supported by the NIH K08NS092895 grant (M.D.).

Acknowledgments: Authors would like to thank Christopher Brown MS for his help with the figure illustrations.

Conflicts of Interest: The authors declare that they have no competing interests.

Abbreviations

APC	Antigen presenting cells
CAR	Chimeric antigen receptor
CMV	Cytomegalovirus
CTL	Cytotoxic T-lymphocytes
DC	Dendritic cells
GBM	Glioblastoma
NK	Natural killer
SA-β-Gal	Senescence-associated-β-galactosidase
TCR	T cell receptor
Th	T helper cells
TME	Tumor microenvironment
Tregs	T- regulatory cells

References

1. Palucka, A.K.; Coussens, L.M. The basis of oncoimmunology. *Cell* **2016**, *164*, 1233–1247. [CrossRef] [PubMed]
2. Kim, N.; Lee, H.H.; Lee, H.J.; Choi, W.S.; Lee, J.; Kim, H.S. Natural killer cells as a promising therapeutic target for cancer immunotherapy. *Arch. Pharm. Res.* **2019**, 1–16. [CrossRef] [PubMed]
3. Spitzer, M.H.; Carmi, Y.; Reticker-Flynn, N.E.; Kwek, S.S.; Madhireddy, D.; Martins, M.M.; Gherardini, P.F.; Prestwood, T.R.; Chabon, J.; Bendall, S.C.; et al. Systemic immunity is required for effective cancer immunotherapy. *Cell* **2017**, *168*, 487–502.e15. [CrossRef] [PubMed]
4. Ostroumov, D.; Fekete-Drimusz, N.; Saborowski, M.; Kuhnel, F.; Woller, N. CD4 and CD8 T lymphocyte interplay in controlling tumor growth. *Cell. Mol. Life Sci.* **2018**, *75*, 689–713. [CrossRef] [PubMed]
5. Sharma, P.; Allison, J.P. The future of immune checkpoint therapy. *Science* **2015**, *348*, 56–61. [CrossRef] [PubMed]
6. Schumacher, T.N.; Schreiber, R.D. Neoantigens in cancer immunotherapy. *Science* **2015**, *348*, 69–74. [CrossRef] [PubMed]
7. Farkona, S.; Diamandis, E.P.; Blasutig, I.M. Cancer immunotherapy: The beginning of the end of cancer? *Bmc Med.* **2016**, *14*, 73. [CrossRef]
8. Moreira, A.; Gross, S.; Kirchberger, M.C.; Erdmann, M.; Schuler, G.; Heinzerling, L. Senescence markers: Predictive for response to checkpoint inhibitors. *Int. J. Cancer* **2019**, *144*, 1147–1150. [CrossRef]
9. Omuro, A.; Vlahovic, G.; Lim, M.; Sahebjam, S.; Baehring, J.; Cloughesy, T.; Voloschin, A.; Ramkissoon, S.H.; Ligon, K.L.; Latek, R.; et al. Nivolumab with or without ipilimumab in patients with recurrent glioblastoma: Results from exploratory phase i cohorts of checkmate 143. *Neuro Oncol.* **2018**, *20*, 674–686. [CrossRef]
10. Reardon, D.A.; Omuro, A.; Brandes, A.A.; Rieger, J.; Wick, A.; Sepulveda, J.; Phuphanich, S.; de Souza, P.; Ahluwalia, M.S.; Lim, M.; et al. OS10.3 randomized phase 3 study evaluating the efficacy and safety of nivolumab vs bevacizumab in patients with recurrent glioblastoma: Checkmate 143. *Neuro Oncol.* **2017**, *19*, iii21. [CrossRef]
11. Powles, T.; Duran, I.; van der Heijden, M.S.; Loriot, Y.; Vogelzang, N.J.; De Giorgi, U.; Oudard, S.; Retz, M.M.; Castellano, D.; Bamias, A.; et al. Atezolizumab versus chemotherapy in patients with platinum-treated locally advanced or metastatic urothelial carcinoma (imvigor211): A multicentre, open-label, phase 3 randomised controlled trial. *Lancet* **2018**, *391*, 748–757. [CrossRef]
12. Brahmer, J.R.; Tykodi, S.S.; Chow, L.Q.; Hwu, W.J.; Topalian, S.L.; Hwu, P.; Drake, C.G.; Camacho, L.H.; Kauh, J.; Odunsi, K.; et al. Safety and activity of anti-PD-l1 antibody in patients with advanced cancer. *N. Engl. J. Med.* **2012**, *366*, 2455–2465. [CrossRef]
13. Topalian, S.L.; Hodi, F.S.; Brahmer, J.R.; Gettinger, S.N.; Smith, D.C.; McDermott, D.F.; Powderly, J.D.; Carvajal, R.D.; Sosman, J.A.; Atkins, M.B.; et al. Safety, activity, and immune correlates of anti-PD-1 antibody in cancer. *N. Engl. J. Med.* **2012**, *366*, 2443–2454. [CrossRef]
14. Crespo, J.; Sun, H.; Welling, T.H.; Tian, Z.; Zou, W. T cell anergy, exhaustion, senescence, and stemness in the tumor microenvironment. *Curr. Opin. Immunol.* **2013**, *25*, 214–221. [CrossRef] [PubMed]

15. Wherry, E.J.; Kurachi, M. Molecular and cellular insights into T cell exhaustion. *Nat. Rev. Immunol.* **2015**, *15*, 486–499. [CrossRef]

16. Thommen, D.S.; Schumacher, T.N. T cell dysfunction in cancer. *Cancer Cell* **2018**, *33*, 547–562. [CrossRef]

17. Woroniecka, K.I.; Rhodin, K.E.; Chongsathidkiet, P.; Keith, K.A.; Fecci, P.E. T-cell dysfunction in glioblastoma: Applying a new framework. *Clin. Cancer Res.* **2018**, *24*, 3792–3802. [CrossRef]

18. Aguilera, M.O.; Delgui, L.R.; Romano, P.S.; Colombo, M.I. Chronic infections: A possible scenario for autophagy and senescence cross-talk. *Cells* **2018**, *7*, 162. [CrossRef]

19. Aberg, J.A. Aging, inflammation, and hiv infection. *Top. Antivir. Med.* **2012**, *20*, 101–105.

20. Childs, B.G.; Gluscevic, M.; Baker, D.J.; Laberge, R.M.; Marquess, D.; Dananberg, J.; van Deursen, J.M. Senescent cells: An emerging target for diseases of ageing. *Nat. Rev. Drug Discov.* **2017**, *16*, 718–735. [CrossRef]

21. Warren, J.A.; Clutton, G.; Goonetilleke, N. Harnessing CD8(+) T cells under hiv antiretroviral therapy. *Front. Immunol.* **2019**, *10*, 291. [CrossRef]

22. Schwartz, R.H. Costimulation of T lymphocytes: The role of CD28, CTLA-4, and B7/BB1 in interleukin-2 production and immunotherapy. *Cell* **1992**, *71*, 1065–1068. [CrossRef]

23. Chen, X.; Liu, Q.; Xiang, A.P. Cd8+CD28- T cells: Not only age-related cells but a subset of regulatory T cells. *Cell. Mol. Immunol.* **2018**, *15*, 734–736. [CrossRef]

24. Filaci, G.; Fenoglio, D.; Fravega, M.; Ansaldo, G.; Borgonovo, G.; Traverso, P.; Villaggio, B.; Ferrera, A.; Kunkl, A.; Rizzi, M.; et al. CD8+ CD28- T regulatory lymphocytes inhibiting T cell proliferative and cytotoxic functions infiltrate human cancers. *J. Immunol.* **2007**, *179*, 4323–4334. [CrossRef]

25. Borthwick, N.J.; Lowdell, M.; Salmon, M.; Akbar, A.N. Loss of CD28 expression on CD8(+) T cells is induced by IL-2 receptor gamma chain signalling cytokines and type i ifn, and increases susceptibility to activation-induced apoptosis. *Int. Immunol.* **2000**, *12*, 1005–1013. [CrossRef]

26. Vallejo, A.N. CD28 extinction in human T cells: Altered functions and the program of T-cell senescence. *Immunol. Rev.* **2005**, *205*, 158–169. [CrossRef]

27. Borst, J.; Ahrends, T.; Babala, N.; Melief, C.J.M.; Kastenmuller, W. CD4(+) T cell help in cancer immunology and immunotherapy. *Nat. Rev. Immunol.* **2018**, *18*, 635–647. [CrossRef]

28. Farhood, B.; Najafi, M.; Mortezaee, K. CD8(+) cytotoxic t lymphocytes in cancer immunotherapy: A review. *J. Cell. Physiol.* **2019**, *234*, 8509–8521. [CrossRef]

29. Singer, A.; Adoro, S.; Park, J.H. Lineage fate and intense debate: Myths, models and mechanisms of CD4- versus CD8-lineage choice. *Nat. Rev. Immunol.* **2008**, *8*, 788–801. [CrossRef]

30. Germain, R.N. T-cell development and the CD4-CD8 lineage decision. *Nat. Rev. Immunol.* **2002**, *2*, 309–322. [CrossRef]

31. Arosa, F.A.; Esgalhado, A.J.; Padrao, C.A.; Cardoso, E.M. Divide, conquer, and sense: Cd8(+)CD28(-) T cells in perspective. *Front. Immunol.* **2016**, *7*, 665. [CrossRef]

32. Fearon, D.T.; Carr, J.M.; Telaranta, A.; Carrasco, M.J.; Thaventhiran, J.E. The rationale for the il-2-independent generation of the self-renewing central memory CD8+ T cells. *Immunol. Rev.* **2006**, *211*, 104–118. [CrossRef]

33. Martinez-Lostao, L.; Anel, A.; Pardo, J. How do cytotoxic lymphocytes kill cancer cells? *Clin. Cancer Res.* **2015**, *21*, 5047–5056. [CrossRef]

34. Janssen, E.M.; Lemmens, E.E.; Wolfe, T.; Christen, U.; von Herrath, M.G.; Schoenberger, S.P. CD4+ T cells are required for secondary expansion and memory in CD8+ T lymphocytes. *Nature* **2003**, *421*, 852–856. [CrossRef]

35. Reading, J.L.; Galvez-Cancino, F.; Swanton, C.; Lladser, A.; Peggs, K.S.; Quezada, S.A. The function and dysfunction of memory CD8(+) T cells in tumor immunity. *Immunol. Rev.* **2018**, *283*, 194–212. [CrossRef]

36. Ara, A.; Ahmed, K.A.; Xiang, J. Multiple effects of CD40-CD40l axis in immunity against infection and cancer. *Immuno Targets Ther.* **2018**, *7*, 55–61. [CrossRef]

37. Kershaw, M.H.; Westwood, J.A.; Darcy, P.K. Gene-engineered T cells for cancer therapy. *Nat. Rev. Cancer* **2013**, *13*, 525–541. [CrossRef]

38. Gottschalk, C.; Mettke, E.; Kurts, C. The role of invariant natural killer T cells in dendritic cell licensing, cross-priming, and memory CD8(+) T cell generation. *Front. Immunol.* **2015**, *6*, 379. [CrossRef]

39. Wu, R.C.; Hwu, P.; Radvanyi, L.G. New insights on the role of CD8(+)CD57(+) T-cells in cancer. *Oncoimmunology* **2012**, *1*, 954–956. [CrossRef]

40. Lohr, J.; Ratliff, T.; Huppertz, A.; Ge, Y.; Dictus, C.; Ahmadi, R.; Grau, S.; Hiraoka, N.; Eckstein, V.; Ecker, R.C.; et al. Effector T-cell infiltration positively impacts survival of glioblastoma patients and is impaired by tumor-derived TGF-β. *Clin. Cancer Res.* **2011**, *17*, 4296–4308. [CrossRef]

41. Mahmoud, S.M.; Paish, E.C.; Powe, D.G.; Macmillan, R.D.; Grainge, M.J.; Lee, A.H.; Ellis, I.O.; Green, A.R. Tumor-infiltrating CD8+ lymphocytes predict clinical outcome in breast cancer. *J. Clin. Oncol.* **2011**, *29*, 1949–1955. [CrossRef]

42. Yang, Z.Q.; Yang, Z.Y.; Zhang, L.D.; Ping, B.; Wang, S.G.; Ma, K.S.; Li, X.W.; Dong, J.H. Increased liver-infiltrating CD8+FOXP3+ regulatory T cells are associated with tumor stage in hepatocellular carcinoma patients. *Hum. Immunol.* **2010**, *71*, 1180–1186. [CrossRef]

43. Sato, E.; Olson, S.H.; Ahn, J.; Bundy, B.; Nishikawa, H.; Qian, F.; Jungbluth, A.A.; Frosina, D.; Gnjatic, S.; Ambrosone, C.; et al. Intraepithelial CD8+ tumor-infiltrating lymphocytes and a high CD8+/regulatory T cell ratio are associated with favorable prognosis in ovarian cancer. *Proc. Natl. Acad. Sci. USA* **2005**, *102*, 18538–18543. [CrossRef]

44. Ling, A.; Edin, S.; Wikberg, M.L.; Oberg, A.; Palmqvist, R. The intratumoural subsite and relation of CD8(+) and FOXP3(+) T lymphocytes in colorectal cancer provide important prognostic clues. *Br. J. Cancer* **2014**, *110*, 2551–2559. [CrossRef]

45. Liu, L.; Zhao, G.; Wu, W.; Rong, Y.; Jin, D.; Wang, D.; Lou, W.; Qin, X. Low intratumoral regulatory T cells and high peritumoral CD8(+) T cells relate to long-term survival in patients with pancreatic ductal adenocarcinoma after pancreatectomy. *Cancer Immunol. Immunother.* **2016**, *65*, 73–82. [CrossRef]

46. Coca, S.; Perez-Piqueras, J.; Martinez, D.; Colmenarejo, A.; Saez, M.A.; Vallejo, C.; Martos, J.A.; Moreno, M. The prognostic significance of intratumoral natural killer cells in patients with colorectal carcinoma. *Cancer* **1997**, *79*, 2320–2328. [CrossRef]

47. Preston, C.C.; Maurer, M.J.; Oberg, A.L.; Visscher, D.W.; Kalli, K.R.; Hartmann, L.C.; Goode, E.L.; Knutson, K.L. The ratios of CD8+ T cells to CD4+CD25+ FOXP3+ and FOXP3- T cells correlate with poor clinical outcome in human serous ovarian cancer. *PLoS ONE* **2013**, *8*, e80063. [CrossRef]

48. Galon, J.; Costes, A.; Sanchez-Cabo, F.; Kirilovsky, A.; Mlecnik, B.; Lagorce-Pages, C.; Tosolini, M.; Camus, M.; Berger, A.; Wind, P.; et al. Type, density, and location of immune cells within human colorectal tumors predict clinical outcome. *Science* **2006**, *313*, 1960–1964. [CrossRef]

49. Huang, Y.; Ma, C.; Zhang, Q.; Ye, J.; Wang, F.; Zhang, Y.; Hunborg, P.; Varvares, M.A.; Hoft, D.F.; Hsueh, E.C.; et al. CD4+ and CD8+ T cells have opposing roles in breast cancer progression and outcome. *Oncotarget* **2015**, *6*, 17462–17478. [CrossRef]

50. Zhu, S.; Lin, J.; Qiao, G.; Xu, Y.; Zou, H. Differential regulation and function of tumor-infiltrating T cells in different stages of breast cancer patients. *Tumour Biol.* **2015**, *36*, 7907–7913. [CrossRef]

51. Watanabe, Y.; Katou, F.; Ohtani, H.; Nakayama, T.; Yoshie, O.; Hashimoto, K. Tumor-infiltrating lymphocytes, particularly the balance between CD8(+) T cells and CCR4(+) regulatory T cells, affect the survival of patients with oral squamous cell carcinoma. *Oral Surg. Oral Med. Oral Pathol. Oral Radiol. Endodontol.* **2010**, *109*, 744–752. [CrossRef]

52. Esensten, J.H.; Helou, Y.A.; Chopra, G.; Weiss, A.; Bluestone, J.A. CD28 costimulation: From mechanism to therapy. *Immunity* **2016**, *44*, 973–988. [CrossRef]

53. Zumerle, S.; Molon, B.; Viola, A. Membrane rafts in T cell activation: A spotlight on CD28 costimulation. *Front. Immunol.* **2017**, *8*, 1467. [CrossRef]

54. Porciello, N.; Grazioli, P.; Campese, A.F.; Kunkl, M.; Caristi, S.; Mastrogiovanni, M.; Muscolini, M.; Spadaro, F.; Favre, C.; Nunès, J.A.; et al. A non-conserved amino acid variant regulates differential signalling between human and mouse CD28. *Nat. Commun.* **2018**, *9*, 1080. [CrossRef]

55. Bour-Jordan, H.; Esensten, J.H.; Martinez-Llordella, M.; Penaranda, C.; Stumpf, M.; Bluestone, J.A. Intrinsic and extrinsic control of peripheral T-cell tolerance by costimulatory molecules of the CD28/B7 family. *Immunol. Rev.* **2011**, *241*, 180–205. [CrossRef]

56. Zhang, Q.; Vignali, D.A. Co-stimulatory and co-inhibitory pathways in autoimmunity. *Immunity* **2016**, *44*, 1034–1051. [CrossRef]

57. Lenschow, D.J.; Zeng, Y.; Thistlethwaite, J.R.; Montag, A.; Brady, W.; Gibson, M.G.; Linsley, P.S.; Bluestone, J.A. Long-term survival of xenogeneic pancreatic islet grafts induced by CTLA4IG. *Science* **1992**, *257*, 789–792. [CrossRef]

58. Blair, H.A.; Deeks, E.D. Abatacept: A review in rheumatoid arthritis. *Drugs* **2017**, *77*, 1221–1233. [CrossRef]

59. Vincenti, F.; Rostaing, L.; Grinyo, J.; Rice, K.; Steinberg, S.; Gaite, L.; Moal, M.C.; Mondragon-Ramirez, G.A.; Kothari, J.; Polinsky, M.S.; et al. Belatacept and long-term outcomes in kidney transplantation. *N. Engl. J. Med.* **2016**, *374*, 333–343. [CrossRef]

60. Ford, M.L. T cell cosignaling molecules in transplantation. *Immunity* **2016**, *44*, 1020–1033. [CrossRef]

61. Attanasio, J.; Wherry, E.J. Costimulatory and coinhibitory receptor pathways in infectious disease. *Immunity* **2016**, *44*, 1052–1068. [CrossRef]

62. Callahan, M.K.; Postow, M.A.; Wolchok, J.D. Targeting T cell co-receptors for cancer therapy. *Immunity* **2016**, *44*, 1069–1078. [CrossRef]

63. Suntharalingam, G.; Perry, M.R.; Ward, S.; Brett, S.J.; Castello-Cortes, A.; Brunner, M.D.; Panoskaltsis, N. Cytokine storm in a phase 1 trial of the anti-CD28 monoclonal antibody TGN1412. *N. Engl. J. Med.* **2006**, *355*, 1018–1028. [CrossRef]

64. Tyrsin, D.; Chuvpilo, S.; Matskevich, A.; Nemenov, D.; Romer, P.S.; Tabares, P.; Hunig, T. From TGN1412 to TAB08: The return of CD28 superagonist therapy to clinical development for the treatment of rheumatoid arthritis. *Clin. Exp. Rheumatol.* **2016**, *34*, 45–48.

65. Hunig, T. The storm has cleared: Lessons from the CD28 superagonist TGN1412 trial. *Nat. Rev. Immunol.* **2012**, *12*, 317–318. [CrossRef]

66. Zhou, J.; Rossi, J. Aptamers as targeted therapeutics: Current potential and challenges. *Nat. Rev. Drug Discov.* **2017**, *16*, 440. [CrossRef]

67. Lozano, T.; Soldevilla, M.M.; Casares, N.; Villanueva, H.; Bendandi, M.; Lasarte, J.J.; Pastor, F. Targeting inhibition of FOXP3 by a CD28 2'-fluro oligonucleotide aptamer conjugated to p60-peptide enhances active cancer immunotherapy. *Biomaterials* **2016**, *91*, 73–80. [CrossRef]

68. Soldevilla, M.M.; Villanueva, H.; Casares, N.; Lasarte, J.J.; Bendandi, M.; Inoges, S.; Lopez-Diaz de Cerio, A.; Pastor, F. MRP1-CD28 bi-specific oligonucleotide aptamers: Target costimulation to drug-resistant melanoma cancer stem cells. *Oncotarget* **2016**, *7*, 23182–23196. [CrossRef]

69. Filley, A.C.; Henriquez, M.; Dey, M. Recurrent glioma clinical trial, checkmate-143: The game is not over yet. *Oncotarget* **2017**, *8*, 91779–91794. [CrossRef]

70. Porter, D.L.; Hwang, W.T.; Frey, N.V.; Lacey, S.F.; Shaw, P.A.; Loren, A.W.; Bagg, A.; Marcucci, K.T.; Shen, A.; Gonzalez, V.; et al. Chimeric antigen receptor T cells persist and induce sustained remissions in relapsed refractory chronic lymphocytic leukemia. *Sci. Transl. Med.* **2015**, *7*, 303ra139. [CrossRef]

71. Tang, X.Y.; Sun, Y.; Zhang, A.; Hu, G.L.; Cao, W.; Wang, D.H.; Zhang, B.; Chen, H. Third-generation CD28/4-1BB chimeric antigen receptor T cells for chemotherapy relapsed or refractory acute lymphoblastic leukaemia: A non-randomised, open-label phase I trial protocol. *BMJ Open* **2016**, *6*, e013904. [CrossRef]

72. Ahmed, N.; Brawley, V.; Hegde, M.; Bielamowicz, K.; Wakefield, A.; Ghazi, A.; Ashoori, A.; Diouf, O.; Gerken, C.; Landi, D.; et al. Autologous HER2 cmv bispecific CAR T cells are safe and demonstrate clinical benefit for glioblastoma in a phase I trial. *J. Immunother. Cancer* **2015**, *3*, O11. [CrossRef]

73. Cabo, M.; Offringa, R.; Zitvogel, L.; Kroemer, G.; Muntasell, A.; Galluzzi, L. Trial watch: Immunostimulatory monoclonal antibodies for oncological indications. *Oncoimmunology* **2017**, *6*, e1371896. [CrossRef]

74. Emerson, D.A.; Redmond, W.L. Overcoming tumor-induced immune suppression: From relieving inhibition to providing costimulation with T cell agonists. *BioDrugs* **2018**, *32*, 221–231. [CrossRef]

75. Sanmamed, M.F.; Pastor, F.; Rodriguez, A.; Perez-Gracia, J.L.; Rodriguez-Ruiz, M.E.; Jure-Kunkel, M.; Melero, I. Agonists of co-stimulation in cancer immunotherapy directed against CD137, OX40, GITR, CD27, CD28, and icos. *Semin. Oncol.* **2015**, *42*, 640–655. [CrossRef]

76. Pratico, E.D.; Sullenger, B.A.; Nair, S.K. Identification and characterization of an agonistic aptamer against the T cell costimulatory receptor, OX40. *Nucleic Acid Ther.* **2013**, *23*, 35–43. [CrossRef]

77. Weigelin, B.; Bolanos, E.; Teijeira, A.; Martinez-Forero, I.; Labiano, S.; Azpilikueta, A.; Morales-Kastresana, A.; Quetglas, J.I.; Wagena, E.; Sanchez-Paulete, A.R.; et al. Focusing and sustaining the antitumor CTL effector killer response by agonist anti-CD137 mAb. *Proc. Natl. Acad. Sci. USA* **2015**, *112*, 7551–7556. [CrossRef]

78. Coe, D.; Begom, S.; Addey, C.; White, M.; Dyson, J.; Chai, J.G. Depletion of regulatory T cells by anti-gitr mab as a novel mechanism for cancer immunotherapy. *Cancer Immunol. Immunother.* **2010**, *59*, 1367–1377. [CrossRef]

79. Schaer, D.A.; Budhu, S.; Liu, C.; Bryson, C.; Malandro, N.; Cohen, A.; Zhong, H.; Yang, X.; Houghton, A.N.; Merghoub, T.; et al. Gitr pathway activation abrogates tumor immune suppression through loss of regulatory T cell lineage stability. *Cancer Immunol. Res.* **2013**, *1*, 320–331. [CrossRef]

80. Lu, L.; Xu, X.; Zhang, B.; Zhang, R.; Ji, H.; Wang, X. Combined PD-1 blockade and GITR triggering induce a potent antitumor immunity in murine cancer models and synergizes with chemotherapeutic drugs. *J. Transl. Med.* **2014**, *12*, 36. [CrossRef]

81. Fan, X.; Quezada, S.A.; Sepulveda, M.A.; Sharma, P.; Allison, J.P. Engagement of the icos pathway markedly enhances efficacy of CTLA-4 blockade in cancer immunotherapy. *J. Exp. Med.* **2014**, *211*, 715–725. [CrossRef]

82. Munoz-Espin, D.; Serrano, M. Cellular senescence: From physiology to pathology. *Nat. Rev. Mol. Cell Biol.* **2014**, *15*, 482–496. [CrossRef]

83. Demaria, M.; Ohtani, N.; Youssef, S.A.; Rodier, F.; Toussaint, W.; Mitchell, J.R.; Laberge, R.M.; Vijg, J.; Van Steeg, H.; Dolle, M.E.; et al. An essential role for senescent cells in optimal wound healing through secretion of PDGF-AA. *Dev. Cell* **2014**, *31*, 722–733. [CrossRef]

84. Rajagopalan, S.; Long, E.O. Cellular senescence induced by CD158D reprograms natural killer cells to promote vascular remodeling. *Proc. Natl. Acad. Sci. USA* **2012**, *109*, 20596–20601. [CrossRef]

85. Howcroft, T.K.; Campisi, J.; Louis, G.B.; Smith, M.T.; Wise, B.; Wyss-Coray, T.; Augustine, A.D.; McElhaney, J.E.; Kohanski, R.; Sierra, F. The role of inflammation in age-related disease. *Aging* **2013**, *5*, 84–93. [CrossRef]

86. Myrianthopoulos, V.; Evangelou, K.; Vasileiou, P.V.S.; Cooks, T.; Vassilakopoulos, T.P.; Pangalis, G.A.; Kouloukoussa, M.; Kittas, C.; Georgakilas, A.G.; Gorgoulis, V.G. Senescence and senotherapeutics: A new field in cancer therapy. *Pharmacol. Ther.* **2019**, *193*, 31–49. [CrossRef]

87. Petrakis, T.G.; Komseli, E.S.; Papaioannou, M.; Vougas, K.; Polyzos, A.; Myrianthopoulos, V.; Mikros, E.; Trougakos, I.P.; Thanos, D.; Branzei, D.; et al. Exploring and exploiting the systemic effects of deregulated replication licensing. *Semin. Cancer Biol.* **2016**, *37–38*, 3–15. [CrossRef]

88. Bartkova, J.; Rezaei, N.; Liontos, M.; Karakaidos, P.; Kletsas, D.; Issaeva, N.; Vassiliou, L.V.; Kolettas, E.; Niforou, K.; Zoumpourlis, V.C.; et al. Oncogene-induced senescence is part of the tumorigenesis barrier imposed by DNA damage checkpoints. *Nature* **2006**, *444*, 633–637. [CrossRef]

89. Rayess, H.; Wang, M.B.; Srivatsan, E.S. Cellular senescence and tumor suppressor gene p16. *Int. J. Cancer* **2012**, *130*, 1715–1725. [CrossRef]

90. Alcorta, D.A.; Xiong, Y.; Phelps, D.; Hannon, G.; Beach, D.; Barrett, J.C. Involvement of the cyclin-dependent kinase inhibitor p16 (INK4A) in replicative senescence of normal human fibroblasts. *Proc. Natl. Acad. Sci. USA* **1996**, *93*, 13742–13747. [CrossRef]

91. Jurk, D.; Wang, C.; Miwa, S.; Maddick, M.; Korolchuk, V.; Tsolou, A.; Gonos, E.S.; Thrasivoulou, C.; Saffrey, M.J.; Cameron, K.; et al. Postmitotic neurons develop a p21-dependent senescence-like phenotype driven by a DNA damage response. *Aging Cell* **2012**, *11*, 996–1004. [CrossRef]

92. Ou, H.L.; Schumacher, B. DNA damage responses and p53 in the aging process. *Blood* **2018**, *131*, 488–495. [CrossRef]

93. Aird, K.M.; Zhang, R. Detection of senescence-associated heterochromatin foci (SAHF). *Methods Mol. Biol.* **2013**, *965*, 185–196.

94. Evangelou, K.; Lougiakis, N.; Rizou, S.V.; Kotsinas, A.; Kletsas, D.; Munoz-Espin, D.; Kastrinakis, N.G.; Pouli, N.; Marakos, P.; Townsend, P.; et al. Robust, universal biomarker assay to detect senescent cells in biological specimens. *Aging Cell* **2017**, *16*, 192–197. [CrossRef]

95. Salminen, A.; Kaarniranta, K.; Kauppinen, A. Immunosenescence: The potential role of myeloid-derived suppressor cells (MDSC) in age-related immune deficiency. *Cell. Mol. Life Sci.* **2019**, *76*, 1901–1918. [CrossRef]

96. Akbar, A.N.; Fletcher, J.M. Memory T cell homeostasis and senescence during aging. *Curr. Opin. Immunol.* **2005**, *17*, 480–485. [CrossRef]

97. Swain, S.; Clise-Dwyer, K.; Haynes, L. Homeostasis and the age-associated defect of CD4 T cells. *Semin. Immunol.* **2005**, *17*, 370–377. [CrossRef]

98. Singhal, S.K.; Roder, J.C.; Duwe, A.K. Suppressor cells in immunosenescence. *Fed. Proc.* **1978**, *37*, 1245–1252.

99. Strioga, M.; Pasukoniene, V.; Characiejus, D. CD8+ CD28- and CD8+ CD57+ T cells and their role in health and disease. *Immunology* **2011**, *134*, 17–32. [CrossRef]

100. Vallejo, A.N.; Weyand, C.M.; Goronzy, J.J. T-cell senescence: A culprit of immune abnormalities in chronic inflammation and persistent infection. *Trends Mol. Med.* **2004**, *10*, 119–124. [CrossRef]

101. Eck, S.C.; Chang, D.; Wells, A.D.; Turka, L.A. Differential down-regulation of CD28 by B7-1 and B7-2 engagement. *Transplantation* **1997**, *64*, 1497–1499. [CrossRef]

102. Laidlaw, B.J.; Craft, J.E.; Kaech, S.M. The multifaceted role of CD4(+) T cells in CD8(+) T cell memory. *Nat. Rev. Immunol.* **2016**, *16*, 102–111. [CrossRef]

103. Weng, N.P.; Akbar, A.N.; Goronzy, J. CD28(-) T cells: Their role in the age-associated decline of immune function. *Trends Immunol.* **2009**, *30*, 306–312. [CrossRef]

104. Qin, L.; Jing, X.; Qiu, Z.; Cao, W.; Jiao, Y.; Routy, J.P.; Li, T. Aging of immune system: Immune signature from peripheral blood lymphocyte subsets in 1068 healthy adults. *Aging* **2016**, *8*, 848–859. [CrossRef]

105. Valenzuela, H.F.; Effros, R.B. Divergent telomerase and CD28 expression patterns in human CD4 and CD8 T cells following repeated encounters with the same antigenic stimulus. *Clin. Immunol.* **2002**, *105*, 117–125. [CrossRef]

106. Khan, N.; Shariff, N.; Cobbold, M.; Bruton, R.; Ainsworth, J.A.; Sinclair, A.J.; Nayak, L.; Moss, P.A. Cytomegalovirus seropositivity drives the CD8 T cell repertoire toward greater clonality in healthy elderly individuals. *J. Immunol.* **2002**, *169*, 1984–1992. [CrossRef]

107. Paillard, F.; Sterkers, G.; Vaquero, C. Transcriptional and post-transcriptional regulation of tcr, CD4 and CD8 gene expression during activation of normal human T lymphocytes. *EMBO J.* **1990**, *9*, 1867–1872. [CrossRef]

108. Mou, D.; Espinosa, J.; Lo, D.J.; Kirk, A.D. CD28 negative T cells: Is their loss our gain? *Am. J. Transpl.* **2014**, *14*, 2460–2466. [CrossRef]

109. Tarazona, R.; DelaRosa, O.; Alonso, C.; Ostos, B.; Espejo, J.; Pena, J.; Solana, R. Increased expression of NK cell markers on T lymphocytes in aging and chronic activation of the immune system reflects the accumulation of effector/senescent T cells. *Mech. Ageing Dev.* **2000**, *121*, 77–88. [CrossRef]

110. Seyda, M.; Elkhal, A.; Quante, M.; Falk, C.S.; Tullius, S.G. T cells going innate. *Trends Immunol.* **2016**, *37*, 546–556. [CrossRef]

111. Bauer, S.; Groh, V.; Wu, J.; Steinle, A.; Phillips, J.H.; Lanier, L.L.; Spies, T. Activation of nk cells and T cells by NKG2D, a receptor for stress-inducible mica. *Science* **1999**, *285*, 727–729. [CrossRef]

112. Verneris, M.R.; Karimi, M.; Baker, J.; Jayaswal, A.; Negrin, R.S. Role of NKG2D signaling in the cytotoxity of activated and expanded CD8+ T cells. *Blood* **2004**, *103*, 3065–3072. [CrossRef]

113. Prajapati, K.; Perez, C.; Rojas, L.B.P.; Burke, B.; Guevara-Patino, J.A. Functions of NKG2D in CD8(+) T cells: An opportunity for immunotherapy. *Cell. Mol. Immunol.* **2018**, *15*, 470–479. [CrossRef]

114. Yi, H.S.; Kim, S.Y.; Kim, J.T.; Lee, Y.S.; Moon, J.S.; Kim, M.; Kang, Y.E.; Joung, K.H.; Lee, J.H.; Kim, H.J.; et al. T-cell senescence contributes to abnormal glucose homeostasis in humans and mice. *Cell Death Dis.* **2019**, *10*, 249. [CrossRef]

115. Onyema, O.O.; Decoster, L.; Njemini, R.; Forti, L.N.; Bautmans, I.; De Waele, M.; Mets, T. Shifts in subsets of CD8+ T-cells as evidence of immunosenescence in patients with cancers affecting the lungs: An observational case-control study. *BMC Cancer* **2015**, *15*, 1016. [CrossRef]

116. Brenchley, J.M.; Karandikar, N.J.; Betts, M.R.; Ambrozak, D.R.; Hill, B.J.; Crotty, L.E.; Casazza, J.P.; Kuruppu, J.; Migueles, S.A.; Connors, M.; et al. Expression of CD57 defines replicative senescence and antigen-induced apoptotic death of CD8+ T cells. *Blood* **2003**, *101*, 2711–2720. [CrossRef]

117. Gunturi, A.; Berg, R.E.; Forman, J. Preferential survival of CD8 T and NK cells expressing high levels of CD94. *J. Immunol.* **2003**, *170*, 1737–1745. [CrossRef]

118. Appay, V.; Nixon, D.F.; Donahoe, S.M.; Gillespie, G.M.; Dong, T.; King, A.; Ogg, G.S.; Spiegel, H.M.; Conlon, C.; Spina, C.A.; et al. Hiv-specific CD8(+) T cells produce antiviral cytokines but are impaired in cytolytic function. *J. Exp. Med.* **2000**, *192*, 63–75. [CrossRef]

119. Kiniry, B.E.; Hunt, P.W.; Hecht, F.M.; Somsouk, M.; Deeks, S.G.; Shacklett, B.L. Differential expression of CD8(+) T cell cytotoxic effector molecules in blood and gastrointestinal mucosa in HIV-1 infection. *J. Immunol.* **2018**, *200*, 1876–1888. [CrossRef]

120. Reuter, M.A.; Del Rio Estrada, P.M.; Buggert, M.; Petrovas, C.; Ferrando-Martinez, S.; Nguyen, S.; Sada Japp, A.; Ablanedo-Terrazas, Y.; Rivero-Arrieta, A.; Kuri-Cervantes, L.; et al. HIV-specific CD8(+) T cells exhibit reduced and differentially regulated cytolytic activity in lymphoid tissue. *Cell Rep.* **2017**, *21*, 3458–3470. [CrossRef]

121. Hodge, G.; Hodge, S. Steroid resistant CD8(+)CD28(null) nkt-like pro-inflammatory cytotoxic cells in chronic obstructive pulmonary disease. *Front. Immunol.* **2016**, *7*, 617. [CrossRef]

122. Cortesini, R.; LeMaoult, J.; Ciubotariu, R.; Cortesini, N.S. Cd8+CD28- t suppressor cells and the induction of antigen-specific, antigen-presenting cell-mediated suppression of th reactivity. *Immunol. Rev.* **2001**, *182*, 201–206. [CrossRef]

123. Plaumann, J.; Engelhardt, M.; Awwad, M.H.S.; Echchannaoui, H.; Amman, E.; Raab, M.S.; Hillengass, J.; Halama, N.; Neuber, B.; Muller-Tidow, C.; et al. IL-10 inducible CD8(+) regulatory T-cells are enriched in patients with multiple myeloma and impact the generation of antigen-specific T-cells. *Cancer Immunol. Immunother.* **2018**, *67*, 1695–1707. [CrossRef]

124. Geng, L.; Liu, J.; Huang, J.; Lin, B.; Yu, S.; Shen, T.; Wang, Z.; Yang, Z.; Zhou, L.; Zheng, S. A high frequency of CD8(+)CD28(-) t-suppressor cells contributes to maintaining stable graft function and reducing immunosuppressant dosage after liver transplantation. *Int. J. Med. Sci.* **2018**, *15*, 892–899. [CrossRef]

125. Vieyra-Lobato, M.R.; Vela-Ojeda, J.; Montiel-Cervantes, L.; Lopez-Santiago, R.; Moreno-Lafont, M.C. Description of CD8(+) regulatory T lymphocytes and their specific intervention in graft-versus-host and infectious diseases, autoimmunity, and cancer. *J. Immunol. Res* **2018**, *2018*, 3758713. [CrossRef]

126. Manavalan, J.S.; Rossi, P.C.; Vlad, G.; Piazza, F.; Yarilina, A.; Cortesini, R.; Mancini, D.; Suciu-Foca, N. High expression of ILT3 and ILT4 is a general feature of tolerogenic dendritic cells. *Transpl. Immunol.* **2003**, *11*, 245–258. [CrossRef]

127. Ye, J.; Huang, X.; Hsueh, E.C.; Zhang, Q.; Ma, C.; Zhang, Y.; Varvares, M.A.; Hoft, D.F.; Peng, G. Human regulatory T cells induce T-lymphocyte senescence. *Blood* **2012**, *120*, 2021–2031. [CrossRef]

128. Sun, Z.; Zhong, W.; Lu, X.; Shi, B.; Zhu, Y.; Chen, L.; Zhang, G.; Zhang, X. Association of graves' disease and prevalence of circulating IFN-γ-producing CD28(-) T cells. *J. Clin. Immunol.* **2008**, *28*, 464–472. [CrossRef]

129. Maly, K.; Schirmer, M. The story of CD4+ CD28- T cells revisited: Solved or still ongoing? *J. Immunol. Res.* **2015**, *2015*, 348746.

130. Tulunay, A.; Yavuz, S.; Direskeneli, H.; Eksioglu-Demiralp, E. CD8+CD28-, suppressive T cells in systemic lupus erythematosus. *Lupus* **2008**, *17*, 630–637. [CrossRef]

131. Najafian, N.; Chitnis, T.; Salama, A.D.; Zhu, B.; Benou, C.; Yuan, X.; Clarkson, M.R.; Sayegh, M.H.; Khoury, S.J. Regulatory functions of CD8+CD28- T cells in an autoimmune disease model. *J. Clin. Investig.* **2003**, *112*, 1037–1048. [CrossRef]

132. Manavalan, J.S.; Kim-Schulze, S.; Scotto, L.; Naiyer, A.J.; Vlad, G.; Colombo, P.C.; Marboe, C.; Mancini, D.; Cortesini, R.; Suciu-Foca, N. Alloantigen specific CD8+CD28- FOXP3+ T suppressor cells induce ILT3+ ILT4+ tolerogenic endothelial cells, inhibiting alloreactivity. *Int. Immunol.* **2004**, *16*, 1055–1068. [CrossRef]

133. Chang, C.C.; Ciubotariu, R.; Manavalan, J.S.; Yuan, J.; Colovai, A.I.; Piazza, F.; Lederman, S.; Colonna, M.; Cortesini, R.; Dalla-Favera, R.; et al. Tolerization of dendritic cells by T(s) cells: The crucial role of inhibitory receptors ILT3 and ILT4. *Nat. Immunol.* **2002**, *3*, 237–243. [CrossRef]

134. Topalian, S.L.; Taube, J.M.; Anders, R.A.; Pardoll, D.M. Mechanism-driven biomarkers to guide immune checkpoint blockade in cancer therapy. *Nat. Rev. Cancer* **2016**, *16*, 275–287. [CrossRef]

135. Tomaszewski, W.; Sanchez-Perez, L.; Gajewski, T.F.; Sampson, J.H. Brain tumor microenvironment and host state: Implications for immunotherapy. *Clin. Cancer Res.* **2019**. [CrossRef]

136. Wargo, J.A.; Reddy, S.M.; Reuben, A.; Sharma, P. Monitoring immune responses in the tumor microenvironment. *Curr. Opin. Immunol.* **2016**, *41*, 23–31. [CrossRef]

137. Woroniecka, K.; Chongsathidkiet, P.; Rhodin, K.; Kemeny, H.; Dechant, C.; Farber, S.H.; Elsamadicy, A.A.; Cui, X.; Koyama, S.; Jackson, C.; et al. T-cell exhaustion signatures vary with tumor type and are severe in glioblastoma. *Clin. Cancer Res.* **2018**, *24*, 4175–4186. [CrossRef]

138. Mirzaei, R.; Sarkar, S.; Yong, V.W. T cell exhaustion in glioblastoma: Intricacies of immune checkpoints. *Trends Immunol.* **2017**, *38*, 104–115. [CrossRef]

139. Nduom, E.K.; Weller, M.; Heimberger, A.B. Immunosuppressive mechanisms in glioblastoma. *Neuro Oncol.* **2015**, *17* (Suppl. 7), vii9–vii14. [CrossRef]

140. Wherry, E.J. T cell exhaustion. *Nat. Immunol.* **2011**, *12*, 492–499. [CrossRef]

141. Pardoll, D.M. The blockade of immune checkpoints in cancer immunotherapy. *Nat. Rev. Cancer* **2012**, *12*, 252–264. [CrossRef]

142. Hodi, F.S.; O'Day, S.J.; McDermott, D.F.; Weber, R.W.; Sosman, J.A.; Haanen, J.B.; Gonzalez, R.; Robert, C.; Schadendorf, D.; Hassel, J.C.; et al. Improved survival with ipilimumab in patients with metastatic melanoma. *N. Engl. J. Med.* **2010**, *363*, 711–723. [CrossRef]

143. Darvin, P.; Toor, S.M.; Sasidharan Nair, V.; Elkord, E. Immune checkpoint inhibitors: Recent progress and potential biomarkers. *Exp. Mol. Med.* **2018**, *50*, 165. [CrossRef]

144. Cheng, W.; Fu, D.; Xu, F.; Zhang, Z. Unwrapping the genomic characteristics of urothelial bladder cancer and successes with immune checkpoint blockade therapy. *Oncogenesis* **2018**, *7*, 2. [CrossRef]

145. Lauko, A.; Thapa, B.; Venur, V.A.; Ahluwalia, M.S. Management of brain metastases in the new era of checkpoint inhibition. *Curr. Neurol. Neurosci. Rep.* **2018**, *18*, 70. [CrossRef]

146. Johnson, D.B.; Sullivan, R.J.; Menzies, A.M. Immune checkpoint inhibitors in challenging populations. *Cancer* **2017**, *123*, 1904–1911. [CrossRef]

147. Meloni, F.; Morosini, M.; Solari, N.; Passadore, I.; Nascimbene, C.; Novo, M.; Ferrari, M.; Cosentino, M.; Marino, F.; Pozzi, E.; et al. Foxp3 expressing CD4+ CD25+ and CD8+CD28- T regulatory cells in the peripheral blood of patients with lung cancer and pleural mesothelioma. *Hum. Immunol.* **2006**, *67*, 1–12. [CrossRef]

148. Chen, C.; Chen, D.; Zhang, Y.; Chen, Z.; Zhu, W.; Zhang, B.; Wang, Z.; Le, H. Changes of CD4+CD25+FOXP3+ and CD8+CD28- regulatory T cells in non-small cell lung cancer patients undergoing surgery. *Int. Immunopharmacol.* **2014**, *18*, 255–261. [CrossRef]

149. Casado, J.G.; Soto, R.; DelaRosa, O.; Peralbo, E.; del Carmen Munoz-Villanueva, M.; Rioja, L.; Pena, J.; Solana, R.; Tarazona, R. CD8 T cells expressing nk associated receptors are increased in melanoma patients and display an effector phenotype. *Cancer Immunol. Immunother.* **2005**, *54*, 1162–1171. [CrossRef]

150. Liu, X.; Mo, W.; Ye, J.; Li, L.; Zhang, Y.; Hsueh, E.C.; Hoft, D.F.; Peng, G. Regulatory T cells trigger effector T cell DNA damage and senescence caused by metabolic competition. *Nat. Commun.* **2018**, *9*, 249. [CrossRef]

151. Ye, J.; Ma, C.; Hsueh, E.C.; Eickhoff, C.S.; Zhang, Y.; Varvares, M.A.; Hoft, D.F.; Peng, G. Tumor-derived gammadelta regulatory T cells suppress innate and adaptive immunity through the induction of immunosenescence. *J. Immunol.* **2013**, *190*, 2403–2414. [CrossRef]

152. Ye, J.; Peng, G. Controlling T cell senescence in the tumor microenvironment for tumor immunotherapy. *Oncoimmunology* **2015**, *4*, e994398. [CrossRef]

153. Lanna, A.; Henson, S.M.; Escors, D.; Akbar, A.N. The kinase p38 activated by the metabolic regulator AMPK and scaffold TAB1 drives the senescence of human T cells. *Nat. Immunol.* **2014**, *15*, 965–972. [CrossRef]

154. Ye, J.; Ma, C.; Hsueh, E.C.; Dou, J.; Mo, W.; Liu, S.; Han, B.; Huang, Y.; Zhang, Y.; Varvares, M.A.; et al. TLR8 signaling enhances tumor immunity by preventing tumor-induced T-cell senescence. *EMBO Mol. Med.* **2014**, *6*, 1294–1311. [CrossRef]

155. Lim, M.; Xia, Y.; Bettegowda, C.; Weller, M. Current state of immunotherapy for glioblastoma. *Nat. Rev. Clin. Oncol.* **2018**, *15*, 422–442. [CrossRef]

156. Broekman, M.L.; Maas, S.L.N.; Abels, E.R.; Mempel, T.R.; Krichevsky, A.M.; Breakefield, X.O. Multidimensional communication in the microenvirons of glioblastoma. *Nat. Rev. Neurol.* **2018**, *14*, 482–495. [CrossRef]

157. Quail, D.F.; Joyce, J.A. The microenvironmental landscape of brain tumors. *Cancer Cell* **2017**, *31*, 326–341. [CrossRef]

158. Sayour, E.J.; McLendon, P.; McLendon, R.; De Leon, G.; Reynolds, R.; Kresak, J.; Sampson, J.H.; Mitchell, D.A. Increased proportion of FoxP3+ regulatory T cells in tumor infiltrating lymphocytes is associated with tumor recurrence and reduced survival in patients with glioblastoma. *Cancer Immunol. Immunother.* **2015**, *64*, 419–427. [CrossRef]

159. Fornara, O.; Odeberg, J.; Wolmer Solberg, N.; Tammik, C.; Skarman, P.; Peredo, I.; Stragliotto, G.; Rahbar, A.; Soderberg-Naucler, C. Poor survival in glioblastoma patients is associated with early signs of immunosenescence in the CD4 T-cell compartment after surgery. *Oncoimmunology* **2015**, *4*, e1036211. [CrossRef]

160. Wang, R.F. CD8+ regulatory T cells, their suppressive mechanisms, and regulation in cancer. *Hum. Immunol.* **2008**, *69*, 811–814. [CrossRef]

161. Kmiecik, J.; Poli, A.; Brons, N.H.; Waha, A.; Eide, G.E.; Enger, P.O.; Zimmer, J.; Chekenya, M. Elevated CD3+ and CD8+ tumor-infiltrating immune cells correlate with prolonged survival in glioblastoma patients despite integrated immunosuppressive mechanisms in the tumor microenvironment and at the systemic level. *J. Neuroimmunol.* **2013**, *264*, 71–83. [CrossRef]

162. Vlad, G.; Cortesini, R.; Suciu-Foca, N. CD8+ T suppressor cells and the ILT3 master switch. *Hum. Immunol.* **2008**, *69*, 681–686. [CrossRef]

163. Ladomersky, E.; Scholtens, D.M.; Kocherginsky, M.; Hibler, E.A.; Bartom, E.T.; Otto-Meyer, S.; Zhai, L.; Lauing, K.L.; Choi, J.; Sosman, J.A.; et al. The coincidence between increasing age, immunosuppression, and the incidence of patients with glioblastoma. *Front. Pharmacol.* **2019**, *10*, 200. [CrossRef]

164. Lamas, A.; Lopez, E.; Carrio, R.; Lopez, D.M. Adipocyte and leptin accumulation in tumor-induced thymic involution. *Int. J. Mol. Med.* **2016**, *37*, 133–138. [CrossRef]

165. Kasakovski, D.; Xu, L.; Li, Y. T cell senescence and CAR-T cell exhaustion in hematological malignancies. *J. Hematol. Oncol.* **2018**, *11*, 91. [CrossRef]

166. Short, S.; Fielder, E.; Miwa, S.; von Zglinicki, T. Senolytics and senostatics as adjuvant tumour therapy. *EBioMedicine* **2019**, *41*, 683–692. [CrossRef]

167. Baar, M.P.; Brandt, R.M.C.; Putavet, D.A.; Klein, J.D.D.; Derks, K.W.J.; Bourgeois, B.R.M.; Stryeck, S.; Rijksen, Y.; van Willigenburg, H.; Feijtel, D.A.; et al. Targeted apoptosis of senescent cells restores tissue homeostasis in response to chemotoxicity and aging. *Cell* **2017**, *169*, 132–147.e16. [CrossRef]

168. Kirkland, J.L.; Tchkonia, T.; Zhu, Y.; Niedernhofer, L.J.; Robbins, P.D. The clinical potential of senolytic drugs. *J. Am. Geriatr. Soc.* **2017**, *65*, 2297–2301. [CrossRef]

169. Rueff, J.; Medinger, M.; Heim, D.; Passweg, J.; Stern, M. Lymphocyte subset recovery and outcome after autologous hematopoietic stem cell transplantation for plasma cell myeloma. *Biol. Blood Marrow Transpl.* **2014**, *20*, 896–899. [CrossRef]

170. Farge, D.; Arruda, L.C.; Brigant, F.; Clave, E.; Douay, C.; Marjanovic, Z.; Deligny, C.; Maki, G.; Gluckman, E.; Toubert, A.; et al. Long-term immune reconstitution and T cell repertoire analysis after autologous hematopoietic stem cell transplantation in systemic sclerosis patients. *J. Hematol. Oncol.* **2017**, *10*, 21. [CrossRef]

171. Zhang, P.; Kishimoto, Y.; Grammatikakis, I.; Gottimukkala, K.; Cutler, R.G.; Zhang, S.; Abdelmohsen, K.; Bohr, V.A.; Misra Sen, J.; Gorospe, M.; et al. Senolytic therapy alleviates abeta-associated oligodendrocyte progenitor cell senescence and cognitive deficits in an alzheimer's disease model. *Nat. Neurosci.* **2019**, *22*, 719–728. [CrossRef]

172. Allsopp, R. Telomere length and ipsc re-programming: Survival of the longest. *Cell Res.* **2012**, *22*, 614–615. [CrossRef]

173. Parish, S.T.; Wu, J.E.; Effros, R.B. Sustained CD28 expression delays multiple features of replicative senescence in human CD8 T lymphocytes. *J. Clin. Immunol.* **2010**, *30*, 798–805. [CrossRef]

174. Le Page, A.; Fortin, C.; Garneau, H.; Allard, N.; Tsvetkova, K.; Tan, C.T.; Larbi, A.; Dupuis, G.; Fulop, T. Downregulation of inhibitory SRC homology 2 domain-containing phosphatase-1 (SHP-1) leads to recovery of T cell responses in elderly. *Cell Commun. Signal.* **2014**, *12*, 2. [CrossRef]

175. Karagiannis, P.; Iriguchi, S.; Kaneko, S. Reprogramming away from the exhausted T cell state. *Semin. Immunol.* **2016**, *28*, 35–44. [CrossRef]

176. Themeli, M.; Kloss, C.C.; Ciriello, G.; Fedorov, V.D.; Perna, F.; Gonen, M.; Sadelain, M. Generation of tumor-targeted human T lymphocytes from induced pluripotent stem cells for cancer therapy. *Nat. Biotechnol.* **2013**, *31*, 928–933. [CrossRef]

177. Kaneko, S. In vitro generation of antigen-specific T cells from induced pluripotent stem cells of antigen-specific T cell origin. *Methods Mol. Biol.* **2016**, *1393*, 67–73.

178. Ramos, C.A.; Rouce, R.; Robertson, C.S.; Reyna, A.; Narala, N.; Vyas, G.; Mehta, B.; Zhang, H.; Dakhova, O.; Carrum, G.; et al. In vivo fate and activity of second- versus third-generation CD19-specific CAR-T cells in B cell non-hodgkin's lymphomas. *Mol. Ther.* **2018**, *26*, 2727–2737. [CrossRef]

179. Petersen, C.T.; Krenciute, G. Next generation CAR T cells for the immunotherapy of high-grade glioma. *Front. Oncol.* **2019**, *9*, 69. [CrossRef]

180. Sahin, A.; Sanchez, C.; Bullain, S.; Waterman, P.; Weissleder, R.; Carter, B.S. Development of third generation anti-egfrviii chimeric T cells and egfrviii-expressing artificial antigen presenting cells for adoptive cell therapy for glioma. *PLoS ONE* **2018**, *13*, e0199414. [CrossRef]

181. June, C.H.; Sadelain, M. Chimeric antigen receptor therapy. *N. Engl. J. Med.* **2018**, *379*, 64–73. [CrossRef]

182. Al-Chami, E.; Tormo, A.; Pasquin, S.; Kanjarawi, R.; Ziouani, S.; Rafei, M. Interleukin-21 administration to aged mice rejuvenates their peripheral T-cell pool by triggering de novo thymopoiesis. *Aging Cell* **2016**, *15*, 349–360. [CrossRef]

183. Fan, Y.; Tajima, A.; Goh, S.K.; Geng, X.; Gualtierotti, G.; Grupillo, M.; Coppola, A.; Bertera, S.; Rudert, W.A.; Banerjee, I.; et al. Bioengineering thymus organoids to restore thymic function and induce donor-specific immune tolerance to allografts. *Mol. Ther.* **2015**, *23*, 1262–1277. [CrossRef]

184. Tajima, A.; Pradhan, I.; Trucco, M.; Fan, Y. Restoration of thymus function with bioengineered thymus organoids. *Curr. Stem Cell Rep.* **2016**, *2*, 128–139. [CrossRef]
185. Akbar, A.N.; Henson, S.M. Are senescence and exhaustion intertwined or unrelated processes that compromise immunity? *Nat. Rev. Immunol.* **2011**, *11*, 289–295. [CrossRef]

International Journal of
Molecular Sciences

Article

A Small Number of HER2 Redirected CAR T Cells Significantly Improves Immune Response of Adoptively Transferred Mouse Lymphocytes against Human Breast Cancer Xenografts

Gábor Tóth [1], János Szöllősi [1,2], Hinrich Abken [3], György Vereb [1,2,4,*] and Árpád Szöőr [1,*]

[1] Faculty of Medicine, Department of Biophysics and Cell Biology, University of Debrecen, 4032 Debrecen, Hungary; tothgab@med.unideb.hu (G.T.); szollo@med.unideb.hu (J.S.)
[2] MTA-DE Cell Biology and Signaling Research Group, Faculty of Medicine, University of Debrecen, 4032 Debrecen, Hungary
[3] Regensburg Center for Interventional Immunology, Dept. Genetic Immunotherapy, and University Hospital Regensburg, D-93053 Regensburg, Germany; Hinrich.Abken@klinik.uni-regensburg.de
[4] Faculty of Pharmacy, University of Debrecen, 4032 Debrecen, Hungary
* Correspondence: gvereb2020@gmail.com (G.V.); akuka@med.unideb.hu (A.S.); Tel.: +36-52-258-603 (G.V. & A.S.)

Received: 30 November 2019; Accepted: 31 January 2020; Published: 4 February 2020

Abstract: HER2 positive JIMT-1 breast tumors are resistant to trastuzumab treatment in vitro and develop resistance to trastuzumab in vivo in SCID mice. We explored whether these resistant tumors could still be eliminated by T cells redirected by a second-generation chimeric antigen receptor (CAR) containing a CD28 costimulatory domain and targeting HER2 with a trastuzumab-derived scFv. In vitro, T cells engineered with this HER2 specific CAR recognized HER2 positive target cells as judged by cytokine production and cytolytic activity. In vivo, the administration of trastuzumab twice weekly had no effect on the growth of JIMT-1 xenografts in SCID mice. At the same time, a single dose of 2.5 million T cells from congenic mice exhibited a moderate xenoimmune response and even stable disease in some cases. In contrast, when the same dose contained 7% (175,000) CAR T cells, complete remission was achieved in 57 days. Even a reduced dose of 250,000 T cells, including only 17,500 CAR T cells, yielded complete remission, although it needed nearly twice the time. We conclude that even a small number of CAR T lymphocytes can evoke a robust anti-tumor response against an antibody resistant xenograft by focusing the activity of xenogenic T cells. This observation may have significance for optimizing the dose of CAR T cells in the therapy of solid tumors.

Keywords: breast cancer; trastuzumab; chimeric antigen receptor; immunotherapy; cell therapy

---

1. Introduction

Human epidermal growth factor receptor 2 (HER2) is overexpressed in 20–25% of breast cancer tumors [1]. HER2 expression is associated with an aggressive disease with a high recurrence rate and increased mortality [2]. Specific monoclonal antibody therapy has revolutionized the treatment of HER2 positive breast cancer since the FDA (U.S. Food and Drug Administration) approval of trastuzumab (Herceptin®) in 1998 [3]. The addition of trastuzumab to chemotherapy results in a lower rate of death after one year (22 percent vs. 33 percent, $P = 0.008$), longer survival (median survival, 25.1 vs. 20.3 months; $P = 0.046$), and a 20 percent reduction in the risk of death [4]. Despite the success, resistance to therapeutic antibodies is a clinical reality that affects the outcome of 60–80% of HER2+ breast cancer patients [5]. One of the underlying mechanisms is epitope masking by components of the tumor microenvironment (TME) such as the MUC4 (mucin 4) or the CD44/Hyaluronan complex [6–11].

The JIMT-1 cell line was established from the pleural metastasis of a breast cancer patient and has recapitulated the trastuzumab resistance of the original tumor in vitro and also in vivo if treatment of JIMT-1 xenografts SCID mice was initiated at a few hundred mm^3 tumor volumes [12–14]. Our recent data indicate that simultaneous targeting of two epitopes on the HER2 molecule with clinical doses of trastuzumab and pertuzumab additionally improves the efficacy of antibody-dependent cellular cytotoxicity and thereby also the anti-tumor response; however, eventually all JIMT-1 xenografts become resistant to antibody treatment at a certain tumor size [15].

In such cases of antibody resistance, Chimeric Antigen Receptor (CAR) engineered T cells [16] represent an appealing option for improving the outcome for patients with advanced breast cancer [17–20]. Several tumor-associated membrane proteins are targeted in clinical trials by CAR T cells, including HER2 (NCT02547961, NCT02713984), CEA (NCT02349724) and mesothelin (NCT02792114). While no results have been disclosed of current HER2 targeting trials, the first reported clinical use of HER2 specific CAR T cells resulted in a serious adverse event following CAR T cell infusion [21]. In this trial, a HER2 positive colon cancer patient was treated with a large number (10^{10}) of CD28-41BB costimulatory (3rd-generation) CAR T cells, which derived their antigen specificity from trastuzumab. The patient developed respiratory distress, followed by multiple cardiac arrests over the course of 5 days, leading to death. The death of this patient may have occurred due to the result of HER2 recognition of highly active and numerous anti-HER2 CAR T cells in the normal lung tissue that caused pulmonary toxicity and edema followed by a cytokine release storm causing multiorgan failure. The immune-mediated recognition of tumor antigens in normal tissues is referred to as "on-target, off-tumor" toxicity. It is thus clear from both preclinical experiments and clinical trials that while CAR T cell-based immune therapy has great potential to improve the outcome for patients with HER2 positive tumors, it still needs plentiful optimization.

Here, we report the generation of mouse T cells that are genetically modified to express a chimeric antigen receptor that consists of a HER2 specific single-chain variable fragment (scFv) derived from trastuzumab, a CD28 costimulatory endodomain, and a CD3z intracellular signaling domain. We demonstrate that these T cells recognize and kill HER2+ tumor cells in vitro and significantly improve the xenogenic immune response against human breast cancer even at very low numbers (17,500), resulting in complete tumor regression, and significant survival advantage.

2. Results

2.1. Generation of Murine HER2 Specific CAR T Cells

To genetically modify mouse T cells (Figure 1), first, we generated VSVG-pseudotyped retroviral particles encoding HER2 specific chimeric antigen receptors (Figure 1A). T cells were isolated from the freshly dissected spleen of congenic Balb/c mice and activated by anti-mouse CD3e and anti-mouse CD28 antibodies. After 24 h, the medium was supplemented with mouse interleukin 2. Finally, activated mouse T cells were retrovirally transduced on RetroNectin-coated plates (Figure 1B).

The CAR contains an scFv obtained from trastuzumab, an IgG1 CH2-CH3 extracellular stalk, a CD28 costimulatory endodomain and a CD3z effector domain (Figure 2A). Using trastuzumab as a recognition domain allowed us to compare the impact of CAR T cells as living drugs with the impact of antibodies. The mean transduction efficiency was 8.7% (range: 5.58–11.84%; *n* = 8) as judged by flow cytometric analysis of the HER2 specific scFv (Figure 2B,C). We confirmed that CARs are stably expressed and re-confirmed CAR expression on day 10.

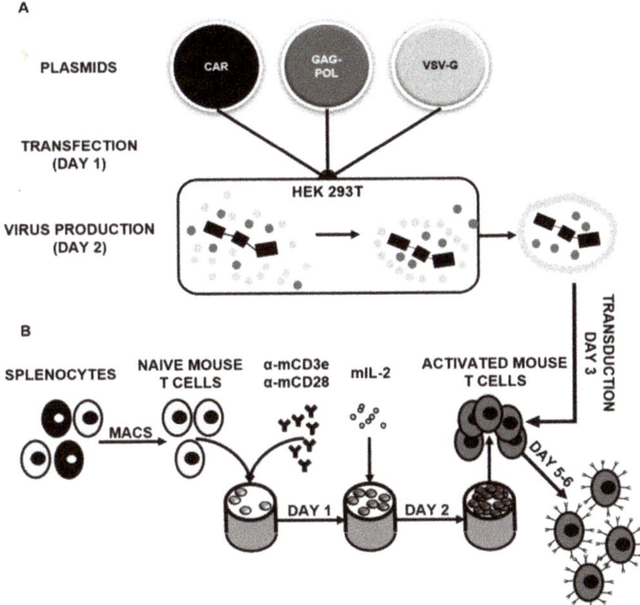

Figure 1. Genetic modification of mouse T cells with chimeric antigen receptors: (**A**) Scheme of retrovirus production. (**B**) Scheme of mouse T cell separation and activation.

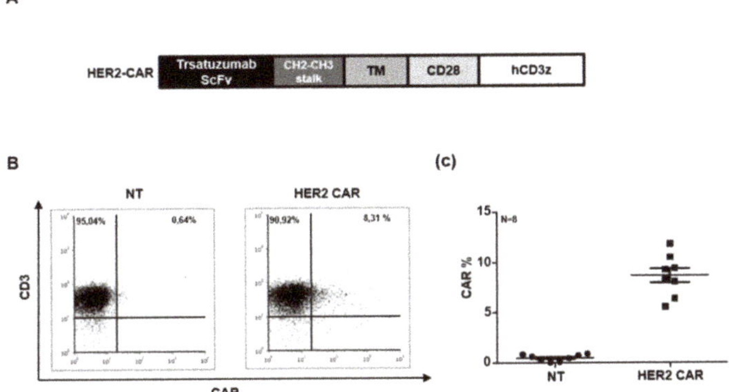

Figure 2. Generation of HER2 specific mouse CAR T cells: (**A**) Schematic diagram illustrating the modular composition of the retroviral vector encoding HER2 specific CAR. (**B,C**) Representative flow cytometry dot-plots and summary data (HER2 CAR mouse T cells (n = 8) and non-transduced (NT) mouse T cells (n = 8)).

2.2. HER2 Specific CARs Redirect Mouse T Cells to HER2 Positive Trastuzumab Resistant Tumor Cells

To demonstrate that the HER2 specific CAR redirects mouse T cells to HER2 positive target cells, we co-cultured HER2 CAR T cells with JIMT-1 cells in various effector to target ratios (from 2.5:1 to 0.01:1). HER2 specific CAR T cells recognized the HER2 positive tumor cells indicated by a significant increase in IFNg secretion ($p < 0.001$). Unmodified (NT) T cells did not induce mIFNg release (Figure 3A). While HER2 specific CAR induced T cells killed JIMT-1 tumor cells, no killing was observed in co-cultures with unmodified T cells ($p < 0.001$; Figure 3B). Taken together, the HER2 specific CAR activates mouse T cells in an antigen-dependent manner and induces antigen-dependent tumor cell killing.

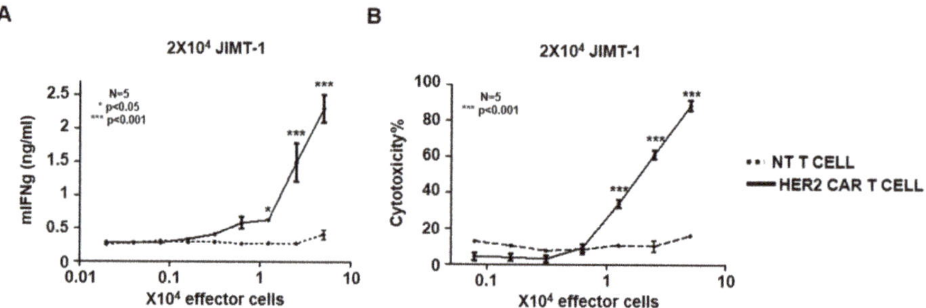

Figure 3. In vitro anti-tumor function of HER2 specific mouse CAR T cells: (**A**) HER2 CAR or non-transduced (NT) mouse T cells were co-cultured with HER2+ JIMT-1 cells at various (2.5:1–0.01:1) T cell to tumor cell ratios. After 48 h, IFNγ release was determined by ELISA ($n = 2$, assay performed in duplicates; HER2 CAR versus non-transduced (NT) T cells: * $p < 0.05$, *** $p < 0.001$). (**B**) XTT-based cytotoxicity assay using HER2 CAR T cells or non-modified mouse T cells and HER2 positive JIMT-1 cells as target at various (2.5:1–0.04:1) T cell to tumor cell ratios ($n = 2$; assay was performed in duplicates; HER2 CAR versus NT T cells: *** $p < 0.001$).

2.3. HER2 Specific CAR T Cells Have Antitumor Activity In Vivo against HER2+ Trastuzumab-Resistant Tumor Xenografts

To compare the anti-tumor function of antibody treatment and HER2-redirected CAR T cells, we established subcutaneous JIMT-1 xenografts (3×10^6 cells) in SCID mice (day −35, Figure 4). Mice were injected with 100 µg trastuzumab intraperitoneally twice weekly from day 0 (35 days after tumor cell injection), when average tumor size reached 800 mm³. Control animals were injected with PBS (Figure 4) and did not show delayed tumor growth and consequently their overall survival was not improved ($p = 0.79$, Figure 5). These data are in line with our previous observations [13].

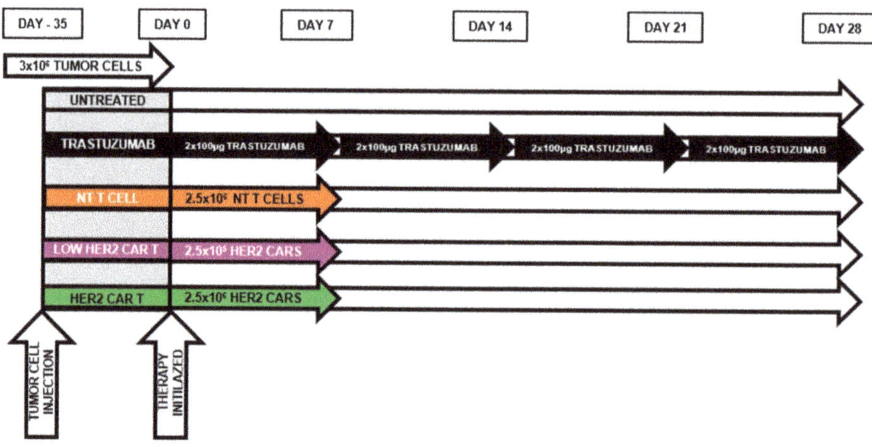

Figure 4. Outline of in vivo animal treatment schedule: Mice were s.c. injected with 3×10^6 JIMT-1 cells. 35 days later (on day 0), mice received 100 µL PBS buffer twice weekly (untreated, $n = 5$), 100 µg trastuzumab in 100 µL PBS buffer twice weekly (trastuzumab, $n = 5$), or an i.v. dose of 2.5×10^5 HER2 CAR mouse T cells (low-dose HER2 CAR, $n = 5$), or an i.v. dose of 2.5×10^6 HER2 CAR mouse T cells (HER2 CAR, $n = 5$). Tumor growth was followed by caliper and was derived as the product of the length, width and height.

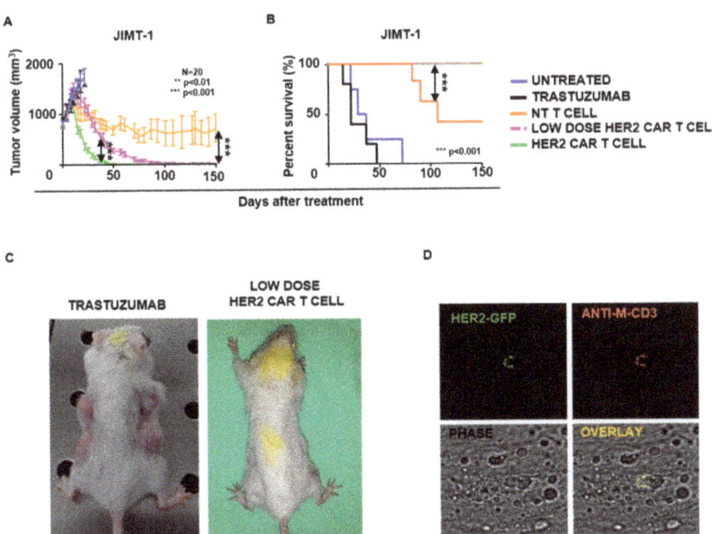

Figure 5. Antitumor activity of HER2-CAR mouse T cells in a xenograft model: (**A**) Quantitative measurement of tumor volumes (volume = mm^3; low dose HER2-CAR versus HER2-CAR groups: *** $p < 0.001$; NT T cell versus HER2-CAR groups *** $p < 0.001$). (**B**) Kaplan-Meier survival curve (NT T cell versus both HER2 CAR groups *** $p < 0.001$). (**C**) Representative images of animals. (**D**) Representative images for the detection of HER2-CAR mouse T cells in JIMT-1 xenografts (field of view: 92 μm × 92 μm). HER2-CAR mouse T cells were visualized by AlexaFluor647 conjugated anti-mouse CD3e and HER2-GFP co-staining.

To evaluate the in vivo efficacy of HER2-CAR T cells against these trastuzumab resistant xenografts, 35 days after JIMT-1 inoculation mice were injected iv. with a single dose of 2.5×10^6 (HER2-CAR T cell group), or 2.5×10^5 (Low Dose HER2-CAR T cell group) congenic mouse T cells, 7% of them expressing the HER2-CAR. Control mice were treated with 2.5×10^6 unmodified T cells (NT T cell group) (Figure 4). In this group, the high number of mouse T cells, in some cases, delayed the progression of the human xenografts resulting in better overall survival in comparison to the untreated group (Figure 5A–C). In contrast, the same dose of mouse T cells, when transduced with the CAR at 7% efficiency, completely eradicated the tumors in 57 days and resulted in long-term tumor-free survival. Moreover, complete tumor regression was also observed, by day 105, in the low dose HER2-CAR T cell group in which animals received only 250,000 T cells, among them 17,500 CAR T cells (Figure 5A–C). There is a cell dose dependence of the rate of tumor regression ($p < 0.001$). Despite the difference in time to complete regression, all CAR T cell treated mice remained tumor free until the termination of the experiment (day 150). To assess on-target off-tumor toxicity, formaldehyde-fixed paraffin-embedded tissue section weremade from the heart and lungs of each sacrificed animal. HE-stained sections were characterized based on morphology by an expert histopathologist and showed no signs of mononuclear infiltration (Figure S1A,B). Visual inspection upon dissection also did not show signs of inflammation.

To demonstrate that HER2 specific CAR T cells penetrated the JIMT-1 xenografts, tumor samples from week 3 after CAR T cell injection were immunostained and analyzed by confocal microscopy. We found T cells positive for the CAR and mouse-CD3e confirming the presence of CAR T cells in tumor xenografts (Figure 5D). Taken together, we conclude that HER2 specific CAR T cells have potent in vivo antitumor activity and penetrate HER2 positive xenografts, which are not eliminated by trastuzumab treatment.

3. Discussion

In this study, we described the generation of HER2 specific CAR-modified mouse T cells that obtained their antigen specificity from trastuzumab, a HER2 specific monoclonal antibody applied in clinical practice. We demonstrated that these cells specifically recognize trastuzumab resistant HER2 positive JIMT-1 target cells in vitro. Moreover, in vivo a low dose of HER2 CAR T cell expressed potent anti-tumor activity in a trastuzumab-resistant mouse xenograft model.

Although HER2 specific CAR T cells are effective against breast cancer cells [19,22], the tumor cells preferentially used as targets were trastuzumab sensitive cell lines (SKBR3 or BT474) leaving the question unanswered whether trastuzumab-resistant xenografts can be successfully treated with trastuzumab-derived CAR T cells.

Incorporation of the trastuzumab scFv into the CAR backbone allowed us to compare the efficacy in tumor eradication by cytolytic CAR T cells versus antibody mediated cytotoxicity.

SCID mice exhibit natural killer cell (NK) activity [23], through which trastuzumab treatment induces antibody dependent cellular cytotocixity against therapy sensitive xenografts (MCF7; BT-474) [24], which makes a direct comparison possible. In vivo, we confirmed that trastuzumab treatment has no potential to delay or revert the growth of established JIMT-1 tumors. which is in line with previous observations [11–14].

Although unmodified mouse T cells in co-cultures with JIMT-1 cells did not release cytokine and did not exhibit cytotoxicity, we wanted to confirm that xenogenic immune response does not reject human tumor xenografts in vivo [23]. A single injection of 2.5×10^6 unmodified mouse T cells on day 35 following JIMT-1 implantation could not eradicate the human tumor xenografts; however, in some cases, it caused tumor regression and resulted in stable disease.

In contrast to the control and trastuzumab-treated groups, 2.5 million mouse T cells with a proportion of 7% CAR-transduced cells (total ~175,000) caused a complete remission in 57. Even a tenth of this dose, 250,000 T cells including ~17,500 CAR T cells, was fully curative, although only in 105 days. The number of CAR T cells yielding complete remission in the latter case is only 0.2–0.3% of the usual 5 to 10 million CAR T cells used in successful mouse CAR-T therapy models [25]. Thus, it is likely that this small fraction of specifically redirected T lymphocytes successfully penetrates the tumor mass and evokes, in addition to direct tumor killing, a focusing effect that concentrates the xenogenic response of non-transduced mouse lymphocytes against the human tumor. By extrapolation, it is possible that in the case of HER2 positive solid tumors, a reduced number of CAR T cells could still maintain therapeutic efficacy through actively penetrating the tumor and enhancing the activity of tumor infiltrating lymphocytes (TILS, [26]). At the same time, while "on target off tumor" toxicity (mainly in the cardiopulmonary system), could be avoided owing to the lower expression of the HER2 target and the lack of TILs in healthy tissues.

Overall, we conclude that even a small number of CAR T lymphocytes can evoke a robust anti-tumor response against an antibody resistant xenograft by focusing the activity of xenogenic T cells. This observation may have significance for optimizing the dose of CAR T cells in the therapy of solid tumors.

4. Materials and Methods

All materials were from Sigma-Aldrich (St. Louis, MO, USA) unless otherwise indicated.

4.1. Cells and Culture Conditions

HEK 293T packaging cells were purchased from the American Type Culture Collection (ATCC, Manassas, VA, USA). Cells were cultured in Dulbecco's Modified Eagle Medium (DMEM) supplemented with 2 mmol/L Glutamax and 10% Fetal Calf Serum (FCS) and antibiotics. The JIMT-1 human breast cancer cell line was established in the laboratory of Cancer Biology, University of Tampere, Finland [12]. These cells were cultured in 1:1 ratio of Ham's F-12 and DMEM supplemented with 20% FCS, 300 U/L insulin, 2 mmol/L GlutaMAX and antibiotics. Primary mouse splenocytes, T cells and CAR T cells were cultured in RPMI 1640 supplemented with 2 mmol/L GlutaMAX, 10% FCS and antibiotics. All of the above-listed cells and cell lines were maintained in a humidified atmosphere containing 5% CO_2 at 37 °C and were routinely checked for the absence of mycoplasma contamination.

4.2. Retrovirus Production and Transduction of T Cells

Retroviral particles were generated by transient transfection of HEK 293T cells with the MSGV retroviral vector containing a trastuzumab derived HER2 specific CAR [17], a Peg-Pam-e plasmid containing the sequence for MoMLV gag-pol, and a pMEVSVg plasmid containing the sequence for VSVg, using jetPrime transfection reagent (Polyplus, Illkirch, France). Supernatants containing the retrovirus were collected after 48 h (Figure 1A).

To generate HER2 specific CAR T cells, T cells of syngenic Balb/c mice were isolated from a freshly dissected spleen by using a mouse-specific T cell isolation MACS sorting kit (130-095-130; Miltenyi Biotech; Bergisch Gladbach, Germany). MACS sorted mouse T cells were plated on non-tissue culture treated 24-well plates (5×10^6 cells/well), which were pre-coated with 1μg/manti-mouse CD3e (ThermoFischer, Waltham, MA, USA) and anti-mouse CD28 (R&D Systems, Min L neapolis, MN, USA) antibodies. After 24 h, mouse interleukin 2 (mIL2; 700 U/ml) was added to cultures. T cells were then transduced with the previously described retroviral particles on RetroNectin-coated (Takara, Kusatsu, Japan) plates on day 3 in the presence of mIL2 (200 U/m L). The expansion of T cells was subsequently supported with mIL2. Anti-mouse CD3e/CD28 activated non-transduced (NT) T cells were expanded in parallel with mIL2. Following 48-72h incubation, cells were collected and used for further experiments (Figure 1B).

4.3. Flow Cytometry

HER2 specific CAR expression was confirmed by a HER2-GFP recombinant protein. T cell purity was determined by Alexa Fluor 647 conjugated anti-mouse CD3 antibody (BD Biosciences, San Jose, CA, USA) staining. Both molecules were used at 10 μg/mL final concentration for 10 minutes on ice. Analysis was performed on at least 10,000 cells per sample using a FACS Calibur (Becton Dickinson, Franklin Lakes, NJ, USA) instrument and FCS Express 6 software (De Novo Software, Glendale, CA, USA).

4.4. CAR-Mediated T Cell Activation

Mouse CAR T cells and non-transduced controls were cultivated in indicator-free RPMI 1640 medium and 10% (*v/v*) FCS, without additional stimuli for 24 h, washed and incubated on 96-well round-bottomed plates in the presence of JIMT-1 target cells for 48 h. Culture supernatants were analyzed for IFN-γ by ELISA (BD Biosciences). IFN-γ was bound to a solid-phase mAb R46A2 and detected by a biotinylated mAb XMG1.2. The reaction product was visualized by a peroxidase-streptavidin conjugate (1:10,000) and ABTS as substrate. To monitor the cytolytic activity, genetically modified and control T cells were co-cultured with JIMT-1 target cells with increasing T cell numbers for 48 h in 96-well round-bottomed plates. Specific cytotoxicity was monitored by a 2,3-bis[2-methoxy-4-nitro-5-sulphophenyl]-2H-tetrazolium-5-carboxanilide salt (XTT)-based colorimetric assay ('Cell Proliferation Kit II', Roche Diagnostics, Risch, Switzerland). Reduction of XTT was determined as OD at 480 nm for treated tumor cells (Tu), for untreated tumor cells (Max) and for T cells only (T). Background (Bg) was measured in complete medium with XTT but no cells. Measurements were run with minimally 3 technical replicates in three independent experiments. Cytotoxicity was calculated as (1).

$$\text{Cytotoxicity (\%)} = \left(1 - \frac{Tu - T}{Max - Bg}\right) \cdot 100\%. \tag{1}$$

4.5. Xenograft Tumors and In Vivo Antibody Treatment

SCID (C.B-17/Icr-Prkdc^{scid}/IcrIcoCrl, Fox-Chase) mice were purchased from Charles River Laboratories, and housed in a specific-pathogen-free environment. All animal experiments were performed in accordance with FELASA guidelines and recommendations and DIN EN ISO 9001 standards. Only non-leaky SCID mice with murine IgG levels below 100 ng/mL were used. Each seven-week-old female SCID mouse participating in the study was given a subcutaneous injection in both flanks, each containing 3×10^6 JIMT-1 cells suspended in 100 μL PBS buffer and mixed with an equal volume of Matrigel (BD Biosciences, San Jose, CA, USA). Tumor volumes were derived as the product of the length, width and height measured with a caliper.

The trastuzumab group was treated with 100 μg trastuzumab intraperitoneally in 100 μL PBS twice weekly from day 35 post tumor cell injection. The untreated control group was injected with 100 μL PBS buffer i.p. twice weekly. In the HER2 CAR T cell and unmodified mouse T cell groups mice received a single dose of 2.5×10^6 effector cells i.v. on day 35 post JIMT-1 inoculation. In the low dose HER2 CAR T cell group mice received 2.5×10^5 cells i.v. (Figure 3A). At the end of the experiment, the animals were euthanized. Experiments were approved by the National Ethical Committee for Animal Research (# 5-1/2018/DEMÁB).

4.6. Tumor Xenograft Sections

At termination, mice were dissected, and fresh tumors were embedded in cryomatrix (Thermo Fischer Scientific, Waltham, MA, USA) and snap-frozen in isopentane submerged in liquid nitrogen. Serial 14 μm thick cryosections were made with a Shandon Cryotome (Thermo Fischer Scientific, Waltham, MA, USA) at −24 °C and air-dried. Staining was carried out at room temperature and all labeling molecules were diluted in PBS buffer supplemented with 1% BSA. After 5 min of rehydration in PBS buffer containing 1% BSA and 0.01% TritonX-100 (Thermo Fischer Scientific, Waltham, MA, USA) HER2 CAR mouse T cells were stained with HER2-GFP recombinant protein and Alexa Fluor 647 conjugated anti-mouse CD3e antibodies. Both molecules were used at 2 μg/mL concentration at 4 °C for 10 h. Sections were washed three times, for 5, 20, and 60 minutes, fixed in formaldehyde, and mounted in Mowiol antifade.

4.7. Haematoxylin and Eosin Stained Sections

The heart and lung were resected from sacrificed mice, fixed in 4% paraformaldehyde for 4–6 h, dehydrated in ethyl alcohol, and embedded in paraffin. Serial sections of 6 μm were cut with a microtome. Deparaffinized sections were HE stained using standard procedures and imaged using a Pannoramic digital histopathology scanner with a 20× objective in transmission mode. A trained histopathologist has examined the digital slides.

4.8. Confocal Laser Scanning Microscopy

Immunofluorescence-labeled tissue sections were analyzed with a confocal laser scanning microscope (LSM 510, Carl Zeiss GmbH, Jena, Germany) using a 40× C-Apochromat water immersion objective (NA = 1.2). GFP was excited at 488 nm and Alexa Fluor 647 at 633 nm. Corresponding fluorescence emissions were separated with an appropriate quad-band dichroic mirror, and detected through 505 to 550 nm bandpass and 650 nm longpass filters, respectively. Pinhole was set for 4 μm thick optical sections.

4.9. Statistical Analysis

GraphPad Prism 5 software (GraphPad software, Inc., La Jolla, CA) was used for statistical analysis. Data were presented as mean ± SD or ± SEM. For comparison between two groups, a two-tailed *t*-test was used. For comparisons of three or more groups, one-way ANOVA with Bonferroni's post-test was used. For the mouse experiments, survival, determined from the time of tumor cell injection, was analyzed by the Kaplan-Meier method and log-rank test. p-values < 0.05 were considered statistically significant.

Supplementary Materials: Supplementary materials can be found at http://www.mdpi.com/1422-0067/21/3/1039/s1.

Author Contributions: G.V. and Á.S. designed the study. G.T and Á.S. performed the experiments. H.A. provided reagents. G.T., J.S., H.A., G.V. and Á.S. analysed data. Á.S. wrote the manuscript, G.V., H.A. and Á.S. revised the manuscript. All authors have read and agreed to the published version of the manuscript.

Funding: We acknowledge the financial support from OTKA K119690 and FK132773 (the National Research, Development and Innovation Office, Hungary), GINOP-2.3.2-15-2016-00050 (co-financed by the European Union and the European Regional Development Fund), and Deutsche Krebshilfe, Bonn, FRG. A.S. was supported by the János Bolyai Research Scholarship of the Hungarian Academy of Sciences and by the UNKP-19- 4-DE-167 New National Excellence Program of the Ministry for Innovation and Technology.

Conflicts of Interest: The authors declare no conflict of interest.

Abbreviations

CAR	Chimeric antigen receptor
GFP	Green fluorescent protein
HER2	Human epidermal growth factor receptor 2
IFN-γ	Interferon gamma
PBS	Phosphate buffered saline
scFv	Single Chain Variable Fragment
SCID	Severe combined immunodeficiency

References

1. Hudis, C.A. Trastuzumab–mechanism of action and use in clinical practice. *N. Engl. J. Med.* **2007**, *357*, 39–51. [CrossRef] [PubMed]
2. Slamon, D.J.; Clark, G.M.; Wong, S.G.; Levin, W.J.; Ullrich, A.; McGuire, W.L. Human breast cancer: correlation of relapse and survival with amplification of the HER-2/neu oncogene. *Science* **1987**, *235*, 177–182. [CrossRef] [PubMed]
3. Friedlander, E.; Barok, M.; Szollosi, J.; Vereb, G. ErbB-directed immunotherapy: antibodies in current practice and promising new agents. *Immunol. Lett.* **2008**, *116*, 126–140. [CrossRef] [PubMed]
4. Slamon, D.J.; Leyland-Jones, B.; Shak, S.; Fuchs, H.; Paton, V.; Bajamonde, A.; Fleming, T.; Eiermann, W.; Wolter, J.; Pegram, M.; et al. Use of chemotherapy plus a monoclonal antibody against HER2 for metastatic breast cancer that overexpresses HER2. *N. Engl. J. Med.* **2001**, *344*, 783–792. [CrossRef] [PubMed]
5. Escriva-de-Romani, S.; Arumi, M.; Bellet, M.; Saura, C. HER2-positive breast cancer: Current and new therapeutic strategies. *Breast* **2018**, *39*, 80–88. [CrossRef] [PubMed]
6. Scott, G.K.; Robles, R.; Park, J.W.; Montgomery, P.A.; Daniel, J.; Holmes, W.E.; Lee, J.; Keller, G.A.; Li, W.L.; Fendly, B.M.; et al. A truncated intracellular HER2/neu receptor produced by alternative RNA processing affects growth of human carcinoma cells. *Mol. Cell Biol.* **1993**, *13*, 2247–2257. [CrossRef]
7. Nagy, P.; Friedlander, E.; Tanner, M.; Kapanen, A.I.; Carraway, K.L.; Isola, J.; Jovin, T.M. Decreased accessibility and lack of activation of ErbB2 in JIMT-1, a herceptin-resistant, MUC4-expressing breast cancer cell line. *Cancer Res.* **2005**, *65*, 473–482.
8. Palyi-Krekk, Z.; Barok, M.; Isola, J.; Tammi, M.; Szollosi, J.; Nagy, P. Hyaluronan-induced masking of ErbB2 and CD44-enhanced trastuzumab internalisation in trastuzumab resistant breast cancer. *Eur. J. Cancer* **2007**, *43*, 2423–2433. [CrossRef]
9. Chen, A.C.; Migliaccio, I.; Rimawi, M.; Lopez-Tarruella, S.; Creighton, C.J.; Massarweh, S.; Huang, C.; Wang, Y.C.; Batra, S.K.; Gutierrez, M.C.; et al. Upregulation of mucin4 in ER-positive/HER2-overexpressing breast cancer xenografts with acquired resistance to endocrine and HER2-targeted therapies. *Breast Cancer Res. Treat* **2012**, *134*, 583–593. [CrossRef]
10. Fiszman, G.L.; Jasnis, M.A. Molecular Mechanisms of Trastuzumab Resistance in HER2 Overexpressing Breast Cancer. *Int. J. Breast Cancer* **2011**, *2011*, 352182. [CrossRef]

11. Singha, N.C.; Nekoroski, T.; Zhao, C.; Symons, R.; Jiang, P.; Frost, G.I.; Huang, Z.; Shepard, H.M. Tumor-associated hyaluronan limits efficacy of monoclonal antibody therapy. *Mol. Cancer Ther.* **2015**, *14*, 523–532. [CrossRef] [PubMed]

12. Tanner, M.; Kapanen, A.I.; Junttila, T.; Raheem, O.; Grenman, S.; Elo, J.; Elenius, K.; Isola, J. Characterization of a novel cell line established from a patient with Herceptin-resistant breast cancer. *Mol. Cancer Ther.* **2004**, *3*, 1585–1592. [PubMed]

13. Barok, M.; Isola, J.; Palyi-Krekk, Z.; Nagy, P.; Juhasz, I.; Vereb, G.; Kauraniemi, P.; Kapanen, A.; Tanner, M.; Vereb, G.; et al. Trastuzumab causes antibody-dependent cellular cytotoxicity-mediated growth inhibition of submacroscopic JIMT-1 breast cancer xenografts despite intrinsic drug resistance. *Mol. Cancer Ther.* **2007**, *6*, 2065–2072. [CrossRef] [PubMed]

14. Barok, M.; Balazs, M.; Nagy, P.; Rakosy, Z.; Treszl, A.; Toth, E.; Juhasz, I.; Park, J.W.; Isola, J.; Vereb, G.; et al. Trastuzumab decreases the number of circulating and disseminated tumor cells despite trastuzumab resistance of the primary tumor. *Cancer Lett* **2008**, *260*, 198–208. [CrossRef] [PubMed]

15. Toth, G.; Szoor, A.; Simon, L.; Yarden, Y.; Szollosi, J.; Vereb, G. The combination of trastuzumab and pertuzumab administered at approved doses may delay development of trastuzumab resistance by additively enhancing antibody-dependent cell-mediated cytotoxicity. *MAbs* **2016**, *8*, 1361–1370. [CrossRef]

16. Holzinger, A.; Abken, H. Advances and Challenges of CAR T Cells in Clinical Trials. *Recent Results Cancer Res.* **2020**, *214*, 93–128. [PubMed]

17. Globerson-Levin, A.; Waks, T.; Eshhar, Z. Elimination of progressive mammary cancer by repeated administrations of chimeric antigen receptor-modified T cells. *Mol. Ther.* **2014**, *22*, 1029–1038. [CrossRef]

18. Williams, A.D.; Payne, K.K.; Posey, A.D., Jr.; Hill, C.; Conejo-Garcia, J.; June, C.H.; Tchou, J. Immunotherapy for Breast Cancer: Current and Future Strategies. *Curr. Surg. Rep.* **2017**, *5*, 31. [CrossRef]

19. Priceman, S.J.; Tilakawardane, D.; Jeang, B.; Aguilar, B.; Murad, J.P.; Park, A.K.; Chang, W.C.; Ostberg, J.R.; Neman, J.; Jandial, R.; et al. Regional Delivery of Chimeric Antigen Receptor-Engineered T Cells Effectively Targets HER2(+) Breast Cancer Metastasis to the Brain. *Clin. Cancer Res.* **2018**, *24*, 95–105. [CrossRef]

20. Wei, J.; Sun, H.; Zhang, A.; Wu, X.; Li, Y.; Liu, J.; Duan, Y.; Xiao, F.; Wang, H.; Lv, M.; et al. A novel AXL chimeric antigen receptor endows T cells with anti-tumor effects against triple negative breast cancers. *Cell Immunol.* **2018**, *331*, 49–58. [CrossRef]

21. Morgan, R.A.; Yang, J.C.; Kitano, M.; Dudley, M.E.; Laurencot, C.M.; Rosenberg, S.A. Case report of a serious adverse event following the administration of T cells transduced with a chimeric antigen receptor recognizing ERBB2. *Mol. Ther.* **2010**, *18*, 843–851. [CrossRef] [PubMed]

22. Sun, M.; Shi, H.; Liu, C.; Liu, J.; Liu, X.; Sun, Y. Construction and evaluation of a novel humanized HER2-specific chimeric receptor. *Breast Cancer Res.* **2014**, *16*, R61. [CrossRef] [PubMed]

23. Morgan, R.A. Human tumor xenografts: the good, the bad, and the ugly. *Mol. Ther.* **2012**, *20*, 882–884. [CrossRef] [PubMed]

24. Ithimakin, S.; Day, K.C.; Malik, F.; Zen, Q.; Dawsey, S.J.; Bersano-Begey, T.F.; Quraishi, A.A.; Ignatoski, K.W.; Daignault, S.; Davis, A.; et al. HER2 drives luminal breast cancer stem cells in the absence of HER2 amplification: implications for efficacy of adjuvant trastuzumab. *Cancer Res.* **2013**, *73*, 1635–1646. [CrossRef]

25. Sampson, J.H.; Choi, B.D.; Sanchez-Perez, L.; Suryadevara, C.M.; Snyder, D.J.; Flores, C.T.; Schmittling, R.J.; Nair, S.K.; Reap, E.A.; Norberg, P.K.; et al. EGFRvIII mCAR-modified T-cell therapy cures mice with established intracerebral glioma and generates host immunity against tumor-antigen loss. *Clin. Cancer Res.* **2014**, *20*, 972–984. [CrossRef]

26. Restifo, N.P.; Dudley, M.E.; Rosenberg, S.A. Adoptive immunotherapy for cancer: harnessing the T cell response. *Nat. Rev. Immunol.* **2012**, *12*, 269–281. [CrossRef] [PubMed]

International Journal of
Molecular Sciences

Article

Flow Cytometry Reveals the Nature of Oncotic Cells

Anna Vossenkamper [1] and Gary Warnes [2,*]

[1] Centre for Immunobiology, The Blizard Institute, Barts and The London School of Medicine and Dentistry, Queen Mary London University, 4 Newark Street, London E1 2AT, UK
[2] Flow Cytometry Core Facility, The Blizard Institute, Barts and The London School of Medicine and Dentistry, Queen Mary London University, 4 Newark Street, London E1 2AT, UK
* Correspondence: g.warnes@qmul.ac.uk; Tel.: +44(0)20-7882-2402

Received: 30 July 2019; Accepted: 4 September 2019; Published: 6 September 2019

Abstract: The term necrosis is commonly applied to cells that have died via a non-specific pathway or mechanism but strictly is the description of the degradation processes involved once the plasma membrane of the cell has lost integrity. The signalling pathways potentially involved in accidental cell death (ACD) or oncosis are under-studied. In this study, the flow cytometric analysis of the intracellular antigens involved in regulated cell death (RCD) revealed the phenotypic nature of cells undergoing oncosis or necrosis. Sodium azide induced oncosis but also classic apoptosis, which was blocked by zVAD (z-Vla-Ala-Asp(OMe)-fluoromethylketone). Oncotic cells were found to be viability^{+ve}/caspase-3^{-ve}/RIP3$^{+ve/-ve}$ (Receptor-interacting serine/threonine protein kinase 3). These two cell populations also displayed a DNA damage response (DDR) phenotype pH2AX^{+ve}/PARP^{-ve}, cleaved PARP induced caspase independent apoptosis H2AX^{-ve}/PARP^{+ve} and hyper-activation or parthanatos H2AX^{+ve}/PARP^{+ve}. Oncotic cells with phenotype cell viability^{+ve}/RIP3^{-ve}/caspase-3^{-ve} showed increased DDR and parthanatos. Necrostatin-1 down-regulated DDR in oncotic cells and increased sodium azide induced apoptosis. This flow cytometric approach to cell death research highlights the link between ACD and the RCD processes of programmed apoptosis and necrosis.

Keywords: accidental cell death; oncosis; DDR; parthanatos; flow cytometry

---

1. Introduction

The recent re-definition of cell death from Type I (programmed cell death by apoptosis), Type II (autophagic cell death), and Type III (programmed necrosis) to programmed cell death (PCD, homeostatic and embryonic), accidental cell death (ACD, oncosis), and regulated cell death (RCD), which includes apoptosis, necroptosis, autophagy, parthanatos or hyper-activation of Poly (ADP-ribose) polymerase (PARP) caused by an excessive DNA damage response (DDR, e.g., by pH2AX phospho H2AX histone) has been advantageous in understanding the complexity of cell death. RIP1-(Receptor-interacting serine/threonine protein kinase 1) dependent apoptosis or RIP1/RIP3/caspase-3 cells not being included highlights that the cell death nomenclature should be reviewed on a regular basis [1–5]. Programmed or regulated necrosis now includes necroptosis and parthanatos, amongst other forms of reported programmed cell death. However, necrosis is also the term commonly used to indicate the presence of dead cells that have lost plasma membrane integrity by any cell death pathway, but is strictly a reference to the degradation of the cell contents and plasma membrane after death [6–10]. The term oncosis or ACD is a better description of cell death induced by mechanical, chemical, and environmental factors that cause a rapid decrease in intracellular ATP leading to the deactivation of Na$^+$ and K$^+$-ATPase, resulting in an influx of Na$^+$, Cl$^-$, and Ca^{2+} ions. The cell then undergoes osmosis and swelling with a bursting of cell organelles and the plasma membrane [11]. Uncoupling protein 2 (UCP-2), located in the inner mitochondrial membrane where protons are pumped by UCP-2 into the mitochondrial matrix or the intermembrane space, which then regulates ATP and superoxide production is modestly up-regulated by oncosis, resulting in a rapid depolarization of the mitochondrial membrane, which has been measured by flow cytometry [12,13]. The signal pathways involved in oncosis are under studied, so little information is known about these mechanisms compared to the knowledge of pathways in RCD [1,2,6].

In contrast to oncosis, classic apoptosis is caspase-3 dependent and these cells form blebs on the surface of the plasma membrane with a gradual loss in mitochondrial membrane potential and hence a gradual lowering of intracellular ATP. The generation of reactive oxygen species (ROS) and loss of cytochrome *c* from the mitochondria to the cytoplasm results in activation of caspases and generation of apoptosomes and DNA fragmentation, accompanied with cell shrinkage and the formation of apoptotic bodies [1,2,8,14].

Until recently, necrotic or oncotic cells were measured via flow cytometry using the Annexin V assay in which such cells are gated as cell viability^{+ve}/AnnexinV^{-ve} and numerous researchers have attempted to understand this dead cell population with varying success [13,15,16]. In our laboratory, mitochondrial and plasma membrane dysfunction were detected by the multiplexing of mitochondrial and plasma membrane probes into the Annexin V assay, leading to a better understanding of the biological processes occurring in this oncotic population [13,15,16]. The recent development of a polychromatic flow cytometric assay in this laboratory, which identifies most RCDs simultaneously and demonstrates pathways affected by use of pan-caspase and RIP1 protein blockers zVAD and necrostatin-1 (Nec-1), led us to re-investigate oncotic cell death for potential pathways by comparison with apoptosis. The markers measured by flow cytometry included a fixable cell viability marker, activated caspase-3 (apoptosis), up-regulated RIP3 (necroptosis, or resting when not), pH2AX (DDR), cleaved PARP (apoptosis), parthanatos, or hyper-activation of cleaved PARP (pH2AX/cleaved PARP; Table 1, Figure S1). Potential modulation of the oncotic response to sodium azide was further investigated by the use of zVAD and necrostatin-1 to evaluate if the oncotic signalling pathways can be modified by these inhibitors before the cell loses plasma membrane permeability and the cell undergoes oncosis [3,4,13,17–19]. This approach may indicate the nature of the oncotic cell phenotype and highlight potential mechanisms that can modify the oncotic cellular response and the ACD connection to RCD processes. This may increase the potential for the use of therapeutic drugs to target the ACD process in the treatment of cancer.

Table 1. Cell description and phenotypes; Figure S1 provides a diagrammatical representation.

Cell Population	Phenotypic Markers
Live resting (or necroptotic)	Caspase-3^{-ve}/Zombie NIR^{-ve}/RIP3^{+ve}
Live double negative (DN)	Caspase-3^{-ve}/Zombie NIR^{-ve}/RIP3^{-ve}
Early apoptosis (EAPO)	Caspase-3^{+ve}/Zombie NIR^{-ve}/RIP3^{-ve}
Live RIP1-dependent apoptosis (RIP1-APO)	Caspase-3^{+ve}/Zombie NIR^{-ve}/RIP3^{+ve}
Late apoptosis (LAPO)	Caspase-3^{+ve}/Zombie NIR^{+ve}/RIP3^{-ve}
Dead/necrotic/oncotic	Caspase-3^{-ve}/Zombie NIR^{+ve}
Dead resting (or necroptotic)	Caspase-3^{-ve}/Zombie NIR^{+ve}/RIP3^{+ve}
Dead double negative (DN)	Caspase-3^{-ve}/Zombie NIR^{+ve}/RIP3^{-ve}
Dead RIP1-dependent apoptosis (RIP1-APO)	Caspase-3^{+ve}/Zombie NIR^{+ve}/RIP3^{+ve}
DNA damage response (DDR)	pH2AX^{+ve}/Cleaved PARP^{-ve}
Hyper-activation of cleaved PARP/parthanatos	pH2AX^{+ve}/Cleaved PARP^{+ve}
Cleaved PARP	pH2AX^{-ve}/Cleaved PARP^{+ve}

2. Results

2.1. Induction of Oncosis

NaN$_3$ induced early apoptosis (28%, Figure 1B, lower right quadrant) and lower levels of late apoptotic (13%, Figure 1B, upper right quadrant) and oncotic cells (17%, Figure 1B, upper left quadrant) compared with untreated cells after 24 h (Figure 1A,B, see Materials and Methods section for details of cell phenotype and gating strategy, Table 1, Figure S1). A lower incidence of live resting cells was observed (RIP3^{+ve}/caspase-3^{-ve}, 44%, Figure 2C, upper left quadrant) but with more early apoptosis after treatment (RIP3^{-ve}/caspase-3^{+ve}, 25%, Figure 2A,C, lower right quadrant). Dead cells arising from NaN$_3$ treatment showed less late apoptosis (25%, Figure 2D, lower right quadrant) than untreated cells (Figure 2B).

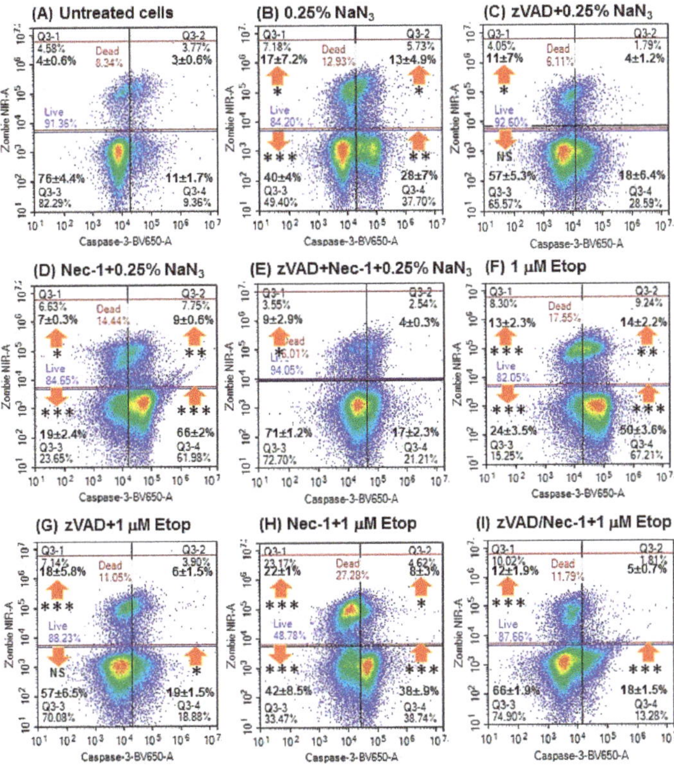

Figure 1. Cell death and caspase-3 activation assay. Cells were (**A**) untreated; (**B**) treated with 0.25% sodium azide (NaN$_3$) for 24 h; (**C**) pre-treated with 20 μM zVAD for 2 h, then with 0.25% NaN$_3$; (**D**) pre-treated with 60 μM necrostatin-1 (Nec-1) for 2 h, then with 0.25% NaN$_3$; (**E**) pre-treated with 20 μM zVAD and 60 μM Nec-1 for 2 h, then with 0.25% NaN$_3$; (**F**) treated with 1 μM Etoposide (Etop) for 24 h; (**G**) pre-treated with 20 μM zVAD for 2 h, then with 1 μM Etop; (**H**) pre-treated with 60 μM Nec-1 for 2 h, then with 1 μM Etop; and (**I**) pre-treated with 20 μM zVAD and 60 μM Nec-1 for 2 h, then with 1 μM Etop. $n = 3$, % Mean ± % SEM; Student's *t*-test: NS (not significant), * $p < 0.05$, ** $p < 0.01$**, *** $p < 0.001$; arrows indicate change compared with untreated cells.

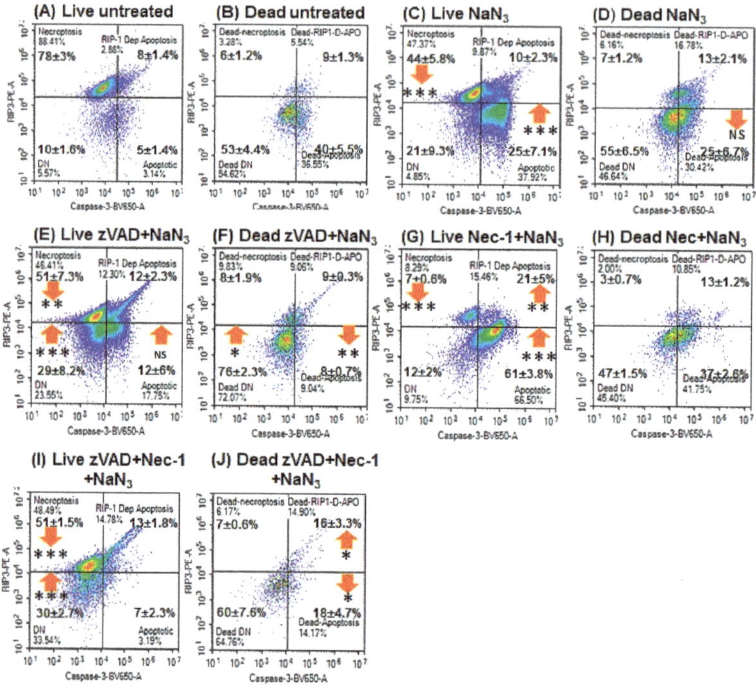

Figure 2. RIP3 and caspase-3 activation analysis of oncosis. After gating on live and dead cells from a Zombie NIR vs. caspase3-BV650 dot-plot (**A**) untreated live and (**B**) dead Jurkat cells were analysed on a RIP3-PE vs. caspase-3-BV650 dot-plot with resting phenotype indicated by RIP3^{+ve}/caspase-3^{-ve}, apoptosis by RIP3^{-ve}/caspase-3^{+ve}, RIP1-dependent apoptosis RIP3^{+ve}/caspase-3^{+ve}, and double negative RIP3^{-ve}/caspase-3^{-ve}. Live and dead cells treated with (**C,D**) 0.25% NaN$_3$ for 24 h; (**E,F**) pre-treated with 20 µM zVAD for 2 h, then treated with 0.25% NaN$_3$; (**G,H**)pre-treated with 60 µM Nec-1 for 2 h, then treated with 0.25% NaN$_3$; and (**I,J**) pre-treated with 20 µM zVAD and 60 µM Nec-1 for 2 h, then treated with 0.25% NaN$_3$, respectively. $n = 3$, % Mean ± % SEM, Student's t–test: NS (not significant), * $p < 0.05$, ** $p < 0.01$**, *** $p < 0.001$; arrows indicate change compared with untreated cells.

The oncotic cells resulting from NaN$_3$ treatment were mainly double negative (55%) for RIP3 and caspase-3 expression (dead resting oncotic cells, <10% caspase-3^{-ve}/RIP3^{+ve}, Figure 2D).

After NaN$_3$ treatment, the two live and dead apoptotic populations showed increased levels of pH2AX hyper-activation of cleaved PARP and a lower degree of apoptosis via cleaved PARP and DDR than untreated cells (Figure 3A–C, Figure S1A,B,E,F,I,J,M,N). Whereas late apoptotic cells showed increased DDR (Figure 3C, Figure S2B,E,M,J). The dead resting oncotic cells (Zombie^{+ve}/caspase-3^{-ve}/RIP3^{+ve}) were, 31% negative for both H2AX and PARP, whereas the dead oncotic DN (Zombie^{+ve}/caspase-3^{-ve}/RIP3^{-ve}) cells were 57% negative for both markers (Figure 3A–C, Figure S2O,P). The live and dead DN populations showed increased levels of parthanatos and DDR (Figure 3A–C, Figure S2D,H,L,P).

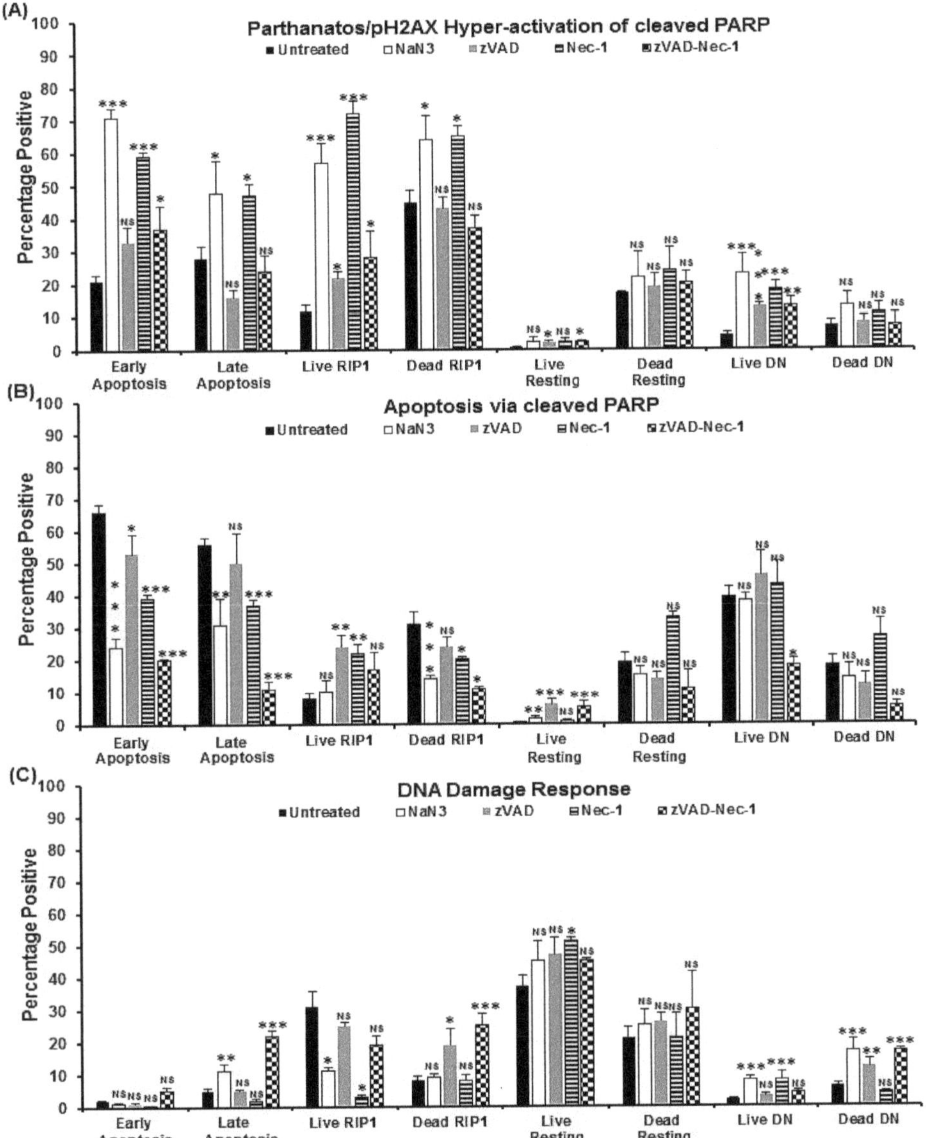

Figure 3. Parthanatos/hyper-activation of cleaved PARP, apoptosis via cleaved PARP, and DDR analysis of oncosis. Untreated Jurkat cells, treated with 0.25% NaN_3 for 24 h, or pre-treated with zVAD (20 µM) and/or Nec-1 (60 µM) for 2 h, then incubated with 0.25% NaN_3. Gating live and dead cells from a Zombie NIR vs. caspase-3-BV650 plot then both were analysed on a RIP3-PE vs. caspase-3-BV650 plot. Next, early and late apoptotic, necroptotic/resting, RIP1-dependent apoptotic, and double negative (DN) populations were analysed for pH2AX and cleaved PARP (Figures S2, S3). The incidence of (**A**) parthanatos/hyper-activation of cleaved PARP, (**B**) apoptosis via cleaved PARP, and (**C**) DDR were determined for all populations listed above. $n = 3$, % Mean, error bars % SEM, Student's *t*-test; NS (not significant), * $p < 0.05$, ** $p < 0.01$**, *** $p < 0.001$ compared with untreated cells.

2.2. Induction of Apoptosis

Induction of apoptosis with Etop showed an increase in early and late apoptosis as well as oncotic cells compared with untreated cells (Figure 1A,F). Live and early apoptotic cells showed increased levels of both types of apoptosis and the DN cells, whereas dead cells showed no such change (Figure 2A,B and Figure 4A,B).

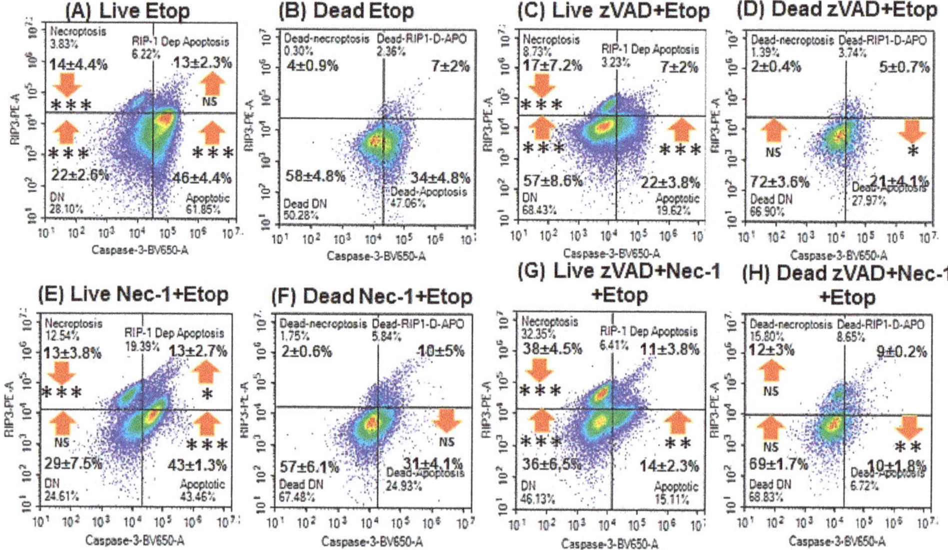

Figure 4. RIP3 and caspase-3 activation analysis of apoptosis. Gating on live and dead cells from a Zombie NIR vs. caspase-3-BV650 plot followed by analysis on a RIP3-PE vs. caspase-3-BV650 plot with resting phenotype indicated by RIP3^{+ve}/caspase-3^{-ve}, apoptosis by RIP3^{-ve}/caspase-3^{+ve}, RIP1-dependent apoptosis by RIP3^{+ve}/caspase-3^{+ve} and double negative by RIP3^{-ve}/caspase-3^{-ve}. (**A,B**) Treated with 1 µM Etop for 24 h; (**C,D**), pre-treated with 20 µM zVAD for 2 h, then treated with 1 µM Etop; (**E,F**) pre-treated with 60 µM Nec-1 for 2 h, then treated with 1 µM Etop; and (**G,H**) pre-treated with 20 µM zVAD and 60 µM Nec-1 for 2 h, then treated with 1 µM Etop. $N = 3$, % Mean ± % SEM, Student's *t*–test: NS (not significant), * $p < 0.05$, ** $p < 0.01$**, *** $p < 0.001$; arrows indicate change compared with untreated cells.

Early and late apoptotic cells after Etop treatment showed increased pH2AX hyper-activation of cleaved PARP with a decrease in apoptosis via cleaved PARP compared with untreated cells (Figure 5A,B and Figure S2Q,U). Live RIP1-dependent cells, however, showed increased pH2AX hyper-activation of cleaved PARP and caspase-3-dependent apoptosis via cleaved PARP with decreased DDR (Figure 5A–C and Figure S2R). However, live and dead oncotic resting and DN phenotypes also showed increased parthanatos and caspase-3-independent apoptosis via cleaved PARP with no increase in DDR, except for an increase observed in the live resting population (Figure 5A–C and Figure S2S,T,V,W,X).

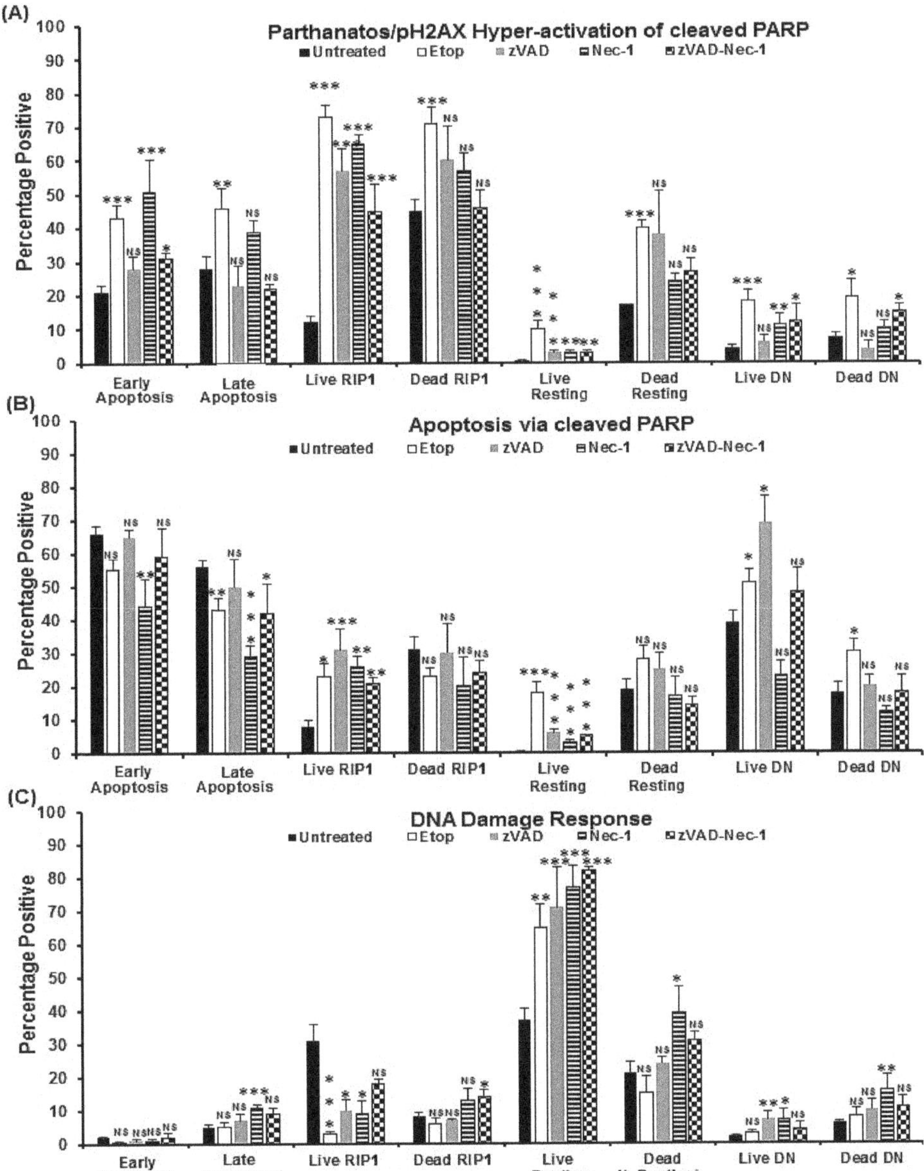

Figure 5. Parthanatos/hyper-activation of cleaved PARP, apoptosis via cleaved PARP, and DDR analysis of apoptosis. Untreated Jurkat, treated with 1 μM Etop or pre-treated with zVAD (20 μM) and/or Nec-1 (60 μM) for 2 h, then incubated with 1 μM Etop for 24 h. Gating on live and dead cells from a Zombie NIR vs. caspase-3-BV650 plot then both were analysed on a RIP3-PE vs. caspase-3-BV650 plot. Early and late apoptotic, necroptotic/resting, RIP1-dependent apoptotic, and double negative (DN) populations were analysed for pH2AX and cleaved PARP (Figures S2,S4 for detailed information). The incidence of (**A**) parthanatos/hyper-activation of cleaved PARP, (**B**) apoptosis via cleaved PARP, and (**C**) DDR were determined for all populations listed above. Mean, error bars % SEM, Student's *t*-test: NS (not significant), * $p < 0.05$, ** $p < 0.01$**, *** $p < 0.001$ compared with untreated cells.

2.3. Blockade of Caspases

Pre-treatment with zVAD to block the activation of caspases by NaN_3 and Etop resulted in lower levels of early apoptosis (<20%) and late apoptosis (<10%) with no change in the incidence of oncotic cells (Figure 1C,G, Figure 2E,F and Figure 4C,D). The proportion of live DN (RIP3^{-ve}/caspase-3^{-ve}) and dead oncotic cells (RIP3^{-ve}/caspase-3^{-ve}) increased after zVAD blockade of both drugs (Figure 2E,F and Figure 4C,D).

After zVAD caspase blockade of NaN_3 and Etop treatments, all live populations showed increased H2AX hyper-activation of cleaved PARP or parthanatos compared with untreated cells but decreased compared to drugs alone (Figure 3A, Figure 5A and Figure S3A–D, S4A–D). The live RIP1-dependent apoptotic, resting, and DN cells from both treatments also showed increased levels of apoptosis via cleaved PARP compared with untreated cells (early apoptosis showed a decrease, Figure 3B, Figure 5B and Figure S3A–D, S4A–D). The live DN population after zVAD/NaN_3 or Etop treatment showed a decrease or increase of DDR compared with drugs alone, respectively (Figure 3C, Figure 5C and Figure S3D, S4D). Live RIP1-dependent apoptotic and resting phenotypes, after both treatments with zVAD, showed increased DDR compared with drugs alone (Figure 3C, Figure 5C and Figure S3B,C, S4B,C). Dead cells from zVAD blockade of NaN_3/Etop treatments returned pH2AX and cleaved PARP expression to that of untreated dead cells (Figure 3, Figure 5 and Figure S3E–H, S4E–H). Except after NaN_3/zVAD treatment, an increase in DDR was observed in the dead RIP1-dependent apoptotic and oncotic DN phenotypes (Figure 5 and Figure S3F,H).

2.4. Blockade with Necrostatin-1

Blockade of NaN_3 with Nec-1 resulted in very high levels of early apoptosis compared with NaN_3, but was lower with Etop treatment (Figure 1D,H). Cell death was lower with Nec-1/NaN_3 but not changed with Etop treatment (Figure 1D,H). The live cells showed a higher incidence of early and RIP1-dependent apoptosis compared with NaN_3 treatment, with no change observed with Etop or in the incidence of dead cells (Figure 2A–D,G,H and Figure 4A,E,F).

pHA2X hyper-activation of cleaved PARP in live and dead cells after Nec-1 showed increased values similar to that observed with only drugs, except for the lower levels in both the DN populations and live resting cells after Nec-1/Etop treatment (Figure 3A, Figure 5A and Figure S3I–P, S4I–P). Apoptosis via cleaved PARP was increased in live and dead cells after Nec-1 blockade of NaN_3 treatment and decreased with Nec-1/Etop treatment compared with drugs alone, except for a decrease in live resting cells (Nec-1/NaN_3) and no change in RIP1-dependent apoptosis after Nec-1/Etop treatment (Figure 3B, Figure 5B and Figure S3I–P, S4I–P). Very low levels of DDR were observed in early and live RIP1-dependent apoptotic cells, but increased in live resting and DN cells (Figure 3C, Figure 5C and Figure S3I–L, S4I–L). Dead cell phenotypes after Nec-1 blockade of NaN_3 showed no increase in DDR compared with untreated cells, but was increased with Nec-1/Etop treatment (Figure 3C, Figure 5C and Figure S3, S4M–P).

2.5. Blockade with zVAD and Necrostatin-1

Pre-treatment with zVAD and Nec-1 to block the activation of caspases and RIP proteins resulted in lower levels of early and late apoptosis (<20%), although oncosis (caspase-3^{-ve}/Zombie^{+ve}) was still maintained (Figure 1E,I, Figure 2I,J and Figure 4G,H). Blocked NaN_3 treated cells had a higher level of live DN cells (and Etop), whereas dead cells showed higher levels of RIP1-dependent apoptosis, which indicated that zVAD did not block caspases in the RIP1-dependent apoptotic pathway in the presence of Nec-1 (Figure 2I,J and Figure 4G,H).

All cell phenotypes after dual blockade showed the same reduced levels of pH2AX hyper-activation of cleaved PARP as that observed after zVAD blockade (no change in Etop DN cells, Figure 3A, Figure 5A and Figure S3Q–X, S4Q–X). Apoptosis via cleaved PARP was lower in the early apoptotic and DN cells after dual blockade of NaN_3 and increased in live resting and RIP1-dependent apoptotic cells (Figure 3B and Figure S3Q,T). In contrast, all live cell (dual-blocked Etop) populations showed no change in apoptosis via cleaved PARP compared with drug treatment, except for the lower levels observed in live resting cells (Figure 5B and Figure S4Q,T). All dead cell populations after dual blockade of both treatments showed lower levels of apoptosis via cleaved PARP compared to any treatment protocol, except no change was observed in Etop-induced late and RIP1-dependent apoptotic cells (Figure 3B, Figure 5B and Figure S3U–X, S4U–X). Lastly, the DDR levels after both treatment protocols showed that all cell populations had higher levels than untreated cells (Figures 3C and 5C), except for the lower levels found in the live RIP1-dependent apoptotic cells compared with both treatments (Figures 3C and 5C).

3. Discussion

The use of oncosis and apoptosis-inducing drugs NaN_3 and Etop, together with caspase and RIP protein blockers, zVAD and Nec-1, has allowed the tracking of the cell death processes involved in ACD and apoptosis a form of RCD using flow cytometry (Figure 6) [17–19]. Induction of oncosis or apoptosis resulted in measurable oncosis (Zombie^{+ve}/caspase-3^{-ve}, commonly termed necrotic) but also early and late apoptosis, which was reduced by zVAD blockade without affecting the degree of oncosis induced by both drugs. Blockade of both drugs with Nec-1 resulted in increased NaN_3-induced early apoptosis while increasing Etop-induced oncosis. Induction of ACD and early apoptosis (by NaN_3) showed that live resting Jurkat cells move the RIP3^{+ve}/caspase-3^{-ve}/Zombie^{-ve} phenotype to the early apoptotic phenotype of RIP3^{-ve}/caspase-3^{+ve}/Zombie^{-ve}, then later to the RIP1-dependent phenotype RIP3^{+ve}/caspase-3^{+ve}/Zombie^{-ve} (Figure 6B), even in the presence of RIP1 inhibitor, Nec-1. This effect has been previously reported [17,19], where it was shown that although Nec-1 inhibited necroptosis by abrogation of the up-regulation of RIP3 (RIP3^{+ve}/caspase-3^{-ve}/Zombie^{-ve}), it did not inhibit cells from undergoing apparent RIP1-dependent apoptosis. The limitation of the current assay is highlighted by the use of RIP3 and caspase-3 antibodies to indirectly identify RIP1-dependent apoptosis due to the lack availability of a fluorescenated RIP1 antibody [17,19]. The interactions of RIP1, RIP3, TRADD (TNFR1-associated death domain), FADD (Fas associated via death domain), and caspase-8 in apoptotic Complex IIa and IIb pathways are not completely understood, so another explanation of the apparent presence (indirectly via RIP3) of RIP1-dependent apoptosis in the presence of Nec-1 is required, which may be elucidated by the use of a fluorescenated RIP1 antibody [5,20,21].

Induction of apoptosis by Etop showed that a high proportion of live resting cells become DN (losing their RIP3), as well as another population expressing caspase-3, possibly indicating that the cells first lose RIP3, become DN, and then express caspase-3 (Figure 6B), but also that early apoptotic cells can also later express RIP3 (Figure 6B) [5,17,19–21]. The route to RIP1-dependent apoptosis may be that these resting cells, rather than lose their RIP3, also start to express caspase-3 (Figure 6B).

Once the cells lose plasma membrane integrity and become Zombie^{+ve}, the cells presumably maintain the late apoptotic phenotype RIP3^{-ve}/caspase-3^{+ve}/Zombie^{+ve} before further degradation resulting in the cells becoming DN (Figure 6C). Oncotic cells (caspase-3^{-ve}/Zombie^{+ve}) induced by NaN_3/Etop, however, can also be divided into those with RIP3^{+ve}/caspase-3^{-ve}/Zombie^{+ve} or the DN phenotype RIP3^{-ve}/caspase-3^{-ve}/Zombie^{+ve} (Figure 6C).

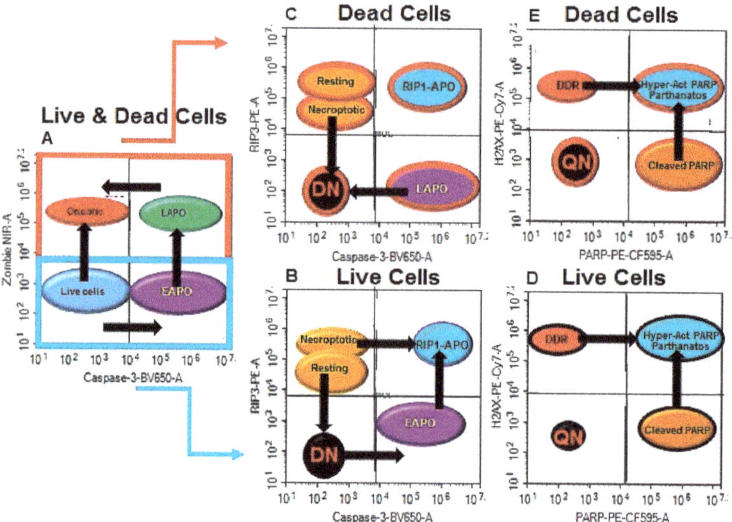

Figure 6. ACD and RCD pathways. (**A**) Live cells can undergo either early apoptosis (EAPO) or oncosis after drug treatment, with early apoptotic cells moving to late apoptotic (LAPO), then later with cell degradation, to the oncotic or necrotic phenotype. (**B**) Live cells may express RIP3^{+ve}/caspase-3^{-ve} when resting, or be RIP3$^{high+ve}$/caspase-3^{-ve} when undergoing necroptosis, or be double negative (DN). EAPO cells lose RIP3 or, if retained, undergo RIP1-dependent apoptosis (RIP1-APO). EAPO cells can also become RIP1^{+ve}. (**C**) Loss of plasma membrane integrity or cell death results in cell phenotypes mirrored in (**B**), with degradation of cells resulting in the DN population. (**B,C**) Live and dead cell phenotypes can also express pH2AX (DDR) or cleaved PARP (apoptosis), both of which can ultimately express both proteins, (**D,E**) resulting in pH2AX hyper-activation of cleaved PARP in the presence of active caspase-3 or parthanatos in the absence of caspase-3. Arrows indicate movement of cell populations.

The expression of pH2AX and cleaved PARP in the four identified phenotypes in live and dead cells are resting, early or late apoptotic, RIP1-dependent apoptosis, and DN, the incidence of which is modified by the action of the two drugs used in this study and can be further manipulated by blockade of ACD and RCD processes by zVAD and Nec-1 (Figure 6D,E) [22–25]. In the first instance live resting cells, the main phenotype of untreated Jurkat cells express RIP3 with a high degree of DDR (37%), whereas resting DN cells showed little DDR (2%) but a high degree of cleaved PARP (39%) in the absence of caspase-3 [25]. Induction of ACD resulted in enhanced levels of pHA2X hyper-activation of cleaved PARP in all live cell phenotypes with consequent reduced levels of DDR in live RIP1-dependent apoptotic cells, but with increased levels in the live DN phenotype (Figure 6D).

Once the cells undergo death, all phenotypes still showed increased pHA2X hyper-activation of cleaved PARP above the levels observed with dead untreated Jurkat cells (Figure 6E), whereas late apoptotic and dead oncotic DN cells showed increased DDR with no change in dead resting cells. Similar results were observed when cells undergo Etop-induced apoptosis, except no increase in DDR was observed in the dead oncotic or resting cells. So, NaN$_3$-derived oncotic cells can express more DDR than Etop-induced oncotic cells, which expressed higher levels of cleaved PARP (in the absence of caspase-3) and parthanatos/pHA2X hyper-activation of cleaved PARP, again in the absence of caspase-3 (Figure 6E).

The main effect of blockade with zVAD was an increased incidence of live and dead DN cells with both treatments, which displayed no change in DDR, increased cleaved PARP and reductions in pHA2X hyper-activation of cleaved PARP, as was the case with most cell phenotypes (Figure 6D, E).

This was with the notable exception of no change in levels of cleaved PARP in dead oncotic resting and DN cells compared to NaN$_3$ treatment.

In contrast, blockade of NaN$_3$ with Nec-1 resulted in no change in the incidence of dead DN and fewer live DN cells (as untreated cells), whereas Etop-induced levels were similar to that observed by drug alone. However, all blocked NaN$_3$ cell phenotypes showed increased cleaved PARP and pH2AX activation of cleaved PARP with reductions in DDR; this observation was especially noteworthy in the dead oncotic resting and DN cells, whereas the opposite was observed in these populations after Nec-1 blockade of Etop.

Dual blockade of both treatments again showed a high incidence of DN cells with reduced levels of pHA2X hyper-activation of cleaved PARP and cleaved PARP expression in the various populations of cells, whereas DDR was generally increased after Nec-1 blockade of both drugs in most populations of cells. So, the blockade of apoptotic and/or necroptotic pathways when the cells are undergoing ACD and RCD processes radically increased the incidence of oncotic cells, especially in the case of blockade of apoptosis alone and necroptosis, which resulted in increased levels of DDR, whereas Nec-1 blockade reduced ACD-related DDR but not in the case of RCD.

For decades, oncotic cells have been an undetermined population of dead cells that have been overlooked due to the difficult in their characterisation rather than being a point of interest. They have been detected by use of the Annexin V assay and classed as cell viability^{+ve}/Annexin V^{-ve} [13,15,16]. Using an active caspase-3 and RIP3 antibodies in tandem with a fixable live/dead dye, we showed that these oncotic cells have two main phenotypes, both of which are cell viability^{+ve}/caspase-3^{-ve}/RIP3$^{+/-ve}$, which can be further divided into DDR, hyper-activation of PARP or parthanatos, apoptosis via PARP or cells undergoing programmed necrosis, as well as double negative oncotic cells. All of them are expressed to different degrees compared with cells derived from untreated cultures. This perhaps reflects their different origins, with differences in expression of these markers when the cells are derived from an ACD or RCD induction process. These differences in origin are further highlighted by their differing oncotic responses to blockade by zVAD and Nec-1 or both.

4. Materials and Methods

4.1. Induction of Oncosis and Apoptosis

Jurkat cells (human acute T cell leukaemia cell line, ECACC, Salisbury, UK) were grown in RPMI (Roswell Park Memorial Institute 1640 Medium) 1640 with 10% FBS (Fetal Bovine Serum, Invitrogen, Paisley, UK) at 37 °C and 5% CO$_2$, either untreated or treated with 0.25% sodium azide or 1 µM etoposide (NaN$_3$, Etop, Sigma, Poole, UK) for 24 h. Cells were pre-treated with pan-caspase blocker zVAD (20 µM, Enzo Life Sciences, Exeter, UK) and/or necroptosis blocker necrostatin-1 (Nec-1, 60 µM Cambridge Bioscience, Cambridge, UK) for 2 h before induction of oncosis with 0.25% sodium azide or apoptosis with 1 µM Etop for 24 h.

4.2. Flow Cytometry Assay

Harvested cells were labelled with fixable live dead stain, Zombie near infra red (NIR; Biolegend, San Diego, CA, USA) at room temperature (RT) for 15 min. Washed cell were fixed in Solution A (CalTag, Little Balmer, UK) then 0.25% Triton X-100 (Sigma, Poole, UK) for 15 min each at RT. Jurkat cells (1×10^6) were incubated for 20 min at RT with anti-RIP3-PE (phycoerythrin clone B-2, Cat. No. sc-374639, Santa Cruz, Dallas, Tx, USA), cleaved PARP-PE-CF-595 (clone F21-852, Becton Dickinson, San Jose, CA, USA), H2A.X-Phospho (ser139)-PE-Cy7 (clone 2F3, Biolegend, San Diego, CA, USA) and anti- active caspase-3-BV650 (clone C92-605, Becton Dickinson, San Jose, CA, USA) for 20 min at RT. Washed cells were resuspended in 400 µL PBS (Phosphate Buffered Saline) and analysed on a ACEA Bioscience Novocyte 3000 flow cytometer (100,000 events, San Diego, CA, USA). Zombie NIR was excited by a 633 nm laser and collected with a 780/60 nm detector. Caspase-3-BV650 was excited by a 405 nm laser and collected at 675/30 nm. RIP3-PE, cleaved PARP-PE-CF-595, and pH2AX-PE-Cy7 were excited by a 488 nm laser and collected at 572/28, 615/20, and 780/60 nm, respectively. Single colour controls were used to determine the colour compensation using the pre-set voltages on the instrument using Novo Express software (ver 1.2.5, ACEA Biosciences, San Diego, CA, USA). Cells were gated on FSC (Forward Scatter) vs. SSC (Side Scatter) with single cells being gated on a FSC-A (Area) vs. FSC-H (Height) plot. Cells were then gated on a plot of caspase-3-BV650 vs. Zombie NIR, with a quadrant placed marking off live cells in the double negative quadrant (Figure S1A, lower left quadrant), with caspase-3-BV650^{+ve}/Zombie NIR^{-ve} (Figure S1A, lower right quadrant) indicating early apoptotic cells (EAPO), and lastly with caspase-3-BV650^{+ve}/Zombie NIR^{+ve} and caspase-3-BV650^{-ve}/Zombie NIR^{+ve} (Figure S1A, upper quadrants) indicating dead late apoptotic (LAPO) (Figure S1A, upper right quadrant) and oncotic cells (Figure S1A, upper left quadrant). Further labelling with RIP3 allowed identification within the live resting and dead oncotic populations of RIP3^{+ve}/caspase-3^{-ve} or necroptotic cells when RIP3 is up-regulated, early or late apoptotic (RIP3^{-ve}/caspase-3^{+ve}), RIP1-dependent apoptosis (RIP3^{+ve}/caspase-3^{+ve}, RIP1APO), and live double negative (DN) or dead oncotic DN cells (Figure S1B,D). Further gating on each of these eight populations for pH2AX and cleaved PARP allowed the identification of DDR (H2AX^{+ve}/PARP^{-ve}), pH2AX hyper-activation of cleaved PARP or parthanatos (H2AX^{+ve}/PARP^{+ve}), apoptotic cell death via cleaved PARP (H2AX^{-ve}/PARP^{+ve}), and DN cells (H2AX^{-ve}/PARP^{-ve}) (Figure S1C,E) [19]. In particular, dead resting oncotic cells (Zombie^{+ve}/caspase-3^{-ve}/RIP3^{+ve}) and dead oncotic DN cells (Zombie^{+ve}/caspase-3^{-ve}/RIP3^{-ve}) were gated for pH2AX and cleaved PARP, revealing the phenotypic nature of these two types of oncotic cells (Figure S1F,G).

4.3. Statistics

For all experiments, $n = 3$ and data are reported as mean ± SEM for percentage positive. Student's *t*-tests were performed in GraphPad software Inc. (San Diego, CA, USA) with $p \geq 0.05$ considered not significant (NS). * denotes $p \leq 0.05$, ** denotes $p \leq 0.01$, and *** denotes $p \leq 0.001$ when treated cells were compared to untreated cells.

Supplementary Materials: Supplementary materials can be found at http://www.mdpi.com/1422-0067/20/18/4379/s1.

Author Contributions: Conceptualization, G.W.; methodology, G.W.; software, G.W., validation, G.W.; formal analysis, G.W. and A.V.; investigation, G.W.; resources, G.W.; data curation, G.W.; writing—original draft preparation, G.W.; writing—review and editing, G.W. and A.V.; visualization, G.W.; supervision, G.W.; project administration, G.W., funding acquisition, G.W.

Funding: APC was funded by Queen Mary University Internal funding.

Conflicts of Interest: A.V. received salary funding from GSK. GSK had no input or role in the conception or undertaking of this study in any manner. Other authors declare no conflict of interest. The funders had no role in the design of the study; in the collection, analyses, or interpretation of data; in the writing of the manuscript, or in the decision to publish the results.

References

1. Galluzzi, L.; Bravo-San Pedro, J.M.; Vitale, I.; Aaronson, S.A.; Abrams, J.M.; Adam, D.; Alnemri, E.S.; Altucci, L.; Andrews, D.; Annicchiarico-Petruzzelli, M.; et al. Essential versus accessory aspects of cell death: Recommendations of the NCCD 2015. *Cell Death Differ.* **2015**, *22*, 58–73. [CrossRef] [PubMed]

2. Galluzzi, L.; Vitale, I.; Aaronson, S.A.; Abrams, J.M.; Adam, D.; Agostinis, P.; Alnemri, E.S.; Altucci, L.; Amelio, I.; Andrews, D.W.; et al. Molecular mechanisms of cell death: Recommendations of the Nomenclature Committee on Cell Death 2018. *Cell Death Differ.* **2018**, *25*, 486–541. [CrossRef] [PubMed]

3. Fatokun, A.A.; Dawson, V.L.; Dawson, T.M. Parthanatos: Mitochondrial-linked mechanisms and therapeutic opportunities. *Br. J. Pharmacol.* **2014**, *171*, 2000–2016. [CrossRef] [PubMed]

4. Morales, J.; Li, L.; Fattah, F.J.; Dong, Y.; Bey, E.A.; Patel, M.; Gao, J.; Boothman, D.A. Review of Poly (ADP-ribose) Polymerase (PARP) Mechanisms of Action and Rationale for Targeting in Cancer and Other Diseases. *Crit. Rev. Eukaryot. Gene Expr.* **2014**, *24*, 15–28. [CrossRef] [PubMed]

5.   Grootjans, S.; Vanden Berghe, T.; Vandenabeele, P. Initiation and execution mechanisms of necroptosis: An overview. *Cell Death Differ.* **2017**, *24*, 1184–1195. [CrossRef] [PubMed]

6.   Weerasinghe, P.; Buja, L.M. Oncosis: An important non-apoptotic mode of cell death. *Exp. Mol. Pathol.* **2012**, *93*, 302–308. [CrossRef] [PubMed]

7.   Majno, G.; Joris, I. Apoptosis, Oncosis, and Necrosis. An Overview of Cell Death. *Am. J. Pathol.* **1995**, *146*, 3–15.

8.   Fink, S.L.; Cookson, B.T. Apoptosis, pyroptosis, and necrosis: Mechanistic description of dead and dying eukaryotic cells. *Infect. Immun.* **2005**, *73*, 1907–1916. [CrossRef] [PubMed]

9.   Darzynkiewicz, Z.; Juan, G.; Li, X.; Gorczyca, W.; Murakami, T.; Traganos, F. Cytometry in Cell Necrobiology: Analysis of Apoptosis and Accidental Cell Death (Necrosis). *Cytom. A* **1997**, *27*, 1–20. [CrossRef]

10. Vanden Berghe, T.; Linkermann, A.; Jouan-Lanhouet, S.; Walczak, H.; Vandenabeele, P. Regulated necrosis: The expanding network of non-apoptotic cell death pathways. *Nat. Rev. Mol. Cell Biol.* **2014**, *15*, 135–147. [CrossRef]

11. Trump, B.E.; Berezesky, I.K.; Chang, S.H.; Phelps, P.C. The Pathways of Cell Death: Oncosis, Apoptosis, and Necrosis. *oxocological Pathol.* **1997**, *25*, 82–88. [CrossRef] [PubMed]

12. Mills, E.M.; Xu, D.; Fergusson, M.M.; Combs, C.A.; Xu, Y.; Finkel, T. Regulation of cellular oncosis by uncoupling protein 2. *J. Biol. Chem.* **2002**, *277*, 27385–27392. [CrossRef] [PubMed]

13. Warnes, G.; Martins, S. Real-time flow cytometry for the kinetic analysis of oncosis. *Cytom. A* **2011**, *79*, 181–191. [CrossRef] [PubMed]

14. Wlodkowic, D.; Skommer, J.; Darzynkiewicz, Z. Cytometry in cell necrobiology revisited. Recent advances and new vistas. *Cytom. A* **2010**, *77A*, 591–606. [CrossRef] [PubMed]

15. Lecoeur, H.; Prévost, M.C.; Gougeon, M.L. Oncosis is associated with exposure of phosphatidylserine residues on the outside layer of the plasma membrane: A reconsideration of the specificity of the annexin V/propidium iodide assay. *Cytom. A* **2001**, *44*, 44–65. [CrossRef]

16. Matteucci, C.; Grelli, S.; De Smaele, E.; Fontana, C.; Mastino, A. identification of nuclei from apoptotic necrotic and viable lymphoid cells by using multiparameter flow cytometry. *Cytom. A* **1999**, *35*, 145–153. [CrossRef]

17. Lee, H.L.; Pike, R.; Chong, M.H.A.; Vossenkamper, A.; Warnes, G. Simultaneous flow cytometric immunophenotyping of necroptosis, apoptosis and RIP1-dependent apoptosis. *Methods* **2018**, *134–135*, 56–66. [CrossRef] [PubMed]

18. Vossenkamper, A.; Warnes, G. A flow cytometric immunophenotyping approach to the detection of regulated cell death processes. *J. Immunol. Sci.* **2018**, *25*, 6–12.

19. Bergamaschi, D.; Vossenkamper, A.; Lee, W.Y.J.; Wang, P.; Bochukova, E.; Warnes, G. Simultaneous polychromatic flow cytometric detection of multiple forms of regulated cell death. *Apoptosis* **2019**, *24*, 453–464. [CrossRef]

20. Cho, Y.; McQuade, T.; Zhang, H.; Zhang, J.; Chan, F.K. RIP1-dependent and independent effects of necrostatin-1 in necrosis and T cell activation. *PLoS ONE* **2011**, *6*, e23209. [CrossRef]

21. Vandenabeele, P.; Grootjans, S.; Callewaert, N.; Takahashi, N. Necrostatin-1 blocks both RIPK1 and IDO: Consequences for the study of cell death in experimental disease models. *Cell Death Differ.* **2013**, *20*, 185–187. [CrossRef] [PubMed]

22. Nikoletopoulou, V.; Markaki, M.; Palikaras, K.; Tavernarakis, N. Crosstalk between apoptosis, necrosis and autophagy. *Biochim. Biophys. Acta* **2013**, *1833*, 3448–3459. [CrossRef] [PubMed]

23. Chen, Y.; Chen, S.; Liang, H.; Yang, H.; Liu, L.; Zhou, K.; Xu, L.; Liu, J.; Yun, L.; Lai, B.; et al. Bcl-2 protects TK6 cells against hydroquinone-induced apoptosis through PARP-1 cytoplasm translocation and stabilizing mitochondrial membrane potential. *Env. Mol. Mutagen.* **2018**, *59*, 49–59. [CrossRef] [PubMed]

24. Henning, R.J.; Bourgeois, M.; Harbison, R.D. Poly(ADP-ribose) Polymerase (PARP) and PARP Inhibitors: Mechanisms of Action and Role in Cardiovascular Disorders. *Cardiovasc. Toxicol.* **2018**, *18*, 493–506. [CrossRef] [PubMed]

25. Jiang, H.Y.; Yang, Y.; Zhang, Y.Y.; Xie, Z.; Zhao, X.Y.; Sun, Y.; Kong, W.J. The dual role of poly(ADP-ribose) polymerase-1 in modulating parthanatos and autophagy under oxidative stress in rat cochlear marginal cells of the stria vascularis. *Redox Biol.* **2018**, *14*, 361–370. [CrossRef] [PubMed]

International Journal of
Molecular Sciences

Article

Single Cell Mass Cytometry Revealed the Immunomodulatory Effect of Cisplatin Via Downregulation of Splenic CD44+, IL-17A+ MDSCs and Promotion of Circulating IFN-γ+ Myeloid Cells in the 4T1 Metastatic Breast Cancer Model

József Á. Balog [1,2], László Hackler Jr. [3], Anita K. Kovács [4], Patrícia Neuperger [1,2], Róbert Alföldi [2,3], Lajos I. Nagy [4], László G. Puskás [1,4] and Gábor J. Szebeni [1,5,*]

[1] Laboratory of Functional Genomics, Biological Research Centre, Temesvári krt. 62, H6726 Szeged, Hungary; balog.jozsef@brc.hu (J.Á.B.); neuperger.patricia@brc.hu (P.N.); laszlo@avidinbiotech.com (L.G.P.)
[2] PhD School in Biology, University of Szeged, H6726 Szeged, Hungary; r.alfoldi@astridbio.com
[3] AstridBio Technologies Ltd., Also kikötő sor 11/D, H6726 Szeged, Hungary; laszlo.hackler@astridbio.com
[4] Avidin Ltd., Also kikötő sor 11/D, H6726 Szeged, Hungary; a.kovacs@avidinbiotech.com (A.K.K.); l.nagy@avidinbiotech.com (L.I.N.)
[5] Department of Physiology, Anatomy and Neuroscience, Faculty of Science and Informatics, University of Szeged, Közép fasor 52, H6726 Szeged, Hungary
* Correspondence: szebeni.gabor@brc.hu

Received: 30 November 2019; Accepted: 24 December 2019; Published: 25 December 2019

Abstract: The treatment of metastatic breast cancer remained a challenge despite the recent breakthrough in the immunotherapy regimens. Here, we addressed the multidimensional immunophenotyping of 4T1 metastatic breast cancer by the state-of-the-art single cell mass cytometry (CyTOF). We determined the dose and time dependent cytotoxicity of cisplatin on 4T1 cells by the xCelligence real-time electronic sensing assay. Cisplatin treatment reduced tumor growth, number of lung metastasis, and the splenomegaly of 4T1 tumor bearing mice. We showed that cisplatin inhibited the tumor stroma formation, the polarization of carcinoma-associated fibroblasts by the diminished proteolytic activity of fibroblast activating protein. The CyTOF analysis revealed the emergence of CD11b+/Gr-1/CD44+ or CD11b+/Gr-1+/IL-17A+ myeloid-derived suppressor cells (MDSCs) and the absence of B220+ or CD62L+ B-cells, the CD62L+/CD4+ and CD62L+/CD8+ T-cells in the spleen of advanced cancer. We could show the immunomodulatory effect of cisplatin via the suppression of splenic MDSCs and via the promotion of peripheral IFN-γ+ myeloid cells. Our data could support the use of low dose chemotherapy with cisplatin as an immunomodulatory agent for metastatic triple negative breast cancer.

Keywords: single cell mass cytometry; metastatic breast cancer; myeloid-derived suppressor cells; immunophenotyping

1. Introduction

The role of the tumor microenvironment, the interaction of cancer cells with the extracellular matrix, endothelial cells, cancer-associated fibroblasts, and leukocytes in the tumor stroma have been increasingly considered as a milestone in cancer development, especially in the last decade [1,2]. The deeper understanding of the disturbances in the regulation and activation of the immune system in cancer resulted in the advancement of anti-cancer therapies, such as the immune checkpoint blockade (ICB) [3]. However, the treatment of poorly immunogenic and metastasizing tumors remained a challenge.

Here, we focus on female breast cancer since it is the most frequent cancer in women and still the deadliest cancer type between the ages of 20–49 years old in contrast to the achievements in early diagnostics and therapeutics [4,5]. In our work the murine mammary carcinoma of the BALB/c mice, the syngeneic 4T1 was studied [6]. The 4T1 model is among the few murine triple negative breast cancer (TNBC) models that spontaneously metastasize to sites affected in human breast cancer (e.g., lung) in an immunocompetent host [7]. Orthotopic transplantation of 4T1 cells offers a relevant tumor model to study efficacy of drug candidates or immune therapy regimens [8]. Previously, we showed the tumor promoting effect of mesenchymal stem cell (cancer associated fibroblast)-derived galectin-1 in the 4T1 model [9], and later on we screened an anti-cancer compound library of imidazo[1-2-b]pyrazole-7-carboxamides in both two- and three-dimensional cell cultures of 4T1 cells [10,11].

We have previously reviewed how cancer-related chronic inflammation can lead to the generation of immature myeloid-derived suppressor cells (MDSCs) and to the alternative polarization of tumor-associated macrophages (TAMs) [12], which manifests autonomously in the 4T1 breast cancer model [13,14]. It has been shown that the granulocytic MDSCs support metastases by suppressing CD8+ T-cells in the 4T1 breast tumor model [15]. It was also recently shown that 4T1 cells shape immune responses with an increase of splenic CD11b+ cells to promote cancer growth in an *Shb* (SRC homology-2 domain protein B) dependent manner [16]. The 4T1 tumor cells are poorly immunogenic and refractory to immune therapies, although the combination of anti-PD-1, anti-CTLA-4 ICB with epigenetic modulators could have a therapeutic benefit curing more than 80% of 4T1 tumor bearing mice via eliminating MDSCs [17]. We have previously reviewed strategies targeting these myeloid-derived suppressors cells or tumor associated macrophages to combat cancer [18]. Here, the traditional chemotherapeutic agent, the DNA crosslinker cisplatin was used, since cisplatin and platinum-based chemoterapeutics are in the clinical routine as first line treatment option in several cancers such as lung, bladder, ovarian and metastatic breast cancer [19]. Recent studies have shown the immune induction by cisplatin in human TNBC (the TONIC trial NCT02499367) [20], or in murine carcinoma models showing enhanced sensitivity to ICB therapy in combination with cisplatin treatment but these studies did not deal with immunophenotyping of the myeloid compartment [21,22]. The beneficial effect of cisplatin on the course of 4T1 tumor development was shown recently in combination with metformin or bromelain [23,24], but these studies also did not address the characterization of the immunophenotype.

To the best of our knowledge our study is the first, where mass cytometry, a multidimensional single cell technology with computational data analysis was carried out in order to reveal the immunophenotype of 4T1 murine triple negative breast carcinoma and the effect of cisplatin treatment on the splenic and circulating immune compartments.

2. Results

2.1. Real-Time Monitoring of 4T1 Cell Viability Hampered by Cisplatin

Determination of the half maximal inhibitory concentration, the IC_{50} of cisplatin on 4T1 cells was carried out using the real-time electronic sensing xCelligence system [25]. The detected impedance is proportional with the percentage of adhered living cells to the gold coated plate and the decline in the normalized cell index corresponds to hampered cell viability (Figure 1). The effect of cisplatin on viability was followed for 120 h after treatment in every 15 min (former studies reported endpoint assays with cisplatin on 4T1 cells). The IC_{50} values were as follows 36.74 µM at 24 h, 7.608 µM at 48 h, 6.962 µM at 72 h, 4.128 µM at 96 h, and 3.995 µM at 120 h (Figure S1).

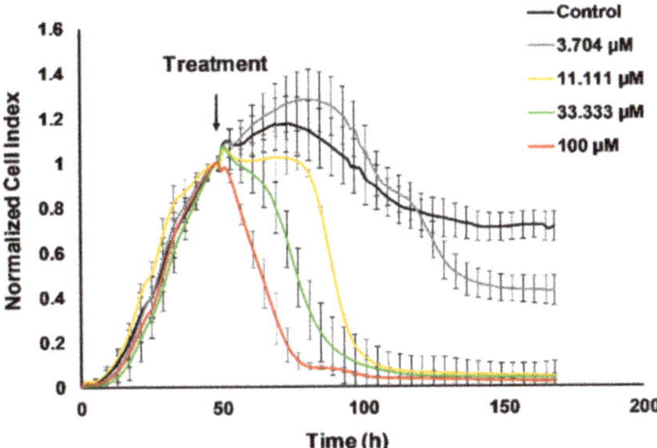

Figure 1. Real-time monitoring of 4T1 cell viability hampered by cisplatin. The 4T1 cells were seeded and the baseline impedance was recorded for 48 h. After that 48 h culturing treatment with 11.111 µM, 33.333 µM, or 100 µM cisplatin reduced viability of 4T1 cells on a time and dose dependent manner. The corresponding dose-response curves with the half maximal inhibitory concentration (IC_{50}) values can be found in Figure S1.

2.2. Cisplatin Treatment Reduced 4T1 Tumor Growth, the Number of Lung Metastatic Nodules and the Weight of the Spleen

The syngeneic BALB/c mice were orthotopically transplanted with 4T1 breast cancer cells in order to establish the animal model for the addressed immunophenotyping. Tumor growth was monitored daily. All mice treated with cisplatin showed markedly reduced tumor growth compared to untreated 4T1 tumor bearing mice represented by the average tumor volume of 102 mm^3 versus (vs.) 1481 mm^3 on the 21st day (Figure 2A). On the 23rd day mice were euthanized for immunophenotyping and the weight of the tumors (Figure 2B), the number of metastatic nodules (macrometastasis) on the lungs (Figure 2C), and the weight of the spleens were measured (Figure 2D).

The average weight of the tumors was reduced by almost 90% due to cisplatin treatment, namely 426.4 ± 110.1 mg (mean ± SEM) in the cisplatin treated 4T1 tumorous mice vs. 3087 ± 356 mg in the untreated 4T1 tumor bearing mice (Figure 2B). The development of metastatic nodules on the surface of the lungs were also inhibited by cisplatin, the average number of lung macrometastasis were as follows: 0.42 ± 0.23 in cisplatin treated vs. 3.25 ± 0.524 in untreated 4T1 tumorous mice (Figure 2C). Splenomegaly is one sign of myeloid cell expansion due to cancer related inflammation in tumor bearing hosts [26]. Cisplatin treatment suppressed the enlargement of the spleen as spleen weights were 143.04 ± 19.25 mg, 843.59 ± 38.69 mg, 123.91 ± 26.08 mg, in the naive, 4T1 tumor bearing and cisplatin treated 4T1 tumor bearing mice, respectively (Figure 2D).

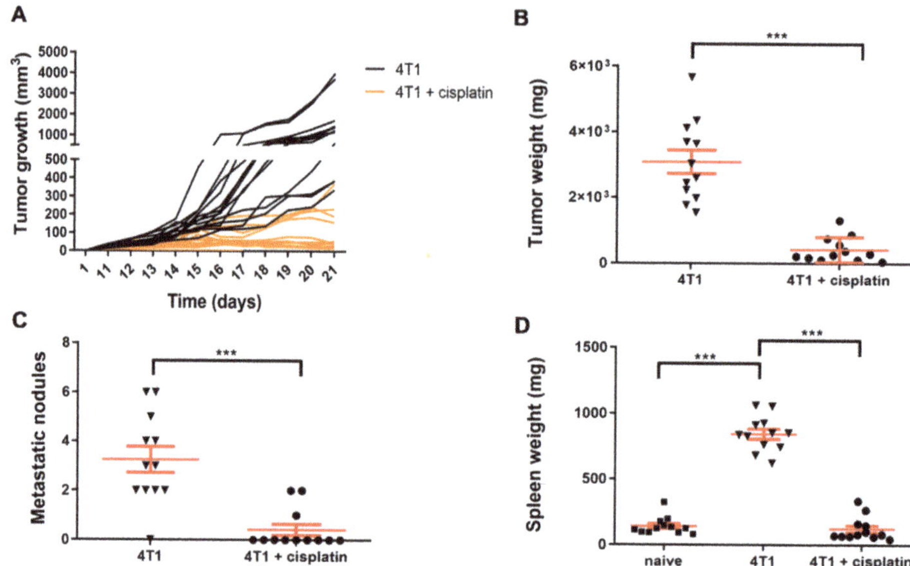

Figure 2. Cisplatin treatment reduced 4T1 tumor growth, the number of lung metastatic nodules and the weight of the spleen. The 4T1 cells (1.2×10^5) were transplanted by the injection into the mammary fat pad of BALB/c mice ($n = 12$). Tumor growth was monitored daily (**A**). On the 23rd day mice were euthanized and the weight of the tumors (**B**), the number of metastatic nodules on the lungs (**C**), and the weight of the spleens were measured (**D**). Individual values and arithmetic mean values of the samples ± standard error of the mean (SEM) are plotted, statistical significance was set to *** $p < 0.001$.

2.3. Cisplatin Reduced the Activity of Fibroblast Activator Protein (FAP)

The increased activity of the prolyl endopeptidase FAP enzyme is a hallmark of the tumor stroma, because of the accumulation and activation of cancer associated fibroblasts (CAFs). The FAP protease activity either membrane bound on CAFs or solubilized, is proportional with the malignancy [27]. The activity of FAP enzyme was investigated, because it has been reported that the accumulation of CAFs in the tumor stroma and their expression of FAP contributes to the chemoresistance to cisplatin in carcinomas [28,29]. We have synthetized a peptide substrate (Fmoc-Gly-Pro-Cysteic acid-Ile-Gly-NH$_2$, Figure S2) in order to measure FAP activity in the plasma of naive, 4T1 tumor bearing and cisplatin treated 4T1 tumorous mice (Figure 3).

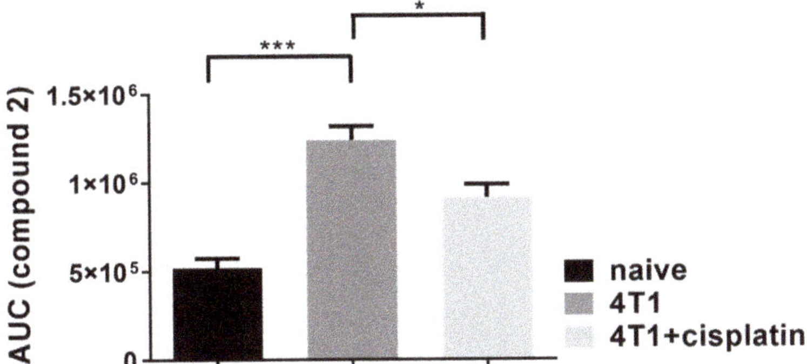

Figure 3. The proteolytic activity of the fibroblast activator protein (FAP) increased by the formation of breast cancer, and was significantly decreased by cisplatin treatment. Area under the curve (AUC) values from high pressure liquid chromatography (HPLC) analysis of the peptide digestion product **2** (Figure S2) were plotted. The results are shown as arithmetic mean values of the samples ± standard error of the mean (SEM), statistical significance was set to *** $p < 0.001$, * $p < 0.05$.

FAP enzyme activity quantification was based on a direct measurement of a digested peptide product with high pressure liquid chromatography (HPLC) (Figure S2). Fmoc-GP-Cox- was synthesized (Avidin Ltd.) and used as a substrate, where "Cox" denotes for oxydized cysteine, namely, cysteic acid. We used oxidized cysteine amino acid at position 3 in the peptide substrate of FAP, instead of unmodified cysteine. In a preliminary experiment, we confirmed that the peptide with the oxidized cysteine residue was the same or better substrate of FAP. Moreover, our improved peptide substrate is more water soluble and after FAP digestion the dipeptide product (Fmoc-GP) could be better separated from the intact substrate during HPLC analysis, then the unmodified version, enabling us more accurate and sensitive quantification. We could detect FAP activity in the plasma of naive, 4T1 tumor bearing and cisplatin treated 4T1 tumorous animals with the following average AUC (± SEM) values from HPLC analysis: $518.5 \pm 55.5 \times 10^3$, $1240.2 \pm 77.9 \times 10^3$, $914.9 \pm 71.7 \times 10^3$, respectively (Figure 3).

2.4. Single Cell Mass Cytometry (SCMC) Revealed the Immunophenotype of Breast Cancer Bearing Mice

2.4.1. Cisplatin Restored the Splenic Immunophenotype of 4T1 Tumor Bearing Mice

In order to investigate the immunophenotype of 4T1 breast cancer single cell mass cytometry was performed with 24 antibodies in one single tube (see Table 1 in Section 4.5). We intended to monitor the draining lymph nodes, bone marrow, spleen, and blood of the naive, 4T1 tumor bearing, and cisplatin treated 4T1 tumorous animals. However, the orthotopic injection of 4T1 cells into the mammary fat pad resulted in the outgrowth of the tumor mass in co-junction with the draining cervical and axillary lymph nodes making them undetectable. Bone marrow staining showed homogenous immature cells (little CD11+ Gr1+ elevation in tumor bearing hosts) excluding these samples from further analysis. Our attention turned toward the spleen since it has been published in the seminal paper of Bronte et al. that tolerance to tumor antigens develops in the spleen [30]. During the SCMC analysis of the spleen CD45+ living singlets were gated and on these leukocytes the unsupervised and multidimensional visualization of stochastic neighbor embedding (viSNE) analysis was performed delineating the separate clouds of different main immune subsets (Figure 4) [31].

Table 1. The list of the antibodies used for mass cytometry.

Target	Clone	Metal Tag
Gr-1 (Ly6C/Ly6G)	RB6-8C5	141_Pr
CD11c	N418	142_Nd
CD69	H1.2F3	145_Nd
CD45	30-F11	147_Sm
CD11b	M1/70	148_Nd
CD19	6D5	149_Sm
CD25	3C7	151_Eu
CD3e	145-2C11	152_Sm
TER-119	TER119	154_Sm
CD62L	MEL-14	160_Gd
CD8a	53-6.7	168_Er
TCRβ	H57-597	169_Tm
NK1.1	PK136	170_Er
CD44	IM7	171_Yb
CD4	RM4-5	172_Yb
B220	Ra3-6B2	176_Yb
IFN-γ	XMG1.2	165_Ho
IL-2	JES6-5H4	144_Nd
IL-4	11B11	166_Er
IL-5	TRFK5	143_Nd
IL-6	MP5-20F3	167_Er
IL-10	JES5-16E3	158_Gd
IL-17A	TC11-18H10.1	174_Yb
TNFα	MP6-XT22	162_Dy

Both the CD4+ and CD8+ T-cells were almost completely absent and the percentage of CD19+ B-cells decreased from 50% to 10% of CD45+ living singlets in the spleen of the 4T1 tumor bearing mice (Figure 4D,E). The percentage of the myeloid CD11b+ cells increased from 10% to 80% at the expense of lymphoid subsets due to breast cancer development (Figure 4D,E). Cisplatin treatment of 4T1 tumorous mice normalized the splenic immunophenotype similar to naive mice with 25% CD4+, 10% CD8+ T-cells, 40% CD19+ B-cells, and 20% CD11b+ cells (Figure 4F).

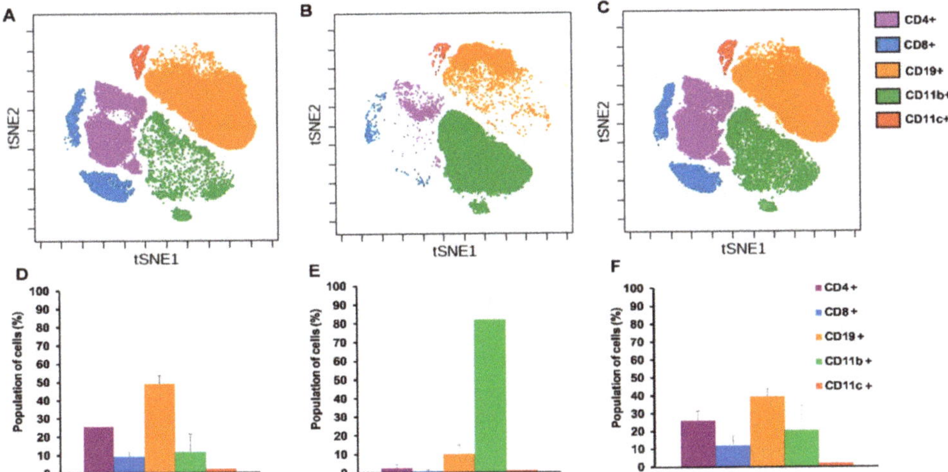

Figure 4. Single cell mass cytometric immunophenotyping showed the accumulation of CD11b+ myeloid cells in 4T1 tumor bearing animals at the expense of lymphoid subsets which was reverted by cisplatin treatment. The visualization of stochastic neighbor embedding (viSNE) analysis was run on the (**A**) naive, (**B**) 4T1 tumor bearing, and (**C**) cisplatin treated 4T1 tumorous animals within the CD45+ living singlets. Quantitative analysis of the main lymphoid subsets: CD4+ T-cells = lilac, CD8+ T-cells = blue, CD19+ B-cells = orange and the myeloid CD11b+ = green, CD11c + = red subsets were performed in the spleen of (**D**) naive, (**E**) 4T1 tumor bearing, and (**F**) cisplatin treated 4T1 tumorous mice within the CD45+ living singlets. Representative viSNE plots and column bars are shown from the pooled samples of 6 mice per group. Data are shown as arithmetic means on the column bars ± SEM. Pairwise comparison of the emergence of CD11b+ population in (**E**) vs. (**D**) and its decrease by cisplatin treatment (**F**) vs. (**E**) has statistical significance at $p < 0.001$.

The expression intensity of certain proteins within the viSNE plots of main subsets showed dramatic changes (Figure 5). The expression pattern of Gr-1+ (Ly6C/Ly6G), CD44+, and IL-17A+ cells within the cloud of CD11b+ myeloid cells were uniquely high in the spleen of 4T1 breast cancer bearing mice (Figure 5) in accordance with the splenomegaly shown in Figure 2D. On the contrary, the B220 and CD62L markers within the cloud of CD19+ B-cells, and the CD62L within the cloud of CD4+ and CD8+ T-cells also were highly reduced in the spleen of 4T1 tumorous mice (Figure 5). Cisplatin treatment reverted the immunophenotype of both splenic CD11b+ myeloid (Gr-1+, CD44+ and IL-17A+) cells and lymphoid (B220+/CD19+, CD62L+/CD19+, CD62L+/CD4+, CD62L+/CD8+ T-cells) cells similar to naive mice (Figure 5).

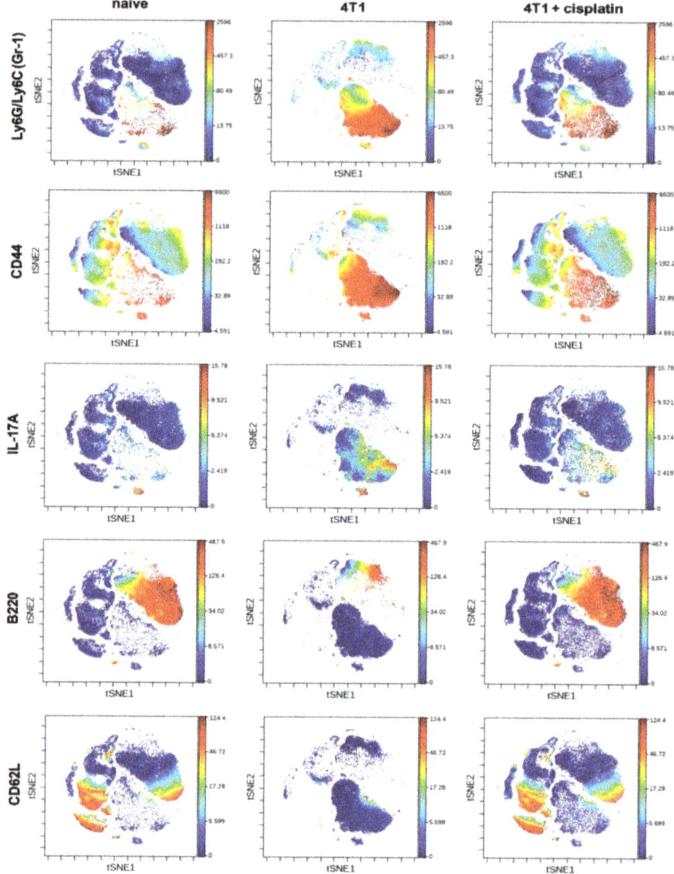

Figure 5. The viSNE plots illustrate the expression intensity of Gr-1, CD44, IL-17A, B220, CD62L markers within the clouds of main subsets defined in Figure 4 in the splenic samples of naive, 4T1 tumor bearing, and cisplatin treated 4T1 tumorous mice. The coloration is proportional to the expression intensity (blue = low, red = high). The list of the antibodies can be found in Table 1 in Section 4.5. Representative viSNE plots are shown from the pooled samples of 6 mice per group. The markers of the panel which were not detected or did not show differential expression are not shown.

Quantitation of the populations with characteristic protein expression was performed by manual gating within the CD45+ living singlets of splenocytes (Figure S3). The trajectories on the radar plots delineate the characteristic marker profile of splenocytes of naive, 4T1 tumor bearing, and cisplatin treated tumorous mice (Figure 6).

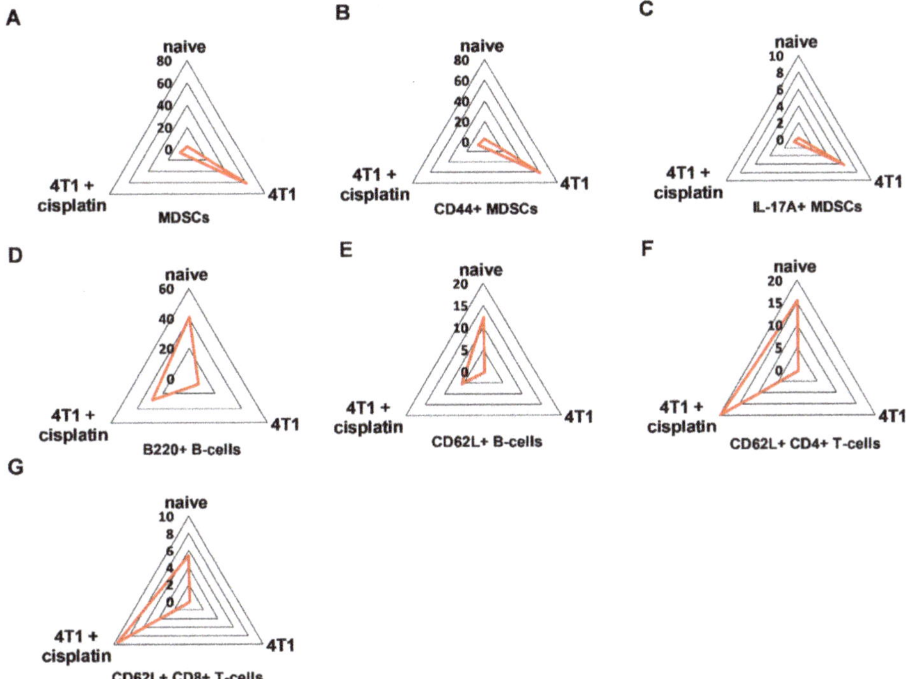

Figure 6. The trajectories on the radar plots delineate the characteristic marker profile of splenocytes in naive, 4T1 tumor bearing, and cisplatin treated tumorous mice. The accumulation of splenic (**A**) CD11b+/Gr-1+ MDSCs, (**B**) CD44+, and in a smaller extent (**C**) IL-17A+ MDSCs is a characteristic of 4T1 breast cancer. Cisplatin restores the percentage of (**D**) B220+ and (**E**) CD62L+ B-cells, (**F**) CD62L+ CD4+ and (**G**) CD8+ T-cells. The percentage of the given populations is demonstrated on the radar plots within the CD45+ living singlets determined by manual gating in Cytobank. The gating hierarchy is explained in the text and in Figure S3. The markers of the panel which were not detected or did not show differential expression are not shown. Representative radar plots are shown from the pooled samples of 6 mice per group as described in Section 4.5.

MDSCs were defined as CD45+/CD3-/CD11+/Gr1+ cells and further evaluated for CD44 and IL-17A staining. These MDSCs, CD44+ MDSCs and IL-17A+ MDSCs represented exclusively 60%, 60% and 6% of CD45+ living singlets in the spleen of 4T1 tumorous mice, respectively. The B-cell marker B220 (except regulatory B-cells, germinal center B-cells, some plasma cells, and certain memory B-cells [32–34]) on CD45+/CD3-/CD19+ B-cells was 40% in the naive, 30% in the cisplatin treated, while only 6% in the untreated 4T1 tumorous mice. The homing receptor L-selectin, CD62L was almost absent (0.2%) on the splenic B-cells of 4T1 tumorous mice but it was restored upon cisplatin treatment to 5.5% (naive mice 12.5%). The CD45+/CD3+/TCRβ+/CD4+ and CD45+/CD3+/TCRβ+/CD8+ T-cells expressed CD62L 15% vs. 20% and 6% vs. 10% in the splenocytes of naive and cisplatin treated 4T1 bearing mice, respectively. Sunburst charts represent the immunocomposition of main subsets of the spleens detected by single cell mass cytometry in Figure S4A and FlowSOM (Flow data using Self-Organizing Map, [35]) analysis demonstrates the identified clusters, minimum spanning trees (MSTs) of splenic subsets in Figure S5.

2.4.2. Cisplatin Could not Completely Restore the Peripheral Immunophenotye of 4T1 Tumor Bearing mice but Increased IFN-γ Production of Myeloid Cells

Immunophenotype of the blood in breast cancer can reflect the state of the peripheral tolerance associated with cancer-related inflammation or the activation of anti-tumor effector cellular responses [36]. The viSNE plots of the main subsets in blood samples show the absence of peripheral CD4+, CD8+ T-, and B-cells and the emergence of CD11b+ cells as a sign of advanced cancer compared to naive blood (Figure 7A,B). Cisplatin treatment at least partially normalized the immunocomposition of the blood of 4T1 tumorous mice (Figure 7C).

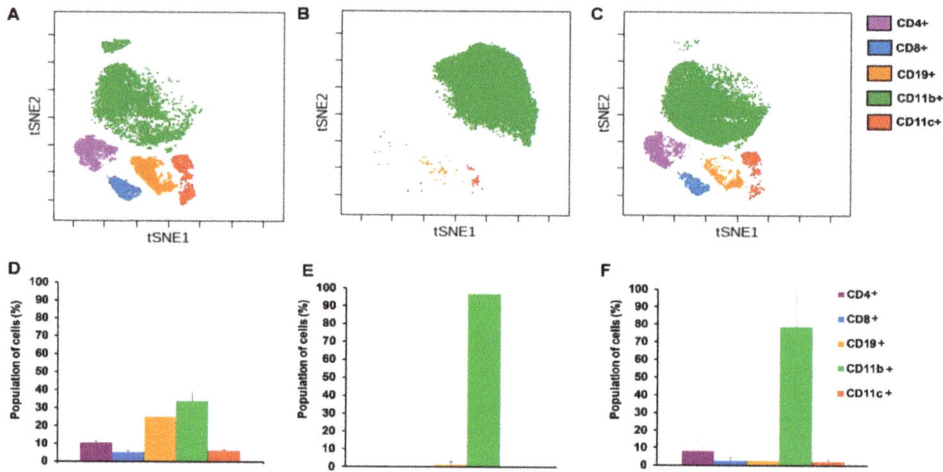

Figure 7. Immunophenotyping of blood showed dramatic expansion of CD11b+ cells in advanced cancer. The viSNE analysis was run on the (**A**) naive, (**B**) 4T1 tumor bearing, and (**C**) cisplatin treated 4T1 tumorous animals within the CD45+ living singlets. Quantitative analysis of the main lymphoid subsets: CD4+ T-cells = lilac, CD8+ T-cells = blue, CD19+ B-cells = orange and the myeloid CD11b+ = green, CD11c + = red subsets was performed in the blood of (**D**) naive, (**E**) 4T1 tumor bearing, and (**F**) cisplatin treated 4T1 tumorous mice within the CD45+ living singlets. Representative viSNE plots and column bars are shown from the pooled samples of 6 mice per group. Data are shown as arithmetic means on the column bars ± SEM. Pairwise comparison of the emergence of CD11b+ population in (**E**) vs. (**D**) has significance at $p < 0.001$.

The percentage of CD11b+ cells of 4T1 tumor bearing mice due to advanced cancer-related myeloid expansion represented around 97% (Figure 7E) in contrast to naive mice with 35% CD11b+ cells (Figure 7D) of CD45+ living singlets in the blood. The treatment of 4T1 breast cancerous mice with cisplatin could not suppress the emergence of CD11b+ cells significantly with an average of 85% CD11b+ cells (Figure 7F). Myeloid expansion led almost to the absence of circulating T- and B-lymphocytes in advanced cancer (Figure 7E). Cisplatin treatment at least partially restored the percentage of these CD4+, CD8+ T- and CD19+ B-lymphocytes with an average 8.1% vs. 10.5%, 2.6% vs. 5.1%, 2.7% vs. 24.8% in naive mice, respectively (Figure 7D,F).

The differential expression intensity of the Gr-1, CD44, IFN-γ markers on blood-derived leukocytes was plotted on the viSNE graphs (Figure 8). Interestingly, the highest expression protein pattern of Gr-1+/CD44+ (Gr-1+ bright/CD44+ bright) within the CD11b+ subset was associated with early cancer disease in the blood of cisplatin treated 4T1 tumorous mice. The Gr-1+ dim/CD44+ dim immature myeloid cells have been reported to suppress T-cells and IFN-γ production [37,38], the dim expression intensity of these markers was characteristic to 4T1 tumor bearing mice (Figure 8). In line with this, the smaller population of IFN-γ producing myeloid cells appeared within the CD11 subtype by cisplatin treatment which was completely absent in the advanced cancer (Figure 8).

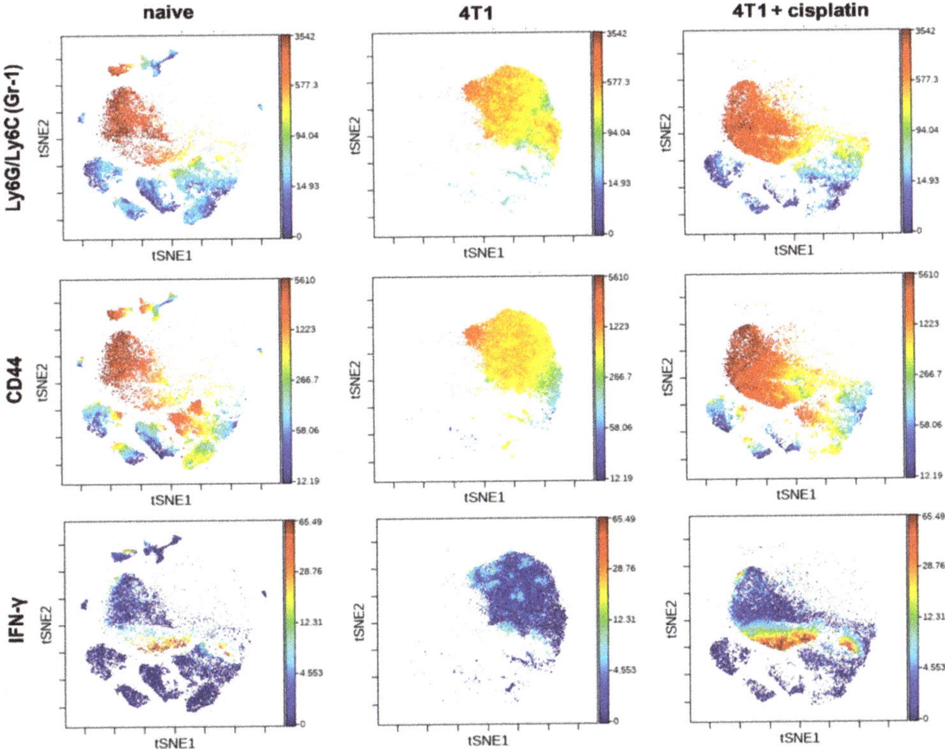

Figure 8. The viSNE plots illustrate the expression intensity of Gr-1, CD44, and IFN-γ markers within the clouds of main subsets defined in Figure 7 in the blood samples of naive, 4T1 tumor bearing, and cisplatin treated 4T1 tumorous mice. The coloration is proportional with the expression intensity (blue = low, red = high). The list of the antibodies can be found in Table 1 in the Section 4.5. Representative viSNE plots are shown from the pooled samples of 6 mice per group. The markers of the panel which were not detected or did not showed different expression are not shown.

Quantitation of the populations with characteristic protein expression was performed by manual gating within the CD45+ living singlets of blood-derived leukocytes. The trajectories on the radar plots delineate the characteristic marker profile of peripheral leukocytes of naive, 4T1 tumor bearing, and cisplatin treated tumorous mice (Figure 9).

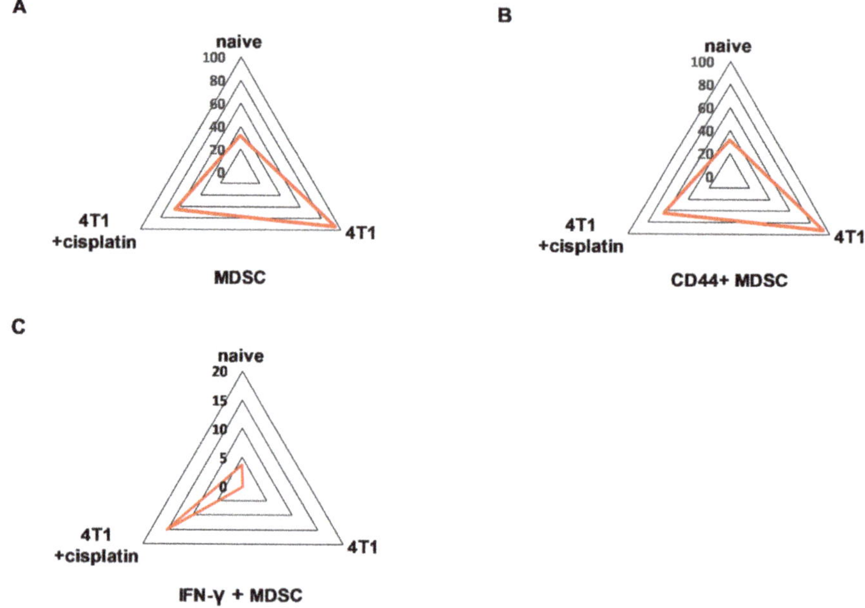

Figure 9. The trajectories on the radar plots delineate the characteristic marker profile of blood-derived leukocytes in naive, 4T1 tumor bearing, and cisplatin treated tumorous mice. The accumulation of peripheral (**A**) CD11b+/Gr-1+ MDSCs and (**B**) CD44+ MDSCs is a characteristic of 4T1 breast cancer. (**C**) Due to cisplatin treatment IFN-γ+ MDSCs were developed at the periphery. The percentage of the given populations is demonstrated on the radar plots within the CD45+ living singlets determined by manual gating in Cytobank. The gating hierarchy is explained in the text. The markers of the panel which were not detected or did not show differential expression are not shown. Representative radar plots are shown from the pooled samples of 6 mice per group as described in Section 4.5.

MDSCs were defined as CD45+/CD3-/CD11+/Gr1+ cells (Figure 9A) and further evaluated for the expression of CD44 (Figure 9B) and IFN-γ (Figure 9C). The total MDSCs, CD44+ MDSCs and IFN-γ + MDSCs represented 92%, 93%, and only 0.15% of CD45+ living singlets in the spleen of 4T1 advanced tumorous mice, respectively. Administration of cisplatin resulted in the decrease of the percentage of both MDSCs, CD44+ MDSCs to 63% (naive blood 32%). Interestingly, IFN-γ which is indispensable for antitumor immune response [39], was increased by cisplatin in 15% of MDSCs (naive 3%). Sunburst charts represent the immunocomposition of main subsets of the blood detected by single cell mass cytometry in Figure S4B and FlowSOM (Flow data using Self-Organizing Map, [35]) analysis demonstrates the identified clusters, minimum spanning trees (MSTs) of peripheral subsets in Figure S5.

3. Discussion

Solid tumors manifest when cancer cells escape from immunosurveillance. During cancer development, malignant cells develop strategies to induce peripheral tolerance and in parallel inflammatory cells change their phenotype to nurse the tumor and suppress anti-tumor effector functions [36,40]. Since metastatic breast cancer is the most frequent and deadliest type of cancer among young adult women [4,5], we focused on the 4T1 murine breast carcinoma model to study the efficacy of anti-cancer drug candidates. Among chemotherapeutic compounds, cisplatin, a well-known DNA crosslinker, a first line option in human carcinomas was investigated (as a reference drug) in the 4T1 murine metastatic breast carcinoma model. This type of breast carcinoma can be a relevant animal model for the human disease because (i) it is orthotopically transplanted into the mammary fat pad, (ii) it grows in immunocompetent mice (BALB/c) and (iii) gives metastasis spontaneously. Here, we studied the immunophenotype of the spleen and blood of mice with advanced cancer and with cisplatin treated cancer compared to naive mice.

We monitored the dose-response curve of cisplatin to 4T1 cells in real-time because previous studies described results with endpoint assays [41–43]. The half maximal inhibitory concentration (IC$_{50}$) values showed time dependence: 36.74 µM at 24 h, 7.608 µM at 48 h, 6.962 µM at 72 h, 4.128 µM at 96 h, and 3.995 µM at 120 h (Figure 1 and Figure S1). In order to investigate the alteration of the tumor stroma or the immunophenotype changes during cancer development the 4T1 cells were orthotopically injected into BALB/c mice. Cisplatin reduced the tumor growth, the number of lung metastatic nodules, and splenomegaly (Figure 2). In addition to malignant cells, the alteration of the stroma of solid tumors, namely the activation of cancer-associated fibroblasts has been described as critical determinant of cancer development and survival, and was also confirmed as therapeutic target [2]. It was shown that the expression of FAP prolyl endopeptidase, even on the cell surface or solubilized in the plasma, correlates with the malignancy and cisplatin resistance of carcinomas [29,44–46]. Therefore, we have developed an assay to measure the activity of FAP cleaving the Fmoc-Gly-Pro-Cysteic acid-Ile-Gly-NH$_2$ peptide substrate (Figure S2). We could show that cisplatin treatment significantly reduced the FAP activity of the plasma of 4T1 tumor bearing mice (Figure 3).

It has been widely known that the alternative polarization of the immune system, the accumulation of immature myeloid cells (MDSCs) with potent immunosuppression contribute to the cancer development [12,47]. The splenic immunophenotye of mice with advanced cancer showed the emergence of the hyaluronic acid receptor CD44+ and in a smaller extent IL-17A+ MDSCs, the loss of B220+ and CD62L+ B-cells, the loss of CD62L+ CD4+ and CD8+T-cells (Figures 5–7). Myeloid expression of IL-17A in the spleen has been published as a sign of advanced cancer [26]. Cisplatin treatment could restore the splenic immunophenotype downregulating CD44+, IL-17A+ MDSCs presumably because of the smaller tumor burden, on the other hand MDSCs are also sensitive to low dose chemotherapy due to their high proliferative potential [30]. The blood showed the expression of Gr-1+^{dim}, more immature MDSCs in advanced cancer (Figure 8), which has been reported to suppress T-cell proliferation and IFN-γ production [37]. The expression intensity of CD44 on circulating MDSCs was also dim suggesting more immature phenotype (Figure 8), however, the percentage of CD44+MDSCs (irrespective of the CD44 marker intensity) was higher in the blood of mice with advanced cancer (Figure 9). Cisplatin partially reduced the accumulation of CD44+/CD11+/Gr1+ myeloid cells in the circulation, but changed their phenotype via induction of INF-γ production (Figure 9). It has been reported that IFN-γ producing immature myeloid cells could exert a potent immune response against sever invasive bacterial infections [48]. Computational tools, unsupervised algorithms were used to summarize our results, such as sunburst population analysis in Figure S4 and the FlowSOM analysis in Figure S5.

The immunomodulatory effect of cisplatin by increasing tumor immunogenicity has been recently published in ovarian cancer [22]. The general overview of chemotherapeutics, especially cisplatin mediated immunomodulation has been reviewed elsewhere [49,50].

To the best of our knowledge our study is the first using single cell mass cytometry to show the immunomodulatory effect of cisplatin in the 4T1 murine model of metastatic triple negative breast cancer via downregulation of immature myeloid cells and upregulation of CD62L+ B- and T-cells and IFN-γ+ peripheral myeloid cells.

4. Materials and Methods

4.1. Real-Time Cell Electronic Sensing (RT-CES) Cytotoxicity Assay

The 4T1 cells were purchased from the ATCC (American Type Culture Collection, Manassas, VA, USA) and maintained as described previously [10]. Briefly, 4T1 were maintained in Roswell Park

Memorial Institute 1640 medium (RPMI-1640) with 10% FCS. The pH of the cell culture media was controlled to be between 7.2–7.4 prior use. The medium was supplemented with 2 mM GlutaMAX, and 100 U/mL penicillin, 100 µg/mL streptomycin (Life Technologies, Carlsbad, CA, USA) before use. Cells were passed every three days and placed in a humidified incubator at 37 °C 5% CO_2 (Sanyo, Osaka, Japan). The xCelligence real-time cell electronic sensing (RT-CES) cytotoxicity assay (Acea Biosciences, San Diego, CA, USA) was performed as previously described with some modifications [25,51]. Briefly, 96-well E-plate (Acea Biosciences) was coated with gelatin solution (0.2% in phosphate buffered saline, PBS) for 20 min at 37 °C, then gelatin was washed twice with PBS solution. Growth media (50 µL) was then gently dispensed into each well of the 96-well E-plate for background readings by the RT-CES system prior to addition of 50 µL of the cell suspension containing 2×10^4 4T1 cells. Plates were kept at room temperature in a tissue culture hood for 30 min prior to insertion into the RT-CES device in the incubator to allow cells to settle. Cell growth was monitored for 48 h by measurements of electrical impedance every 15 min. Continuous recording of impedance in cells was reflected by cell index value. 48 h later cells were treated with an increasing concentration (134 nM–100 µM) of cisplatin (Selleckchem, Houston, TX, USA), treatments are demonstrated between 3.704–100 µM. Treated and control wells were dynamically monitored over 120 h by measurements of electrical impedance every 15 min. The raw plate reads for each titration point were normalized relative to the cell index status right before treatment. Each treatment was repeated in 3 wells per plate during the experiments. The half maximal inhibitory concentration (IC_{50}) was calculated with relation to untreated control cells (1 corresponds to 100% viability on the y axis), and blank wells containing media without cells. IC_{50} values (50% inhibiting concentration) were calculated by GraphPad Prism® (version 5.01, La Jolla, CA, USA).

4.2. The 4T1 in Vivo Breast Carcinoma Model

The animal experiments were performed in accordance with animal experimentation and ethics guidelines of the EU (2010/63/EU). Experimental protocols were approved by the responsible governmental agency (National Food Chain Safety Office) in possession of an ethical clearance XXIX./128/2013.

Female Charles River-derivative BALB/c mice (8–10 week old) were purchased from Kobay Ltd., (Ankara, Turkey) and were injected orthotopically with 4T1 breast carcinoma cells (1.2×10^5 cells) as described previously [9]. The animals had free access to food and water. Six mice were included into each experimental group. Treatment by cisplatin (Ebewe Pharma, Unterach am Attersee, Austria) was started after 10 days of inoculation and followed in every 5th day in 5 mg/kg dose administered intraperitoneally twice on the day of the treatment. The experiments were repeated independently two times under the same conditions, the pooled results have been presented in the paper ($n = 12$). Tumors were evaluated macroscopically by the following parameters: 1) incidence of palpable tumors was determined by the daily monitoring of animals in each experimental group; 2) tumor size was measured with a precision caliper and calculated according to the formula: d2 × D × 0.5, where d and D are the minor and major diameters, respectively; 3) after euthanizing the animals, weights of the excised primary tumors, spleens, and lungs were measured. Spleen and blood were processed freshly in order to isolate leukocytes. Mice showing signs of suffering (lost 15% of body weight and/or loss of the righting reflex and/or enable to eat, drink) due to (ethical) legislation were sacrificed.

4.3. Synthesis of Fmoc-Gly-Pro-Cysteic Acid-Ile-Gly-NH$_2$ Peptide

In order to obtain the target peptide, Fmoc-Gly-Pro-Cysteic acid-Ile-Gly-NH$_2$, first Fmoc-Gly-Pro-Cys-Ile-Gly-NH$_2$ peptide was synthesized (Fmoc, 9-fluorenylmethoxycarbonyl, Avidin Ltd., Szeged, Hungary). Reagents, otherwise not stated, were purchased from Sigma-Aldrich (St. Louis, MI, USA). Fmoc-strategy synthesis was carried out manually in a solid-phase vessel on Rink Amide ChemMatrix resin, the protected Fmoc-amino acids (3 equiv.) were coupled using DCC (dicyclohexylcarbodiimide 3 equiv.) and HOBt (1-hydroxybenzotriazole, 3 equiv.) in DMF

(*N*,*N*-dimethylformamide) for 2 h at room temperature. The deprotection of the Fmoc-group was achieved with the treatment of the resin with 5% piperazine/DMF (1 × 5 min, 1 × 20 min). Washings between coupling and deprotection were performed with DMF (3 × 1 min), MeOH (methanol) (1 × 1 min) and DMF (3 × 1 min). The completion of the coupling was monitored using the Kaiser test. Following the final coupling, the resin was washed with DMF (3 × 1 min) and MeOH (3 × 1 min), and dried under a stream of air. The dry resin was treated with TFA/H_2O (trifluoroacetic acid/H_2O, 95:5) for 3 h at room temperature. The cleavage mixture was lyophilized, and the pellet was redissolved in MeOH/AcN (methanol/acetonitrile, 1:1) for Liquid chromatography-mass spectrometry (LC-MS) analysis (Agilent, Santa Clara, CA, USA). The LC-MS analysis found that Fmoc-Gly-Pro-Cys-Ile-Gly-NH_2 was obtained in 90% purity (linear gradient from 0% to 100% AcN over 30 min, t_R: 21.77 min). LC-MS observed $[M + H]^+$ 667.2, required $[M + H]^+$ 667.8. The obtained Fmoc-Gly-Pro-Cys-Ile-Gly-NH_2 was transformed into Fmoc-Gly-Pro-Cysteic acid-Ile-Gly-NH_2 peptide with no further purification. It was dissolved in acetone, cooled to 1 °C in a mixture of ice and water. First 30% aqueous H_2O_2 was added, then $Na_2WO_4.2H_2O$ in catalytic amount. After 3 h, we did not detect the starting material (Fmoc-Gly-Pro-Cys-Ile-Gly-NH_2), it completely transformed into the desired Fmoc-Gly-Pro-Cysteic acid-Ile-Gly-NH_2 peptide. The LC-MS analysis found that Fmoc-Gly-Pro-Cysteic acid-Ile-Gly-NH_2 was obtained in 75% purity (linear gradient from 0% to 100% AcN over 30 min, t_R: 16.90 min). LC-MS observed $[M + H]^+$ 715.2, required $[M + H]^+$ 715.8. To purify the crude peptide, it was dissolved in AcN/MeOH/H_2O, then filtered, using a 0.45 μm nylon filter. Gradient elution was used, 70–100% AcN in 60 min at a 3 mL min-1 flow rate with detection at 220 nm. Pure fractions were collected and lyophilized to give a fluffy white material.

4.4. FAP Activity Assay

Blood from control and tumor bearing (untreated and cisplatin treated) BALB/c mice was drawn (200 μL) into EDTA containing tubes (separately from 3 animals from each group). Reagents, otherwise not stated, were purchased from Sigma-Aldrich. Blood samples were centrifuged at 12,000 RPM for 10 min. Supernatant (plasma) was removed and transferred to fresh tubes and were immediately used. Digestion reactions were set up by combining 36 μL plasma and 9 μL solution of cysteine acid containing peptide solution (5 mg/mL) for a final peptide concentration of 1 mg/mL. Reactions were incubated at 37 °C for 60 min. After incubation, 75 μL of acetonitrile containing 0.2% trifluoro-acetic (TFA) acid was added and samples were centrifuged at 12,000 RPM for 5 min. 80 μL of supernatant was removed and transferred to fresh tubes and were stored at −20 °C until analysis. Area under the curve (AUC) values of the peptide digestion product **2** were plotted.

4.5. Mass Cytometry

Single cell mass cytometry (CyTOF, Fluidigm, San Francisco, CA, USA)) was performed as described previously with some modifications [52]. Briefly, naive, 4T1 breast tumor bearing and cisplatin treated 4T1 tumor bearing mice were euthanized on the 23rd day after 4T1 injection. Spleen and blood were processed freshly. Withdrawal of the blood was carried by cardiac puncture using 50 μL EDTA (30 mg/mL) per syringe (Beckton Dickinson, Franklin Lakes, NY, USA). Spleen was smashed on 100 μm cell strainer (VWR, Radnoe, PA, USA), washed with PBS and centrifuged at 1400 rpm 5 min. Blood was centrifuged at 2000 rpm for 10 min, plasma was harvested and stored at −80 °C. Both the pellet of spleen and blood were resuspended. Red blood cell lysis was carried out by the incubation of cells with 5 mL ACK (0.155 M NH_4Cl, 10 mM $KHCO_3$, 0.1 mM Na_2EDTA, pH 7.3, Sigma-Aldrich) solution for 5 min. Samples were loaded on cell strainer (70 μm in pore size,) and washed by 20 mL PBS. Cells were counted using Bürker chamber and trypan blue viability dye. Three million cells were pooled from six mice per group for mass cytometry (Helios, Fluidigm, San Francisco, CA, USA). The in vivo experiment and CyTOF were repeated twice. Cells viability was determined by cisplatin (5 μM 195Pt, Fluidigm) staining for 3 min on ice in 300 μL PBS. Sample was diluted by 1500 μL Maxpar Cell Staining Buffer (MCSB, Fluidigm) and centrifuged at 350 g for 5 min. Cells were suspended in

50 µL MCSB and the antibody mix (Table 1) was added in 50 µL (each antibody diluted finally 1:100). Two commercially available antibody panel was combined, the Maxpar® Mouse Sp/LN Phenotyping Panel kit (Fluidigm, cat. numbers 201306) and Maxpar® Mouse Intracellular I Cytokine Panel kit (Fluidigm, cat. number 201310).

Samples after 60 min incubation at 4 °C, antibodies were washed by 2 mL MCSB and centrifuged at 300 g 5 min, two times. The pellet was suspended in the residual volume. Cells were fixed in 1.6% formaldehyde (freshly diluted from 16% Pierce formaldehyde with PBS, Thermo Fisher Scientific, Waltham, MA, USA) and incubated for 10 min at room temperature. Cells were centrifuged at 800 g for 5 min. Cell ID DNA intercalator (125 µM, 191/193 Iridium, Fluidigm) was added in 1000× dilution in Maxpar Fix and Perm for overnight at 4 °C. Cells for the acquisition were centrifuged at 800g for 5 min then were washed by 2 mL MCSB and centrifuged at 800 g for 5 min. Cells were suspended in 1 mL PBS (for WB injector) and counted in Bürker-chamber during centrifugation. For the acquisition the concentration of cells was set to 0.5×10^6/mL in cell acquisition solution (CAS, Fluidigm) containing 10% EQ Calibration Beads. Cells were filtered through 30 µm Celltrics gravity filter (Sysmex, Görlitz, Germany) and acquired freshly. Mass cytometry data were analyzed in Cytobank (Beckman Coulter, Brea, CA, USA). Single living cells were determined. The viSNE analysis was carried-out (iterations = 1000, perplexity = 30, theta = 0.5) on 5×10^4 for the spleen and 2×10^4 for blood of CD45+ living singlets. Reduction of dimensionality was performed by FlowSOM also, by an algorithm creating self-organizing maps during automated clustering in Cytobank [35].

4.6. Statistical Analysis

Results are shown as arithmetic mean ± standard error of the mean (SEM); statistical comparisons were performed by two-tailed Student's t-test as pairwise comparisons as described in the figure legends. In all statistical comparisons, probability "p" was set as the level of significance (set at * $p < 0.05$, ** $p < 0.01$, ***, $p < 0.001$). Data were processed and analyzed using Microsoft Excel (Microsoft Office 2016, Redmond, WA, USA), and visualized using GraphPad Prism or Cytobank.

5. Conclusions

Our findings showed that cisplatin treatment reduced tumor growth, number of lung metastasis and the splenomegaly of 4T1 tumor bearing mice. Cisplatin inhibited the tumor stroma formation, the activation of carcinoma-associated fibroblasts by the diminished proteolytic activity of fibroblast activating protein. Single cell mass cytometry revealed that cisplatin could exert a potent immunomodulatory effect via inhibiting the accumulation of splenic MDSCs in a murine model of metastatic triple negative breast cancer. Emergence of certain myeloid subsets in the spleen, such as CD44+MDSCs and IL-17A+MDSC were associated with advanced cancer, while within the lymphoid subsets, the absence the B220+ B-cells, CD62L+ B-cells, CD62L+CD4+ T-cells, and CD62L+ CD8+ T-cells was shown in the untreated tumor bearing mice. However, cisplatin treatment could restore the splenic immunophenotype similar to naive mice. Peripheral MDSCs in the circulation were not completely eliminated by cisplatin but myeloid-derived IFN-γ production was increased. Thus, our study highlights the use of low-dose chemotherapy, such as cisplatin in combination with immunotherapies to treat triple negative breast cancer.

Supplementary Materials: Supplementary Materials can be found at http://www.mdpi.com/1422-0067/21/1/170/s1.

Author Contributions: Conceptualization, L.G.P. and G.J.S.; methodology, J.A.B., L.H.J., A.K.K., P.N., R.A., L.I.N., and G.J.S.; software, J.Á.B.; validation, J.Á.B.; formal analysis, J.Á.B., L.H.J., and G.J.S.; investigation, J.A.B, L.H.J., A.K.K., P.N., R.A., L.I.N., and G.J.S.; resources, L.G.P.; data curation, J.A.B., L.H.J., L.G.P., and G.J.S.; writing—original draft preparation, L.H.J., L.G.P., and G.J.S.; writing—review and editing, L.H.J., L.G.P., and G.J.S.; visualization, J.Á.B. and G.J.S.; supervision, L.G.P. and G.J.S., project administration, L.G.P.; funding acquisition, L.G.P. All authors have read and agreed to the published version of the manuscript.

Funding: This research was funded by the following grants: GINOP-2.3.2-15-2016-00001 (BRC) by the National Research, Development and Innovation Office, Hungary; and EUREKA Network Program (H2020; E12655

MITOME) (Avidin Ltd.). Gábor J. Szebeni was supported by János Bolyai Research Scholarship of the Hungarian Academy of Sciences (BO/00139/17/8) and by the UNKP-19-4-SZTE-36 New National Excellence Program of the Ministry for Innovation and Technology.

Conflicts of Interest: The authors declare no conflict of interest.

References

1. Marx, J. Cancer biology. All in the stroma: cancer's Cosa Nostra. *Science* **2008**, *320*, 38–41. [CrossRef]
2. Hanahan, D.; Weinberg, R.A. Hallmarks of cancer: The next generation. *Cell* **2011**, *144*, 646–674. [CrossRef]
3. Adams, S.; Gatti-Mays, M.E.; Kalinsky, K.; Korde, L.A.; Sharon, E.; Amiri-Kordestani, L.; Bear, H.; McArthur, H.L.; Frank, E.; Perlmutter, J.; et al. Current Landscape of Immunotherapy in Breast Cancer: A Review. *JAMA Oncol.* **2019**. [CrossRef]
4. Harbeck, N.; Penault-Llorca, F.; Cortes, J.; Gnant, M.; Houssami, N.; Poortmans, P.; Ruddy, K.; Tsang, J.; Cardoso, F. Breast cancer. *Nat. Rev. Dis. Primers* **2019**, *5*, 66. [CrossRef] [PubMed]
5. Anders, C.K.; Johnson, R.; Litton, J.; Phillips, M.; Bleyer, A. Breast cancer before age 40 years. *Semin. Oncol.* **2009**, *36*, 237–249. [CrossRef] [PubMed]
6. Aslakson, C.J.; Miller, F.R. Selective events in the metastatic process defined by analysis of the sequential dissemination of subpopulations of a mouse mammary tumor. *Cancer Res.* **1992**, *52*, 1399–1405. [PubMed]
7. Tao, K.; Fang, M.; Alroy, J.; Sahagian, G.G. Imagable 4T1 model for the study of late stage breast cancer. *BMC Cancer* **2008**, *8*, 228. [CrossRef] [PubMed]
8. Lasso, P.; Llano Murcia, M.; Sandoval, T.A.; Uruena, C.; Barreto, A.; Fiorentino, S. Breast Tumor Cells Highly Resistant to Drugs Are Controlled Only by the Immune Response Induced in an Immunocompetent Mouse Model. *Integr. Cancer Ther.* **2019**, *18*. [CrossRef]
9. Szebeni, G.J.; Kriston-Pal, E.; Blazso, P.; Katona, R.L.; Novak, J.; Szabo, E.; Czibula, A.; Fajka-Boja, R.; Hegyi, B.; Uher, F.; et al. Identification of galectin-1 as a critical factor in function of mouse mesenchymal stromal cell-mediated tumor promotion. *PLoS ONE* **2012**, *7*, e41372. [CrossRef]
10. Szebeni, G.J.; Balog, J.A.; Demjen, A.; Alfoldi, R.; Vegi, V.L.; Feher, L.Z.; Man, I.; Kotogany, E.; Guban, B.; Batar, P.; et al. Imidazo [1,2-b] pyrazole-7-carboxamides Induce Apoptosis in Human Leukemia Cells at Nanomolar Concentrations. *Molecules* **2018**, *23*, 2845. [CrossRef]
11. Demjen, A.; Alfoldi, R.; Angyal, A.; Gyuris, M.; Hackler, L., Jr.; Szebeni, G.J.; Wolfling, J.; Puskas, L.G.; Kanizsai, I. Synthesis, cytotoxic characterization, and SAR study of imidazo [1,2-b] pyrazole-7-carboxamides. *Arch. Pharm. (Weinheim)* **2018**, *351*. [CrossRef] [PubMed]
12. Szebeni, G.J.; Vizler, C.; Kitajka, K.; Puskas, L.G. Inflammation and Cancer: Extra- and Intracellular Determinants of Tumor-Associated Macrophages as Tumor Promoters. *Mediat. Inflamm.* **2017**, *2017*. [CrossRef] [PubMed]
13. Ouzounova, M.; Lee, E.; Piranlioglu, R.; El Andaloussi, A.; Kolhe, R.; Demirci, M.F.; Marasco, D.; Asm, I.; Chadli, A.; Hassan, K.A.; et al. Monocytic and granulocytic myeloid derived suppressor cells differentially regulate spatiotemporal tumour plasticity during metastatic cascade. *Nat. Commun.* **2017**, *8*. [CrossRef] [PubMed]
14. Dos Reis, D.C.; Damasceno, K.A.; de Campos, C.B.; Veloso, E.S.; Pegas, G.R.A.; Kraemer, L.R.; Rodrigues, M.A.; Mattos, M.S.; Gomes, D.A.; Campos, P.P.; et al. Versican and Tumor-Associated Macrophages Promotes Tumor Progression and Metastasis in Canine and Murine Models of Breast Carcinoma. *Front. Oncol.* **2019**, *9*, 577. [CrossRef]
15. Piranlioglu, R.; Lee, E.; Ouzounova, M.; Bollag, R.J.; Vinyard, A.H.; Arbab, A.S.; Marasco, D.; Guzel, M.; Cowell, J.K.; Thangaraju, M.; et al. Primary tumor-induced immunity eradicates disseminated tumor cells in syngeneic mouse model. *Nat. Commun.* **2019**, *10*, 1430. [CrossRef]
16. Li, X.; Singh, K.; Luo, Z.; Mejia-Cordova, M.; Jamalpour, M.; Lindahl, B.; Zhang, G.; Sandler, S.; Welsh, M. Pro-tumoral immune cell alterations in wild type and Shb-deficient mice in response to 4T1 breast carcinomas. *Oncotarget* **2018**, *9*, 18720–18733. [CrossRef]
17. Kim, K.; Skora, A.D.; Li, Z.; Liu, Q.; Tam, A.J.; Blosser, R.L.; Diaz, L.A., Jr.; Papadopoulos, N.; Kinzler, K.W.; Vogelstein, B.; et al. Eradication of metastatic mouse cancers resistant to immune checkpoint blockade by suppression of myeloid-derived cells. *Proc. Natl. Acad. Sci. USA* **2014**, *111*, 11774–11779. [CrossRef]
18. Szebeni, G.J.; Vizler, C.; Nagy, L.I.; Kitajka, K.; Puskas, L.G. Pro-Tumoral Inflammatory Myeloid Cells as Emerging Therapeutic Targets. *Int. J. Mol. Sci.* **2016**, *17*, 1958. [CrossRef]

19. Dasari, S.; Tchounwou, P.B. Cisplatin in cancer therapy: Molecular mechanisms of action. *Eur. J. Pharmacol.* **2014**, *740*, 364–378. [CrossRef]

20. Voorwerk, L.; Slagter, M.; Horlings, H.M.; Sikorska, K.; van de Vijver, K.K.; de Maaker, M.; Nederlof, I.; Kluin, R.J.C.; Warren, S.; Ong, S.; et al. Immune induction strategies in metastatic triple-negative breast cancer to enhance the sensitivity to PD-1 blockade: The TONIC trial. *Nat. Med.* **2019**, *25*, 920–928. [CrossRef]

21. Wakita, D.; Iwai, T.; Harada, S.; Suzuki, M.; Yamamoto, K.; Sugimoto, M. Cisplatin Augments Antitumor T-Cell Responses Leading to a Potent Therapeutic Effect in Combination With PD-L1 Blockade. *Anticancer Res.* **2019**, *39*, 1749–1760. [CrossRef] [PubMed]

22. Grabosch, S.; Bulatovic, M.; Zeng, F.; Ma, T.; Zhang, L.; Ross, M.; Brozick, J.; Fang, Y.; Tseng, G.; Kim, E.; et al. Cisplatin-induced immune modulation in ovarian cancer mouse models with distinct inflammation profiles. *Oncogene* **2019**, *38*, 2380–2393. [CrossRef] [PubMed]

23. Lee, J.O.; Kang, M.J.; Byun, W.S.; Kim, S.A.; Seo, I.H.; Han, J.A.; Moon, J.W.; Kim, J.H.; Kim, S.J.; Lee, E.J.; et al. Metformin overcomes resistance to cisplatin in triple-negative breast cancer (TNBC) cells by targeting RAD51. *Breast Cancer Res.* **2019**, *21*, 115. [CrossRef] [PubMed]

24. Mohamad, N.E.; Abu, N.; Yeap, S.K.; Alitheen, N.B. Bromelain Enhances the Anti-tumor Effects of Cisplatin on 4T1 Breast Tumor Model In Vivo. *Integr. Cancer Ther.* **2019**, *18*. [CrossRef] [PubMed]

25. Ozsvari, B.; Puskas, L.G.; Nagy, L.I.; Kanizsai, I.; Gyuris, M.; Madacsi, R.; Feher, L.Z.; Gero, D.; Szabo, C. A cell-microelectronic sensing technique for the screening of cytoprotective compounds. *Int. J. Mol. Med.* **2010**, *25*, 525–530. [CrossRef] [PubMed]

26. Strauss, L.; Sangaletti, S.; Consonni, F.M.; Szebeni, G.; Morlacchi, S.; Totaro, M.G.; Porta, C.; Anselmo, A.; Tartari, S.; Doni, A.; et al. RORC1 Regulates Tumor-Promoting "Emergency" Granulo-Monocytopoiesis. *Cancer Cell* **2015**, *28*, 253–269. [CrossRef]

27. Hamson, E.J.; Keane, F.M.; Tholen, S.; Schilling, O.; Gorrell, M.D. Understanding fibroblast activation protein (FAP): Substrates, activities, expression and targeting for cancer therapy. *Proteom. Clin. Appl.* **2014**, *8*, 454–463. [CrossRef]

28. Long, X.; Xiong, W.; Zeng, X.; Qi, L.; Cai, Y.; Mo, M.; Jiang, H.; Zhu, B.; Chen, Z.; Li, Y. Cancer-associated fibroblasts promote cisplatin resistance in bladder cancer cells by increasing IGF-1/ERbeta/Bcl-2 signalling. *Cell Death Dis.* **2019**, *10*, 375. [CrossRef]

29. Mhawech-Fauceglia, P.; Yan, L.; Sharifian, M.; Ren, X.; Liu, S.; Kim, G.; Gayther, S.A.; Pejovic, T.; Lawrenson, K. Stromal Expression of Fibroblast Activation Protein Alpha (FAP) Predicts Platinum Resistance and Shorter Recurrence in patients with Epithelial Ovarian Cancer. *Cancer Microenviron.* **2015**, *8*, 23–31. [CrossRef]

30. Ugel, S.; Peranzoni, E.; Desantis, G.; Chioda, M.; Walter, S.; Weinschenk, T.; Ochando, J.C.; Cabrelle, A.; Mandruzzato, S.; Bronte, V. Immune tolerance to tumor antigens occurs in a specialized environment of the spleen. *Cell Rep.* **2012**, *2*, 628–639. [CrossRef]

31. Amir el, A.D.; Davis, K.L.; Tadmor, M.D.; Simonds, E.F.; Levine, J.H.; Bendall, S.C.; Shenfeld, D.K.; Krishnaswamy, S.; Nolan, G.P.; Pe'er, D. viSNE enables visualization of high dimensional single-cell data and reveals phenotypic heterogeneity of leukemia. *Nat. Biotechnol.* **2013**, *31*, 545–552. [CrossRef] [PubMed]

32. Rodig, S.J.; Shahsafaei, A.; Li, B.; Dorfman, D.M. The CD45 isoform B220 identifies select subsets of human B cells and B-cell lymphoproliferative disorders. *Hum. Pathol.* **2005**, *36*, 51–57. [CrossRef] [PubMed]

33. Driver, D.J.; McHeyzer-Williams, L.J.; Cool, M.; Stetson, D.B.; McHeyzer-Williams, M.G. Development and maintenance of a B220- memory B cell compartment. *J. Immunol.* **2001**, *167*, 1393–1405. [CrossRef] [PubMed]

34. Mauri, C.; Menon, M. The expanding family of regulatory B cells. *Int. Immunol.* **2015**, *27*, 479–486. [CrossRef]

35. Van Gassen, S.; Callebaut, B.; Van Helden, M.J.; Lambrecht, B.N.; Demeester, P.; Dhaene, T.; Saeys, Y. FlowSOM: Using self-organizing maps for visualization and interpretation of cytometry data. *Cytom. A* **2015**, *87*, 636–645. [CrossRef]

36. Gonzalez, H.; Hagerling, C.; Werb, Z. Roles of the immune system in cancer: From tumor initiation to metastatic progression. *Genes Dev.* **2018**, *32*, 1267–1284. [CrossRef]

37. Tsiganov, E.N.; Verbina, E.M.; Radaeva, T.V.; Sosunov, V.V.; Kosmiadi, G.A.; Nikitina, I.Y.; Lyadova, I.V. Gr-1dimCD11b+ immature myeloid-derived suppressor cells but not neutrophils are markers of lethal tuberculosis infection in mice. *J. Immunol.* **2014**, *192*, 4718–4727. [CrossRef]

38. Sendo, S.; Saegusa, J.; Okano, T.; Takahashi, S.; Akashi, K.; Morinobu, A. CD11b+Gr-1(dim) Tolerogenic Dendritic Cell-Like Cells Are Expanded in Interstitial Lung Disease in SKG Mice. *Arthritis Rheumatol.* **2017**, *69*, 2314–2327. [CrossRef]

39. Lin, C.F.; Lin, C.M.; Lee, K.Y.; Wu, S.Y.; Feng, P.H.; Chen, K.Y.; Chuang, H.C.; Chen, C.L.; Wang, Y.C.; Tseng, P.C.; et al. Escape from IFN-gamma-dependent immunosurveillance in tumorigenesis. *J. Biomed. Sci.* **2017**, *24*, 10. [CrossRef]

40. Katoh, H.; Watanabe, M. Myeloid-Derived Suppressor Cells and Therapeutic Strategies in Cancer. *Mediat. Inflamm.* **2015**, *2015*. [CrossRef]

41. Yerlikaya, A.; Altikat, S.; Irmak, R.; Cavga, F.Z.; Kocacan, S.A.; Boyaci, I. Effect of bortezomib in combination with cisplatin and 5-fluorouracil on 4T1 breast cancer cells. *Mol. Med. Rep.* **2013**, *8*, 277–281. [CrossRef] [PubMed]

42. Paraskar, A.S.; Soni, S.; Chin, K.T.; Chaudhuri, P.; Muto, K.W.; Berkowitz, J.; Handlogten, M.W.; Alves, N.J.; Bilgicer, B.; Dinulescu, D.M.; et al. Harnessing structure-activity relationship to engineer a cisplatin nanoparticle for enhanced antitumor efficacy. *Proc. Natl. Acad. Sci. USA* **2010**, *107*, 12435–12440. [CrossRef] [PubMed]

43. Shiassi Arani, F.; Karimzadeh, L.; Ghafoori, S.M.; Nabiuni, M. Antimutagenic and Synergistic Cytotoxic Effect of Cisplatin and Honey Bee Venom on 4T1 Invasive Mammary Carcinoma Cell Line. *Adv. Pharmacol. Sci.* **2019**, *2019*. [CrossRef] [PubMed]

44. Liu, R.; Li, H.; Liu, L.; Yu, J.; Ren, X. Fibroblast activation protein: A potential therapeutic target in cancer. *Cancer Biol. Ther.* **2012**, *13*, 123–129. [CrossRef] [PubMed]

45. Cremasco, V.; Astarita, J.L.; Grauel, A.L.; Keerthivasan, S.; MacIsaac, K.; Woodruff, M.C.; Wu, M.; Spel, L.; Santoro, S.; Amoozgar, Z.; et al. FAP Delineates Heterogeneous and Functionally Divergent Stromal Cells in Immune-Excluded Breast Tumors. *Cancer Immunol. Res.* **2018**, *6*, 1472–1485. [CrossRef] [PubMed]

46. Sandberg, T.P.; Stuart, M.; Oosting, J.; Tollenaar, R.; Sier, C.F.M.; Mesker, W.E. Increased expression of cancer-associated fibroblast markers at the invasive front and its association with tumor-stroma ratio in colorectal cancer. *BMC Cancer* **2019**, *19*, 284. [CrossRef] [PubMed]

47. Ostrand-Rosenberg, S.; Fenselau, C. Myeloid-Derived Suppressor Cells: Immune-Suppressive Cells That Impair Antitumor Immunity and Are Sculpted by Their Environment. *J. Immunol.* **2018**, *200*, 422–431. [CrossRef]

48. Matsumura, T.; Ato, M.; Ikebe, T.; Ohnishi, M.; Watanabe, H.; Kobayashi, K. Interferon-gamma-producing immature myeloid cells confer protection against severe invasive group A Streptococcus infections. *Nat. Commun.* **2012**, *3*, 678. [CrossRef]

49. Kersten, K.; Salvagno, C.; de Visser, K.E. Exploiting the Immunomodulatory Properties of Chemotherapeutic Drugs to Improve the Success of Cancer Immunotherapy. *Front. Immunol.* **2015**, *6*, 516. [CrossRef]

50. De Biasi, A.R.; Villena-Vargas, J.; Adusumilli, P.S. Cisplatin-induced antitumor immunomodulation: A review of preclinical and clinical evidence. *Clin. Cancer Res.* **2014**, *20*, 5384–5391. [CrossRef]

51. Man, I.; Szebeni, G.J.; Plangar, I.; Szabo, E.R.; Tokes, T.; Szabo, Z.; Nagy, Z.; Fekete, G.; Fajka-Boja, R.; Puskas, L.G.; et al. Novel real-time cell analysis platform for the dynamic monitoring of ionizing radiation effects on human tumor cell lines and primary fibroblasts. *Mol. Med. Rep.* **2015**, *12*, 4610–4619. [CrossRef] [PubMed]

52. Alfoldi, R.; Balog, J.A.; Farago, N.; Halmai, M.; Kotogany, E.; Neuperger, P.; Nagy, L.I.; Feher, L.Z.; Szebeni, G.J.; Puskas, L.G. Single Cell Mass Cytometry of Non-Small Cell Lung Cancer Cells Reveals Complexity of In vivo And Three-Dimensional Models over the Petri-dish. *Cells* **2019**, *8*, 1903. [CrossRef] [PubMed]

International Journal of
Molecular Sciences

Article

Syntaxin 1: A Novel Robust Immunophenotypic Marker of Neuroendocrine Tumors

Bence Kővári [1,*,†], Sándor Turkevi-Nagy [1,†], Ágnes Báthori [1], Zoltán Fekete [1] and László Krenács [2]

1   Department of Pathology, University of Szeged, Szeged 6725, Hungary;
    turkevi-nagy.sandor@med.u-szeged.hu (S.T.-N.); bathori.agnes87@gmail.com (Á.B.);
    z0li950912@gmail.com (Z.F.)
2   Laboratory of Tumor Pathology and Molecular Diagnostics, Szeged 6726, Hungary; krenacsl@vipmail.hu
*   Correspondence: kovari.bence.p@gmail.com
†   These authors contributed equally to this work.

Received: 4 December 2019; Accepted: 10 February 2020; Published: 12 February 2020

Abstract: Considering the specific clinical management of neuroendocrine (NE) neoplasms (NENs), immunohistochemistry (IHC) is required to confirm their diagnosis. Nowadays, synaptophysin (SYP), chromogranin A (CHGA), and CD56 are the most frequently used NE immunohistochemical markers; however, their sensitivity and specificity are less than optimal. Syntaxin 1 (STX1) is a member of a membrane-integrated protein family involved in neuromediator release, and its expression has been reported to be restricted to neuronal and NE tissues. In this study, we evaluated STX1 as an immunohistochemical marker of NE differentiation. STX1, SYP, CHGA, and CD56 expression was analyzed in a diverse series of NE tumors (NETs), NE carcinomas (NECs), and non-NE tumors. All but one (64/65; 98%) NETs and all (54/54; 100%) NECs revealed STX1 positivity in at least 50% of the tumor cells. STX1 showed the highest sensitivity both in NETs (99%) and NECs (100%) compared to CHGA (98% and 91%), SYP (96% and 89%), and CD56 (70% and 93%), respectively. A wide variety of non-NE tumors were tested and found to be uniformly negative, yielding a perfect specificity. We established that STX1 is a robust NE marker with an outstanding sensitivity and specificity. Its expression is independent of the site and grade of the NENs.

Keywords: neuroendocrine neoplasia; neuroendocrine tumor; neuroendocrine carcinoma; immunohistochemistry; syntaxin 1

---

1. Introduction

Neuroendocrine (NE) cells comprise a cellular network integrating the nervous and endocrine systems. They are found in virtually all organs, and the most common are in the gastrointestinal and lower respiratory tracts. The NE cells secrete biogenic amines and peptide hormones regulating a wide variety of functions into the bloodstream. Tumorous growths of NE cells are collectively referred to as NE neoplasms (NENs). The bioactive substances secreted by neoplastic NE cells can lead to distinct clinical syndromes. Although the NENs may follow an indolent clinical course, a significant number of patients are diagnosed with advanced disease. Although there are many organ-specific differences in tumor biology and prognostic factors among NENs of different localizations, according to the recommendations of the International Agency for Research on Cancer (IARC) and World Health Organization (WHO) expert consensus proposal, the terminology of NENs should be uniformized in the future. Therefore, the currently substantially differing organ-specific classification schemes of NENs will be potentially revised and harmonized within the next edition of each WHO Blue Book [1,2]. This proposal separates well-differentiated and poorly differentiated NENs. The well-differentiated NENs are also referred to as NE tumors (NETs), while poorly differentiated highly malignant NENs are also designated as NE carcinomas (NECs) ([2,3] (pp. 210–214), [4] (pp. 18–19)). The histopathologic diagnosis of NENs is based on the typical cytomorphological and architectural features of these tumors; nevertheless, in atypical and especially poorly differentiated cases, immunophenotyping is inevitable [5–7]. Rare tumors with two distinct NE and non-NE cell populations, in which either component represents at least 30%, are defined as a mixed neuroendocrine-non-neuroendocrine neoplasm (MiNEN) [4] (pp. 16–20). These cases may be difficult to recognize, particularly when they are poorly differentiated, and their accurate classification also requires immunohistochemistry (IHC) evaluation.

NE cells share common antigens characteristic of NE differentiation, which can be utilized as immunophenotypic markers. Some of these are rather obsolete, such as the neuron-specific enolase (NSE) and protein gene product 9.5 (PGP9.5), which are very sensitive, but seriously lack specificity [6,8]; others, such as synaptophysin (SYP), chromogranin-A (CHGA), CD56, and insulinoma-associated protein 1 (INSM1), are more reliable, but still show disadvantages when used as individual markers [8–15].

Syntaxin 1 (STX1) is a member of a membrane-integrated nervous system-specific protein superfamily involved in the neuromediator release from synaptic vesicles [16–18]. STX1 plays a crucial role in ion channel regulation and synaptic exocytosis [16–18]. Two STX1 isoforms, HPC-1/syntaxin 1A and syntaxin 1B, are thought to have similar functions in the exocytosis of synaptic vesicles and show a very high homology [19,20]. Additionally, it has been reported that STX1 is associated with chromaffin granules in the adrenal medulla [21] and expressed in alpha, beta, and delta cells of pancreatic islets [22,23]. Using a Web-based in silico biomarker analysis, syntaxin 1A protein expression has been found to be restricted to NE cells [24]; however, this finding was not further evaluated.

Since STX1 represents a promising NE marker and has not yet been comprehensively tested in diagnostic pathology, we performed a retro- and prospective IHC study on a diverse series of benign and malignant tumors, in order to establish its utility in the immunophenotyping of NENs.

2. Results

Virtually all normal NE cells and hyperplastic NE lesions showed strong membranous and weak to moderate cytoplasmic STX1 staining (Figure 1). Thyroid follicular cells, adrenocortical tissue, and non-NE cells, including gastrointestinal mucosa, exocrine pancreas, liver, kidney, lymphoid tissue, and skin, showed no STX1 positivity. Neuronal tissue in the brain and peripheral nerves, including hypertrophic myenteric plexus in appendicitis specimens, revealed consistent STX1 staining. The results obtained with STX1 in NE and neuroepithelial neoplastic samples are summarized in Table 1.

Table 1. STX1, chromogranin A (CHGA), synaptophysin (SYP), and CD56 immunoreactivity of NE neoplasms (NENs) and neuroectodermal/neuroepithelial neoplasms.

Gastrointestinal NETs	STX1		CHGA		SYP		CD56	
Gastric enterochromaffin-like (ECL) cell NET	5/5	100.0%	4/4	100.0%	3/3	100.0%	0/1	0.0%
Duodenal non-functioning NET	2/2	100.0%	2/2	100.0%	2/2	100.0%	na	na
Ampullary NET	1/2	50.0%	2/2	100.0%	1/2	50.0%	1/2	50.0%
Small intestinal enterochromaffin (EC) cell NET	12/12	100.0%	12/12	100.0%	12/12	100.0%	8/12	66.7%
Appendix NET	9/9	100.0%	8/8	100.0%	7/7	100.0%	9/9	100.0%
Rectal EC cell NET	2/2	100.0%	2/2	100.0%	1/1	100.0%	0/2	0.0%
Rectal L cell NET	5/5	100.0%	2/5	33.3%	4/4	100.0%	5/5	100.0%
Metastatic NET. gastrointestinal origin	4/4	100.0%	4/4	100.0%	1/2	50.0%	0/2	0.0%
Pancreatic NETs								
Pancreatic functioning NET *	3/3	100.0%	2/2	100.0%	2/2	100.0%	1/1	100.0%
Pancreatic non-functioning NET	16/16	100.0%	13/14	92.9%	12/13	92.3%	10/11	90.9%
Pulmonary carcinoids								
Pulmonary carcinoid. typical	3/3	100.0%	3/3	100.0%	2/2	100.0%	2/2	100.0%
Pulmonary carcinoid. atypical	2/2	100.0%	na	na	na	na	na	na

Table 1. *Cont.*

Gastrointestinal NETs	STX1		CHGA		SYP		CD56	
Gastroenteropancreatic NECs								
Stomach small cell NEC	2/2	100.0%	1/1	100.0%	1/1	100.0%	1/1	100.0%
Duodenal large cell NEC	1/1	100.0%	1/1	100.0%	1/1	100.0%	na	na
Pancreatic large cell NEC	2/2	100.0%	1/1	100.0%	1/1	100.0%	2/2	100.0%
Ampullary small cell NEC	2/2	100.0%	1/1	100.0%	na	na	2/2	100.0%
Pulmonary NECs								
Pulmonary small cell NEC	19/19	100.0%	9/10	90.0%	2/2	100.0%	16/16	100.0%
Pulmonary large cell NEC	4/4	100.0%	1/1	100.0%	na		2/2	100.0%
Etc NECs								
Merkel cell carcinoma	5/5	100.0%	1/1	100.0%	1/1	100.0%	2/2	100.0%
Medullary thyroid carcinoma	10/10	100.0%	5/5	100.0%	2/2	100.0%	3/3	100.0%
Metastatic small cell NEC with unknown primary	3/3	100.0%	na	na	na	na	2/2	100.0%
Metastatic large cell NEC with unknown primary	3/3	100.0%	3/3	100.0%	na	na	1/3	33.3%
Small cell NEC of the prostate	1/1	100.0%	1/1	100.0%	1/1	100.0%	1/1	100.0%
Small cell NEC of the uterine cervix	1/1	100.0%	na	na	na	na	na	na
Small cell NEC of the breast	1/1	100.0%	na	na	na	na	na	na
Tumors with mixed NE and non-NE components **								
MiNEN **	2/2	100.0%	0/1	0	2/2	100.0%	0/2	0.0%
Mixed tumor with minor NE component **	4/4	100.0%	na	na	2/2	100.0%	0/2	0.0%
Etc NENs								
Pituitary adenoma	19/19	100.0%	4/4	100.0%	1/1	100.0%	na	na
Pheochromocytoma/paraganglioma	16/18	88.9%	15/16	93.8%	9/10	90.0%	2/2	100.0%
Neuroectodermal/neuroepithelial neoplasia								
Medulloblastoma	8/9	88.9%	na	na	8/9	88.9%	2/2	100.0%
Neuroblastoma	8/8	100.0%	5/5	100.0%	5/5	100.0%	2/2	100.0%
Ganglioneuroma	2/2	100.0%	na	na	na	na	na	na
Ewing sarcoma/PNET	2/2	100.0%	na	na	na	na	na	na

*: 2 insulinomas, 1 gastrinoma. **: the component showing an NE morphology was positive, and the non-NE component was negative for STX1.

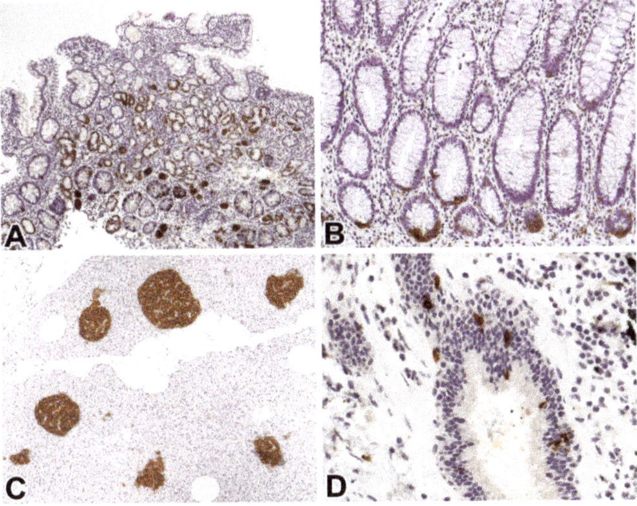

Figure 1. Syntaxin 1 (STX1) immunoreactivity in non-neoplastic tissues. (**A**) Nodular neuroendocrine (NE) cell hyperplasia in autoimmune metaplastic atrophic gastritis, 10x; (**B**) NE cells in normal colonic mucosa, 20x; (**C**) Langerhans islets in normal pancreatic tissue, 10x; (**D**) NE cells in normal bronchial mucosa, 40x.

All but one (99/100, 99%) cases of NENs, including metastatic NETs, proved to be STX1-positive (Figure 2). Regarding specific subsets of gastrointestinal NETs, vesicular monoamine transporter 1 (VMAT1)-positive EC cell NETs; VMAT1-negative, autoimmune gastritis-associated NETs consistent with ECL cell NETs; and even rectal glucagon-like peptide 1 (GLP1)-positive L cell NETs, were consistently positive (Figure 3). At least 50% of the tumor cells were labeled with a moderate to strong intensity, but in 92% of cases, the positivity rate was more than 85%. The STX1 staining intensity showed no correlation with the mitotic activity, Ki-67 labeling index, or tumor grade. The single STX1-negative case represented a grade 2 CHGA-positive NET of the major duodenal papilla. All small and large cell NECs were consistently STX1-positive, mostly with intense diffuse staining, regardless of the anatomical site (Figure 4). Furthermore, special types of NECs, such as Merkel cell and medullary thyroid carcinomas, similarly showed uniform strong diffuse immunoreactivity for STX1 (Figure 5).

The staining pattern was usually both membranous and cytoplasmic. The ratio of these two patterns varied considerably, from predominantly cytoplasmic to complete diffuse membranous. Membrane staining was typically complete in NETs, although in some cases with an acinar architecture, a basolateral membranous staining pattern was revealed. In some NECs, predominantly aberrant incomplete membrane staining was noticed. In comparison to common NE IHC markers, STX1 showed the highest sensitivity both in NETs (99%) and NECs (100%), which was followed by CHGA (98% and 91%), SYP (96% and 89%), and CD56 (70% and 93%), respectively (Table 2). The four applied markers detected a significantly different proportion of positive cases regarding gastrointestinal NETs ($p < 0.000001$), gastrointestinal NECs ($p = 0.01$), and NECs in general ($p = 0.007$). In terms of pancreatic NETs, no statistically significant level of such an association was seen.

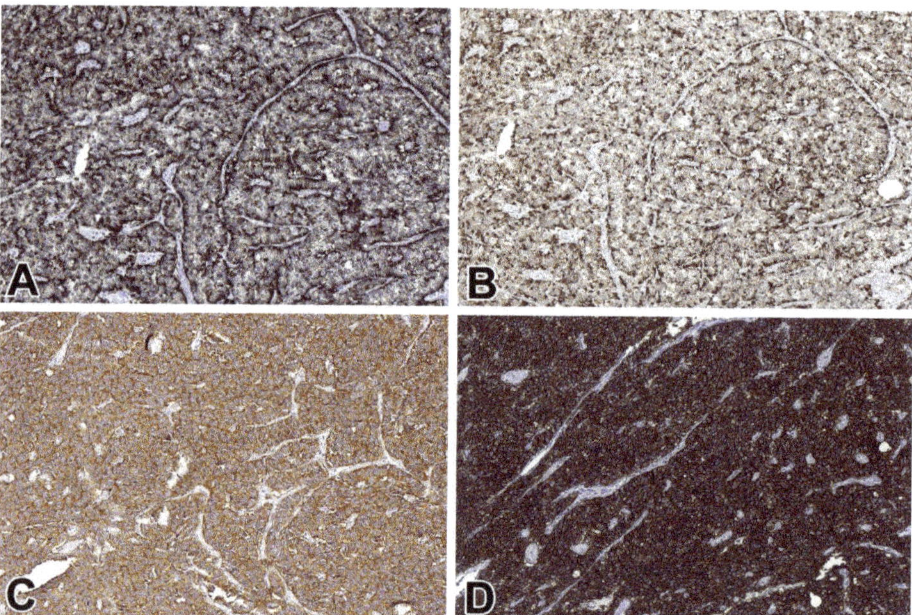

Figure 2. CD56, CHGA, SYP, and STX1 immunoreactivity in a pancreatic non-functioning NE tumor (NET) grade 2. (**A**) CD56, 20x; (**B**) CHGA, 20x; (**C**) SYP, 20x; (**D**) STX1, 20x.

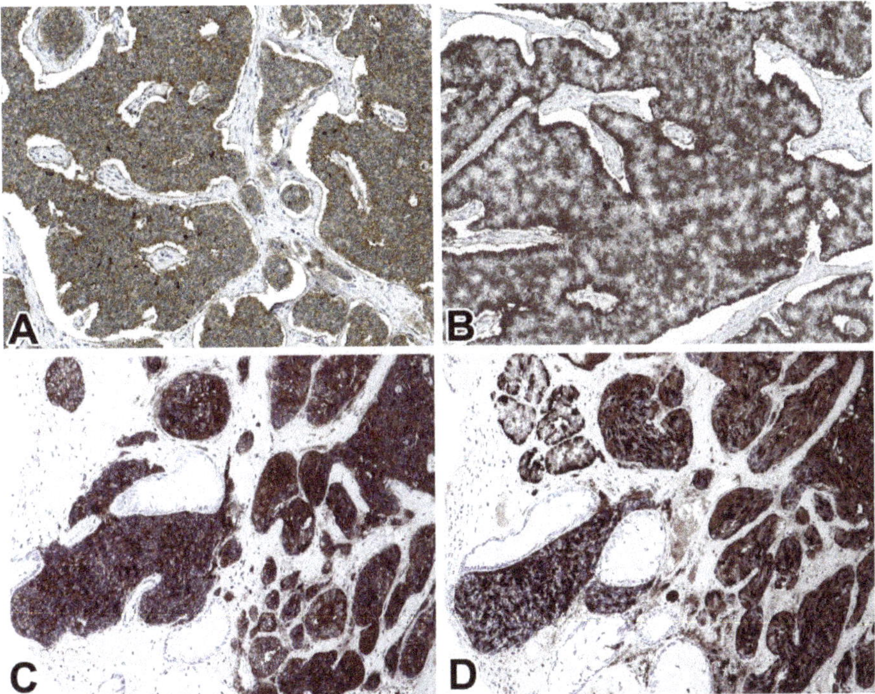

Figure 3. STX1 (**A,C**) vesicular monoamine transporter 1 (VMAT1), (**B**) and glucagon-like peptide 1 (GLP1) (**D**) immunoreactivity in a small intestinal enterochromaffin (EC) cell NET grade 2 (**A,B**) and a rectal L cell NET grade 1 (**C,D**) 20x.

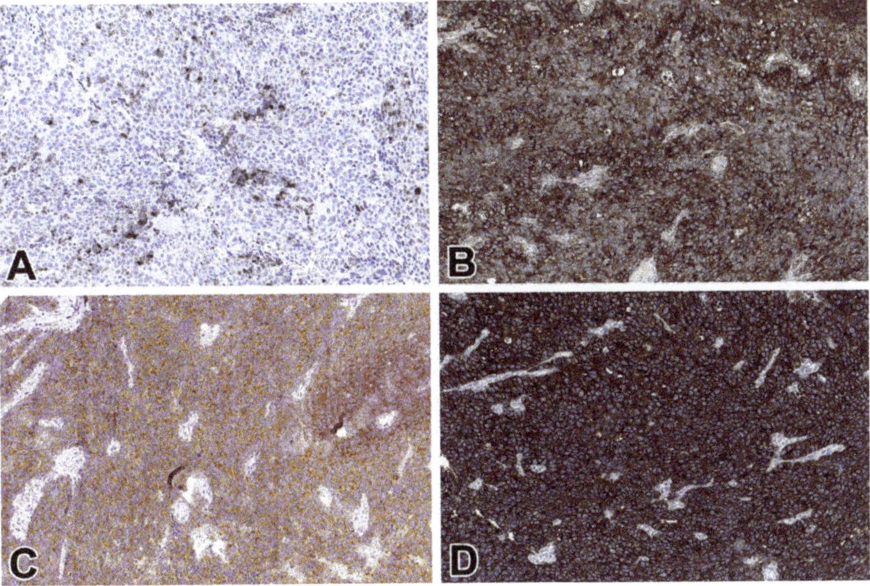

Figure 4. CD56, CHGA, SYP, and STX1 immunoreactivity in a pancreatic large cell NE carcinoma (NEC). (**A**) CD56, 20x; (**B**) CHGA, 20x; (**C**) SYP, 20x; (**D**) STX1, 20x.

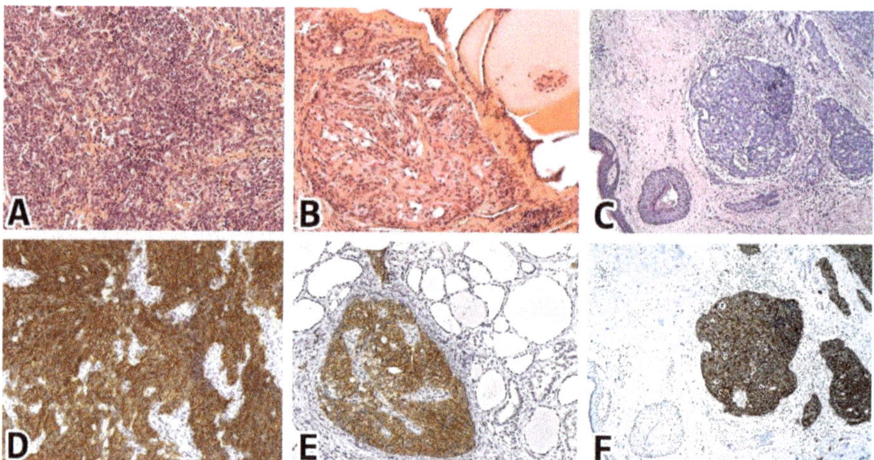

Figure 5. STX1 immunoreactivity in various types of NECs. (**A–C**) Hematoxylin-eosin stainings; (**D–F**) STX1 immunohistochemistry (IHC) reactions; (**A,D**) small cell NEC of the lung, 20x; (**B,E**) medullary thyroid carcinoma, 20x; (**C,F**) Merkel cell carcinoma of the skin, 10x.

Table 2. Sensitivity of STX1, CHGA1, SYP, and CD56 in NETs and NECs.

	Sensitivity—NET	Sensitivity—NEC
STX1	99%	100%
CHGA	98%	91%
SYP	96%	89%
CD56	70%	93%

In six tumors with mixed NE and non-NE components, including those with only minor NE components and those qualifying as MiNENs, the STX1 expression topographically always colocalized with the morphological features of NE differentiation (Table 1), while the exocrine component showed no STX1 positivity.

STX1 expression was generally absent in conventional carcinomas. In a subset (16/215, 7%) of conventional (non-NE) neoplasia, a discrete (<10%) and scattered intratumoral STX1-positive cell population was noticeable without a definite NE morphology. The expression of at least one further (CD56, SYP, or CHGA) NE marker was also detected in all these cases. The majority (10/13, 77%) of the hypercellular variant (type B) of mucinous breast carcinoma cases and one case of ductal carcinoma demonstrated STX1 positivity, usually in more than 60% of the tumor cells (Table 3).

All 20 (100%) pituitary adenoma cases and 14/16 (88%) of pheochromocytoma cases showed STX1 positivity (Table 1, Figure 6). The latter tumor group included 12 cases with diffuse staining and two cases with 30% tumor cell positivity. The two STX1-negative medullary chromaffin cell-derived tumors represented one benign and one malignant pheochromocytoma case. In three pheochromocytomas, faint and focal STX1 positivity was also exhibited by the sustentacular cells. Endocrine neoplasias, including adrenocortical adenoma and adrenocortical carcinoma, parathyroid adenoma, papillary thyroid carcinoma, and follicular thyroid adenoma cases, were consistently STX1-negative (Table 3).

All but one neuroectodermal/neuroepithelial tumor samples revealed strong diffuse STX1 positivity in neuroblasts and ganglion cells (Table 1), whereas less than 10% of tumor cells were positive in one medulloblastoma case. Where it was present, the neuropil component also showed a moderate expression. The Schwann cell component of the ganglioneuromas, and all peripheral nerve sheet tumor cases, proved to be STX1-negative (Table 3). Scattered STX1-positive tumor cells were found in 2/2 Ewing family tumor cases studied.

Table 3. STX1 immunoreactivity in conventional (non-NE) tumors.

Tumor Type	STX1 (Positive/Total No. of Cases; %)		No. of Cases with Scarcely Scattered STX1 Positive Cells
Colorectal adenocarcinoma	0/8	0.0%	5
Gastric adenocarcinoma	0/12	0.0%	9
Hepatobiliary and pancreatic carcinoma	0/11	0.0%	1
Lung squamous cell carcinoma	0/5	0.0%	
Lung adenocarcinoma	0/8	0.0%	
Head and neck carcinoma	0/7	0.0%	
Basal cell carcinoma of the skin	0/2	0.0%	
Cervical and ovarian carcinoma	0/8	0.0%	1
Prostatic adenocarcinoma	0/7	0.0%	
Melanocytic tumor	0/9	0.0%	
Soft tissue tumor	0/13	0.0%	
Lymphoma and myeloid neoplasm	0/31	0.0%	
Genital germ cell tumor	0/9	0.0%	
Gonadal sex-cord stromal tumor	0/5	0.0%	
Solid pseudopapillary neoplasm (pancreas)	0/1	0.0%	
Adrenocortical adenoma	0/11	0.0%	
Adrenocortical carcinoma	0/12	0.0%	
Parathyroid adenoma	0/19	0.0%	
Papillary thyroid carcinoma	0/5	0.0%	
Thyroid follicular adenoma	0/1	0.0%	
Carcinomas of the breast			
No special type	1/18	5.6%	1
Invasive lobular carcinoma	0/2	0.0%	
Mucinous carcinoma, hypocellular type	0/6	0.0%	
Mucinous carcinoma, hypercellular type	7/10	70.0%	
Metaplastic carcinoma	0/1	0.0%	

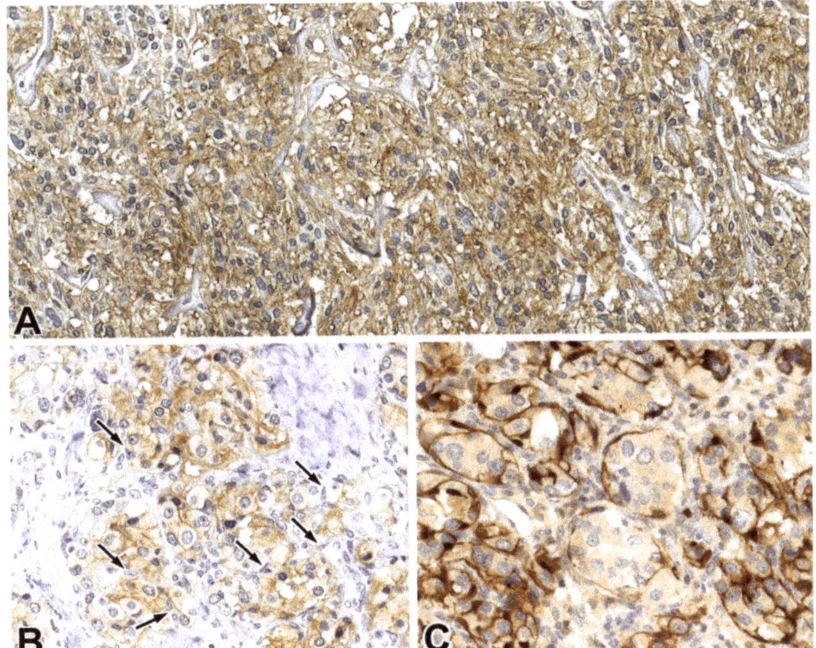

Figure 6. STX1 (**A,B**) and S100 (**C**) immunoreactivity in pheochromocytomas, for which the arrows indicate sustentacular cells (**B**) The latter cells were visualized with an S100 IHC reaction. (**A**): 20x, (**B,C**): 40x.

3. Discussion

There are several IHC NE markers used in diagnostic pathology. To use these markers responsibly, one should be aware of their advantages and limitations [12–15]. Currently, CD56, CHGA, and SYP represent the most widely used NE markers. CD56 is highly sensitive to NE cells [4]; however, it is also detectable in a heterogeneous group of non-NE neoplasms, including nephroblastoma, neuroblastoma, myeloid, lymphoid, and plasma cell tumors, as well as in hepatocellular, renal cell, ovarian, endometrial, and thyroid carcinomas [25,26]. Therefore, it is unreliable if used as a single marker. CHGA constitutes a secretory granule protein of the NE, adenohypophyseal, and parathyroid cells [27,28]. As an IHC NE marker, CHGA shows a good sensitivity; nevertheless, some NE cells, such as Merkel cells and hindgut-derived NE cells [29], may demonstrate significantly weaker staining or can lack CHGA immunoreactivity. Furthermore, the strong cytoplasmic expression seen in most NE cells may be weak or absent in NECs, due to the low density of mature secretory granules [30,31]. SYP is an integral membrane calcium-binding glycoprotein of synaptic vesicles. It is a sensitive marker usually expressed in the cytoplasmic microvesicles of neuronal tumors, NENs, and neuroectoderm-derived small blue cell tumors, as well as in many endocrine tissues, such as normal and neoplastic pituitary and parathyroid glands and adrenocortical tissue [32–34].

In this study, we performed a comprehensive IHC analysis in an extensive series of benign and malignant tumors to evaluate STX1 as an IHC NE marker. Using a well-characterized monoclonal mouse antibody, we found that STX1 represents an impressive robust NE marker, with a sensitivity of 99% in NETs and 100% in NECs, outperforming other common NE markers, such as SYP (96% and 89%), CHGA (93% and 91%), and CD56 (70% and 93%), which was proven to be statistically significant regarding gastrointestinal NETs and NECs. Normal NE cells in different organs, as well as pulmonary, gastrointestinal, and pancreatic NETs and NECs, were likewise positive. Therefore, the STX1 expression of NENs seems to be unrelated to the anatomical site. In contrast to the frequently negative CHGA staining in rectal and appendiceal L-cell NETs [9–11], STX1 was consistently positive in all such cases studied. Pheochromocytomas and paragangliomas were also almost consistently positive. STX1 was uniformly expressed in all NECs, regardless of the morphological or clinicopathological subtype, including small and large cell NECs, as well as in Merkel cell carcinomas and medullary thyroid carcinomas. In contrast to the sometimes faint or dot-like cytoplasmic expression of SYP and CHGA, the STX1 revealed crisp membranous and strong cytoplasmic staining in the majority of cases studied, which makes the evaluation straightforward. Concerning the specificity of STX1, many endocrine tumors known to express either CHGA or SYP, such as parathyroid and adrenocortical neoplasms [35], were consistently negative for STX1. As further evidence of the excellent specificity of STX1, a broad spectrum of non-NE tumors, including various types of carcinomas, were consistently negative. In MiNENs and carcinomas with focal NE components, the STX1 immunostaining was restricted to the NE areas, as was confirmed by other NE markers. A caveat is that in 7% of conventional (non-NE) neoplasia, a discrete (<10%) and scarcely scattered STX1-positive cell population was present without morphological features of NE differentiation. In our opinion, by correlating the IHC stains to the histomorphology, these infrequent positive cells should not cause any diagnostic problem, and such reactions should be interpreted as negative. Although STX1 was also expressed in neural tumors, considering the rather distinct presentation of these neoplasms, the differentiation appears to be straightforward.

INSM1 is a recently described IHC NE marker [36], demonstrating nuclear localization. It has been studied in the NENs of many anatomical regions, including the lung, head and neck, central nervous system, prostate, skin, and pancreas, but, according to the reported series, the INSM1 reveals a lower sensitivity [13,36–41] than STX1 in our series. Although a direct comparison of the two markers is warranted in future studies, based on the literature data and results of the present study, STX1 outperforms INSM1 in terms of its specificity and sensitivity.

In keeping with the frequent NE differentiation of hypercellular (type B) mucinous breast carcinomas, a small cohort of these tumors were also stained positive for STX1, while breast cancer types infrequently expressing NE markers, such as hypocellular (type A) mucinous, conventional invasive ductal, and lobular carcinomas, were negative. Due to the limited number of breast tumors included in this study, additional investigations of breast cancer are warranted.

4. Materials and Methods

4.1. Samples Studied

All samples were assigned by the Department of Pathology, University of Szeged, and the Laboratory of Tumor Pathology and Molecular Diagnostics, Szeged, Hungary, and represented formalin-fixed paraffin-embedded (FFPE) tissues.

To study STX1 in non-neoplastic tissues, normal thyroid (8 samples), parathyroid (11 samples), skin (2 samples), pancreas (5 samples), adrenal gland (9 samples), and brain tissue (5 samples), as well as appendectomy specimens from acute appendicitis (5 cases), were evaluated. Hyperplastic NE lesions, such as linear and nodular NE cell hyperplasia in autoimmune metaplastic atrophic gastritis (10 cases) morphologically consistent with ECL cell hyperplasia [42] and one case of pancreatic nesidioblastosis, were also included. In tumor samples, peritumoral non-neoplastic NE cells, including gastric antral and oxyntic mucosal, as well as intestinal NE cells, pancreatic Langerhans islets, bronchial NE cells, and cutaneous Merkel cells, were also assessed for STX1 expression, where they were present.

To evaluate the specificity and sensitivity of STX1 in neoplastic conditions, altogether, 398 cases of various non-NE and NE neoplasms were studied (Table 1) in either whole tissue sections or tissue microarrays (TMAs). Cases with potential diagnostic pitfalls, for example, frequently CHGA-negative L cell NETs (three rectal and one appendiceal), were also included. In 6/215 cases of predominantly non-neuroendocrine carcinomas, a component showing morphological features of neuroendocrine differentiation was also present. In four (one colonic and one diffuse gastric adenocarcinoma, one poorly differentiated squamous cell, and one adenocarcinoma of the lung) cases, the neuroendocrine component represented 10%–30% of the tumor, while in two (one non-mucinous breast and one poorly differentiated gastric adenocarcinoma) cases, the NE component exceeded the MiNEN defining 30% [5] (pp. 18–19). TMA blocks were constructed with the manual TMA builder instrument (Histopathology Ltd., Pécs, Hungary), as previously published [43]. Each tumor case in TMA blocks was represented with at least two cores (central and peripheral regions) of 2 mm. To compare the proportion of positive cases by the applied markers, Fisher's exact test was performed.

All NENs were diagnosed and graded according to the WHO Blue Book [1], corresponding to their anatomical location.

The study was performed in agreement with the guidelines of the Declaration of Helsinki for human medical research and was ethically approved by the Clinical Research Coordination Office of the University of Szeged (4430/2018) 7 January 2019.

4.2. Immunohistochemistry and Evaluation of Staining Patterns

The IHC reactions were uniformly performed in FFPE sections. Briefly, 2–5 μm-thick paraffin sections were routinely de-waxed, blocked for endogenic peroxidase activities in ethanol containing 1.5% (v/v) H2O2, and heat-treated in 10 mM Sodium citrate (0.05 % Tween-20, pH 10.0) antigen retrieval buffer using a household electric pressure cooker. After protein blocking in 50 mM Tris-buffered saline (TBS, pH 7.4) containing 5% (w/v) low-fat milk powder, the sections were incubated with the primary antibodies at room temperature for 70 min. Detection was performed using the Novolink polymer kit (Leica Biosystems/Novocastra, Newcastle Upon Tyne, United Kingdom), and nuclear staining was carried out with Mayer's hematoxylin. The IHC stains were executed by a four-channel TECAN Freedom Evo liquid handling platform (TECAN, Mannedorf, Switzerland).

For STX1, we utilized a well-characterized mouse monoclonal antibody HPC-1 (sc-12736; 1:200; Santa Cruz Biotechnology, Dallas, TX, USA), which detects both STX1 A and B isoforms. A tumor sample was considered to be STX1-positive if more than 50% of the neoplastic cells showed either membranous or cytoplasmic staining. The staining intensity was categorized as weak, moderate, or strong. Samples were independently assessed by three of the authors (S.T-N., B.K., and L.K.). In a case with discordant results, a consensus was reached by a second-look evaluation made jointly.

In NENs investigated using the TMA technique, immunostainings for the most common NE markers, such as SYP (27G12; 1:100; Leica Biosystems), CHGA (5H7; Leica Biosystems/Novocastra, Newcastle Upon Tyne, United Kingdom), and CD56 (MRQ-42; 1:500; Cell Marque, Rocklin, CA, USA), were also performed to compare the results with STX1 expression. In cases where STX1 immunostainings were performed on whole tissue sections, the SYP, CHGA, and CD56 expression was not tested systematically; however, the SYP, CHGA, and CD56 immunostainings performed for the original pathology report were re-evaluated, if available. Cytoplasmic staining for SYP and CHGA or membranous staining for CD56 were considered positive. To identify the various subtypes of gastrointestinal NETs, IHC reactions were performed against VMAT1 (RMT77; 1:100; Leica Biosystems/Novocastra, Newcastle Upon Tyne, United Kingdom) as a marker of EC cells and GLP1 (sc-57166; HYB 147-06; 1:100; Santa Cruz Biotechnology, Dallas, TX, USA) as a marker of L-cells. The sustentacular cell population in paragangliomas and pheochromocytomas was identified with S100.

5. Conclusions

In conclusion, we showed that STX1 outperforms common NE IHC markers in terms of its specificity and sensitivity and appears to be the most advantageous immunophenotypic marker of NE cells and NENs. STX1 demonstrated a near-perfect specificity and an outstanding sensitivity, even in NECs. We recommend that STX1 be added to the IHC panel of NE differentiation in routine diagnostic histopathology. The consistent STX1 expression in all NETs, regardless of the anatomical site or subtype, makes it a reliable marker, even in the hands of routine pathologists who are less experienced with NENs and unaware of the specific expression patterns and possible pitfalls of classic NE markers.

Author Contributions: Conceptualization: L.K.; data curation: B.K., S.T.-N., Á.B., Z.F., and L.K.; formal analysis: B.K., S.T.-N., and L.K.; funding acquisition: B.K.; investigation: B.K., S.T.-N., L.K., and Z.F.; methodology: B.K. and L.K.; project administration: L.K. and B.K.; resources: H.G. Research Fund; supervision: L.K. and B.K.; Visualization: B.K. and S.T.-N.; writing—original draft: B.K. and S.T.-N.; writing—review and editing: L.K. All authors have read and agreed to the published version of the manuscript.

Funding: This research was funded by the University of Szeged, Faculty of medicine Research Fund—Hetényi Géza grant, grant number 5S582.

Acknowledgments: The authors thank photographer Mihály Dezső for his assistance with the figures.

Conflicts of Interest: The authors declare no conflicts of interest.

Abbreviations

NE	Neuroendocrine
NEN	Neuroendocrine neoplasm
IHC	Immunohistochemistry
SYP	Synaptophysin
CHGA	Chromogranin-A
STX1	Syntaxin 1
NET	Neuroendocrine tumor
NEC	Neuroendocrine carcinoma
IARC	International Agency for Research on Cancer
WHO	World Health Organization
MiNEN	Mixed neuroendocrine-non neuroendocrine neoplasm
NSE	Neuron-specific enolase
PGP9.5	Protein gene product 9.5
INSM1	Insulinoma-associated protein 1
HPC-1	Syntaxin 1

ECL	Enterochromaffin-like
EC	Enterochromaffin
VMAT1	Vesicular monoamine transporter 1
GLP1	Glucagon-like peptide-1
FFPE	Formalin-fixed paraffin-embedded
TMA	Tissue microarray
TBS	Tris-buffered saline

References

1. WHO Classification of Tumours. Available online: http://whobluebooks.iarc.fr (accessed on 31 January 2020).
2. Rindi, G.; Klimstra, D.S.; Abedi-Ardekani, B.; Asa, S.L.; Bosman, F.T.; Brambilla, E.; Busam, K.J.; de Krijger, R.R.; Dietel, M.; El-Naggar, A.K.; et al. A common classification framework for neuroendocrine neoplasms: An International Agency for Research on Cancer (IARC) and World Health Organization (WHO) expert consensus proposal. *Mod. Pathol.* **2018**, *31*, 1770–1786. [CrossRef] [PubMed]
3. Lloyd, R.V.; Osamura, R.Y.; Klöppel, G.; Rosai, J. *WHO Classification of Tumours of Endocrine Organs*, 4th ed.; International Agency for Research on Cancer: Lyon, France, 2017; pp. 210–214.
4. WHO Classification of Tumours Editorial Board. *WHO Classification of Tumours, Digestive System Tumours*, 5th ed.; International Agency for Research on Cancer: Lyon, France, 2019; pp. 16–20.
5. Kyriakopoulos, G.; Mavroeidi, V.; Chatzellis, E.; Kaltsas, G.A.; Alexandraki, K.I. Histopathological, immunohistochemical, genetic and molecular markers of neuroendocrine neoplasms. *Ann. Transl. Med.* **2018**, *6*, 252. [CrossRef] [PubMed]
6. Lloyd, R.V. Practical markers used in the diagnosis of neuroendocrine tumors. *Endocr. Pathol.* **2003**, *14*, 293–301. [CrossRef] [PubMed]
7. Uccella, S.; La Rosa, S.; Volante, M.; Papotti, M. Immunohistochemical Biomarkers of Gastrointestinal, Pancreatic, Pulmonary, and Thymic Neuroendocrine Neoplasms. *Endocr. Pathol.* **2018**, *29*, 150–168. [CrossRef] [PubMed]
8. Kosuke, F.; Kazuhiro, Y.; Shinji, K.; Yamato, M.; Yonosuke, S.; Joeji, W.; Ichiro, K.; Makoto, S.; Takaaki, I. INSM1 is the best marker for the diagnosis of neuroendocrine tumors: Comparison with chromogranin-A, SYP and CD56. *Int. J. Clin. Exp. Pathol.* **2017**, *10*, 5393–5405.
9. Weiler, R.; Feichtinger, H.; Schmid, K.W.; Fischer-Colbrie, R.; Grimelius, L.; Cedermark, B.; Papotti, M.; Bussolati, G.; Winkler, H. Chromogranin A and B and secretogranin II in bronchial and intestinal carcinoids. *Virchows Arch. A Pathol. Anat. Histopathol.* **1987**, *412*, 103–109. [CrossRef]
10. Fahrenkamp, A.G.; Wibbeke, C.; Winde, G.; Ofner, D.; Bocker, W.; Fischer-Colbrie, R.; Schmid, K.W. Immunohistochemical distribution of chromogranins A and B and secretogranin II in neuroendocrine tumours of the gastrointestinal tract. *Virchows Arch.* **1995**, *426*, 361–367. [CrossRef]
11. Al-Khafaji, B.; Noffsinger, A.E.; Miller, M.A.; DeVoe, G.; Stemmermann, G.N.; Fenoglio-Preiser, C. Immunohistologic analysis of gastrointestinal and pulmonary carcinoid tumors. *Hum. Pathol.* **1998**, *29*, 992–999. [CrossRef]
12. Washington, M.K.; Tang, L.H.; Berlin, J.; Branton, P.A.; Burgart, L.J.; Carter, D.K.; Compton, C.C.; Fitzgibbons, P.L.; Frankel, W.L.; Jessup, J.M.; et al. Protocol for the Examination of Specimens From Patients With Neuroendocrine Tumors (Carcinoid Tumors) of the Small Intestine and Ampulla. *Arch. Pathol. Lab. Med.* **2010**, *134*, 181–186. [CrossRef]
13. Rooper, L.M.; Sharma, R.; Li, Q.K.; Illei, P.B.; Westra, W.H. INSM1 Demonstrates Superior Performance to the Individual and Combined Use of Synaptophysin, Chromogranin and CD56 for Diagnosing Neuroendocrine Tumors of the Thoracic Cavity. *Am. J. Surg. Pathol.* **2017**, *41*, 1561–1569. [CrossRef]
14. Jirásek, T.; Hozák, P.; Mandys, V. Different patterns of chromogranin A and Leu-7 (CD57) expression in gastrointestinal carcinoids: Immunohistochemical and confocal laser scanning microscopy study. *Neoplasma* **2003**, *50*, 1–7. [PubMed]
15. Tartaglia, A.; Portela-Gomes, G.M.; Oberg, K.; Vezzadini, P.; Foschini, M.P.; Stridsberg, M. Chromogranin A in gastric neuroendocrine tumours: An immunohistochemical and biochemical study with region-specific antibodies. *Virchows Arch.* **2006**, *448*, 399–406. [CrossRef] [PubMed]

16. Rizo, J. Mechanism of neurotransmitter release coming into focus. *Protein Sci.* **2018**, *27*, 1364–1391. [CrossRef] [PubMed]

17. Xie, Z.; Long, J.; Liu, J.; Chai, Z.; Kang, X.; Wang, C. Molecular Mechanisms for the Coupling of Endocytosis to Exocytosis in Neurons. *Front. Mol. Neurosci.* **2017**, *10*, 47. [CrossRef]

18. Han, J.; Pluhackova, K.; Böckmann, R.A. The ceted Role of SNARE Proteins in Membrane Fusion. *Front. Physiol.* **2017**, *8*, 5. [CrossRef]

19. Inoue, A.; Obata, K.; Akagawa, K. Cloning and sequence analysis of cDNA for a neuronal cell membrane antigen, HPC-1. *J. Biol. Chem.* **1992**, *267*, 10613–10619.

20. Kushima, Y.; Fujiwara, T.; Sanada, M.; Akagawa, K. Characterization of HPC-1 antigen, an isoform of syntaxin-1, with the isoform-specific monoclonal antibody, 14D8. *J. Mol. Neurosci.* **1997**, *8*, 19–27. [CrossRef]

21. Yoo, S.H.; You, S.H.; Huh, Y.H. Presence of syntaxin 1A in secretory granules of chromaffin cells and interaction with chromogranins A and B. *FEBS Lett.* **2005**, *579*, 222–228. [CrossRef]

22. Jacobsson, G.; Beant, A.J.; Schellert, R.H.; Juntti-Berggrent, L.; Deeneyt, J.T.; Berggrent, P.-O.; Meister, B. Identification of synaptic proteins and their isoform mRNAs in compartments of pancreatic endocrine cells (exocytosis/secretion/insulin/diabetes). *Proc. Natl. Acad. Sci. USA* **1994**, *91*, 12487–12491. [CrossRef]

23. Xia, F.; Leung, Y.M.; Gaisano, G.; Gao, X.; Chen, Y.; Fox, J.E.; Bhattacharjee, A.; Wheeler, M.B.; Gaisano, H.Y.; Tsushima, R.G. Targeting of voltage-gated K+ and Ca2+ channels and soluble N-ethylmaleimide-sensitive factor attachment protein receptor proteins to cholesterol-rich lipid rafts in pancreatic alpha-cells: Effects on glucagon stimulus-secretion coupling. *Endocrinology* **2007**, *148*, 2157–2167. [CrossRef]

24. Bjorling, E.; Lindskog, C.; Oksvold, P.; Linne, J.; Kampf, C.; Hober, S.; Uhlen, M.; Ponten, F. A web-based tool for in silico biomarker discovery based on tissue-specific protein profiles in normal and cancer tissues. *Mol. Cell. Proteom.* **2008**, *7*, 825–844. [CrossRef] [PubMed]

25. Kurokawa, M.; Nabeshima, K.; Akiyama, Y.; Maeda, S.; Nishida, T.; Nakayama, F.; Amano, M.; Ogata, K.; Setoyama, M. CD56: A useful marker for diagnosing Merkel cell carcinoma. *J. Derm. Sci.* **2003**, *31*, 219–224. [CrossRef]

26. Bahrami, A.; Gown, A.M.; Baird, G.S.; Hicks, M.J.; Folpe, A.L. Aberrant expression of epithelial and neuroendocrine markers in alveolar rhabdomyosarcoma: A potentially serious diagnostic pitfall. *Mod. Pathol.* **2008**, *21*, 795–806. [CrossRef] [PubMed]

27. Nobels, F.R.; Kwekkeboom, D.J.; Bouillon, R.; Lamberts, S.W. Chromogranin A: Its clinical value as marker of neuroendocrine tumours. *Eur. J. Clin. Investig.* **1998**, *28*, 431–440. [CrossRef]

28. Deftos, L.J.; Woloszczuk, W.; Krisch, I.; Horvat, G.; Ulrich, W.; Neuhold, N.; Braun, O.; Reiner, A.; Srikanta, S.; Krisch, K. Medullary Thyroid Carcinomas Express Chromogranin A and a Novel Neuroendocrine Protein Recognized by Monoclonal Antibody HISL-19. *Am. J. Med.* **1988**, *85*, 780–784. [CrossRef]

29. Portela, G.M.; Stridsberg, M. Chromogranin A in the Human Gastrointestinal Tract: An Immunocytochemical Study with Region-specific Antibodies. *J. Histochem. Cytochem.* **2002**, *50*, 1487–1492. [CrossRef]

30. Williams, G.T. Endocrine tumours of the gastrointestinal tract–selected topics. *Histopathology* **2007**, *50*, 30–41. [CrossRef]

31. Washington, M.K.; Tang, L.H.; Berlin, J.; Branton, P.A.; Burgart, L.J.; Carter, D.K.; Compton, C.C.; Fitzgibbons, P.L.; Frankel, W.L.; Jessup, J.M.; et al. Protocol for the examination of specimens from patients with neuroendocrine tumors (carcinoid tumors) of the colon and rectum. *Arch. Pathol. Lab. Med.* **2010**, *134*, 176–180. [CrossRef]

32. Gould, V.E.; Lee, I.; Wiedenmann, B.; Moll, R.; Chejfec, G.; Franke, W.W. Synaptophysin: A novel marker for neurons, certain neuroendocrine cells, and their neoplasms. *Hum. Pathol.* **1986**, *17*, 979–983. [CrossRef]

33. Wiedenmann, B.; Franke, W.W.; Kuhn, C.; Moll, R.; Gould, V.E. Synaptophysin: A marker protein for neuroendocrine cells and neoplasms. *Proc. Natl. Acad. Sci. USA* **1986**, *83*, 3500–3504. [CrossRef]

34. Wiedenmann, B.; Franke, W.W. Identification and localization of synaptophysin, an integral membrane glycoprotein of Mr 38,000 characteristic of presynaptic vesicles. *Cell* **1985**, *41*, 1017–1028. [CrossRef]

35. Erickson, L.A.; Mete, O. Immunohistochemistry in Diagnostic Parathyroid Pathology. *Endocr. Pathol.* **2018**, *29*, 113–129. [CrossRef]

36. Rosenbaum, J.N.; Guo, Z.; Baus, R.M.; Werner, H.; Rehrauer, W.M.; Lloyd, R.V. INSM1: A Novel Immunohistochemical and Molecular Marker for Neuroendocrine and Neuroepithelial Neoplasms. *Am. J. Clin. Pathol.* **2015**, *144*, 579–591. [CrossRef]

37. Gonzalez, I.; Lu, H.C.; Sninsky, J.; Yang, C.; Bishnupuri, K.; Dieckgraefe, B.; Cao, D.; Chatterjee, D. Insulinoma-associated protein 1 expression in primary and metastatic neuroendocrine neoplasms of the gastrointestinal and pancreaticobiliary tracts. *Histopathology* **2019**, *75*, 568–577. [CrossRef] [PubMed]
38. Kriegsmann, K.; Zgorzelski, C.; Kazdal, D.; Cremer, M.; Muley, T.; Winter, H.; Longuespee, R.; Kriegsmann, J.; Warth, A.; Kriegsmann, M. Insulinoma-associated Protein 1 (INSM1) in Thoracic Tumors is Less Sensitive but More Specific Compared With Synaptophysin, Chromogranin A, and CD56. *Appl. Immunohistochem. Mol. Morphol.* **2018**. [CrossRef]
39. Xin, Z.; Zhang, Y.; Jiang, Z.; Zhao, L.; Fan, L.; Wang, Y.; Xie, S.; Shangguan, X.; Zhu, Y.; Pan, J.; et al. Insulinoma-associated protein 1 is a novel sensitive and specific marker for small cell carcinoma of the prostate. *Hum. Pathol.* **2018**, *79*, 151–159. [CrossRef] [PubMed]
40. Rooper, L.M.; Bishop, J.A.; Westra, W.H. INSM1 is a Sensitive and Specific Marker of Neuroendocrine Differentiation in Head and Neck Tumors. *Am. J. Surg. Pathol.* **2018**, *42*, 665–671. [CrossRef]
41. Lilo, M.T.; Chen, Y.; LeBlanc, R.E. INSM1 Is More Sensitive and Interpretable than Conventional Immunohistochemical Stains Used to Diagnose Merkel Cell Carcinoma. *Am. J. Surg. Pathol.* **2018**, *42*, 1541–1548. [CrossRef]
42. Bordi, C.; Pilato, F.; Carfagna, G.; Ferrari, C.; DAdda, T.; Sivelli, R.; Bertelé, A.; Missale, G. Argyrophil Cell Hyperplasia of Fundic Mucosa in Patients with Chronic Atrophic Gastritis. *Digestion* **1986**, *35*, 130–143. [CrossRef]
43. Toth-Liptak, J.; Piukovics, K.; Borbenyi, Z.; Demeter, J.; Bagdi, E.; Krenacs, L. A comprehensive immunophenotypic marker analysis of hairy cell leukemia in paraffin-embedded bone marrow trephine biopsies-a tissue microarray study. *Pathol. Oncol. Res.* **2015**, *21*, 203–211. [CrossRef]

Article

Dysregulation of Placental Functions and Immune Pathways in Complete Hydatidiform Moles

Jennifer R. King [1,†], Melissa L. Wilson [2,†], Szabolcs Hetey [3], Peter Kiraly [3], Koji Matsuo [1], Antonio V. Castaneda [4], Eszter Toth [3], Tibor Krenacs [5], Petronella Hupuczi [6], Paulette Mhawech-Fauceglia [4], Andrea Balogh [3], Andras Szilagyi [3], Janos Matko [7], Zoltan Papp [6,8], Lynda D. Roman [1], Victoria K. Cortessis [1,2,*] and Nandor Gabor Than [1,3,5,6,*]

[1] Department of Obstetrics and Gynecology, Keck School of Medicine, University of Southern California, Los Angeles, CA 90033, USA; jennyrenaeking@gmail.com (J.R.K.); koji.matsuo@med.usc.edu (K.M.); lroman@usc.edu (L.D.R.)

[2] Department of Preventive Medicine, University of Southern California, Los Angeles, CA 90033, USA; melisslw@usc.edu

[3] Systems Biology of Reproduction Research Group, Institute of Enzymology, Research Centre for Natural Sciences, H-1117 Budapest, Hungary; heteysz@gmail.com (S.H.); peter0kiraly@gmail.com (P.K.); toth.eszter@ttk.hu (E.T.); balogh.andrea@ttk.hu (A.B.); szilagyi.andras@ttk.hu (A.S.)

[4] Department of Pathology, Keck School of Medicine, University of Southern California, Los Angeles, CA 90033, USA; antonio.castaneda@osumc.edu (A.V.C.); pfauceglia@hotmail.com (P.M.-F.)

[5] First Department of Pathology and Experimental Cancer Research, Semmelweis University, H-1085 Budapest, Hungary; krenacst@gmail.com

[6] Maternity Private Clinic of Obstetrics and Gynecology, H-1126 Budapest, Hungary; hupuczi.petronella@maternity.hu (P.H.); pzorvosihetilap@maternity.hu (Z.P.)

[7] Department of Immunology, Institute of Biology, Eotvos Lorand University, H-1117 Budapest, Hungary; matko@elte.hu

[8] Department of Obstetrics and Gynecology, Semmelweis University, H-1088 Budapest, Hungary

* Correspondence: victoria.cortessis@med.usc.edu (V.K.C.); than.gabor@ttk.hu (N.G.T.); Tel.: +1-(323)-865-0544 (V.K.C.); +36-(1)-382-6788 (N.G.T.)

† These authors contributed equally to this work.

Received: 17 September 2019; Accepted: 30 September 2019; Published: 10 October 2019

Abstract: Gene expression studies of molar pregnancy have been limited to a small number of candidate loci. We analyzed high-dimensional RNA and protein data to characterize molecular features of complete hydatidiform moles (CHMs) and corresponding pathologic pathways. CHMs and first trimester placentas were collected, histopathologically examined, then flash-frozen or paraffin-embedded. Frozen CHMs and control placentas were subjected to RNA-Seq, with resulting data and published placental RNA-Seq data subjected to bioinformatics analyses. Paraffin-embedded tissues from CHMs and control placentas were used for tissue microarray (TMA) construction, immunohistochemistry, and immunoscoring for galectin-14. Of the 14,022 protein-coding genes expressed in all samples, 3,729 were differentially expressed (DE) in CHMs, of which 72% were up-regulated. DE genes were enriched in placenta-specific genes (OR = 1.88, p = 0.0001), of which 79% were down-regulated, imprinted genes (OR = 2.38, p = 1.54 × 10^{-6}), and immune genes (OR = 1.82, p = 7.34 × 10^{-18}), of which 73% were up-regulated. DNA methylation-related enzymes and histone demethylases were dysregulated. "Cytokine–cytokine receptor interaction" was the most impacted of 38 dysregulated pathways, among which 17 were immune-related pathways. TMA-based immunoscoring validated the lower expression of galectin-14 in CHM. In conclusion, placental functions were down-regulated, imprinted gene expression was altered, and immune pathways were activated, indicating complex dysregulation of placental developmental and immune processes in CHMs.

Keywords: choriocarcinoma; hydatidiform mole; galectin; gestational trophoblastic disease; placental-specific gene; systems biology; trophoblast differentiation

1. Introduction

Gestational trophoblastic disease (GTD) is characterized by abnormal trophoblastic proliferation and includes hydatidiform mole (complete and partial) and gestational trophoblastic neoplasia (invasive mole, choriocarcinoma, placental site trophoblastic tumor, and epithelioid trophoblastic tumor) [1,2]. Diagnosis relies on clinical presentation and histologic assessment, and treatment is aimed at uterine evacuation with chemotherapy typically reserved for gestational trophoblastic neoplasia (GTN) [2,3]. Post-molar human chorionic gonadotropin (hCG) monitoring is an essential part of management for evaluating the development of GTN, which follows complete hydatidiform moles (CHMs) in ~15–20% of cases [4]. Although GTN is considered among the most curable of all solid tumors with cure rates over 90%, unrecognized and misdiagnosed GTD can result in unnecessarily increased maternal morbidity and mortality [3].

The reported incidence of GTD varies by geographic location, race or ethnicity, maternal age, and histopathologic subtype. Hydatidiform mole reportedly complicates up to two per 1000 pregnancies in Southeast Asia and Japan, nearly twice that proportion reported in North America, Australia, New Zealand, and Europe [5]. This geographic variation has been partially attributed to racial and ethnic differences, as the prevalence of GTD is elevated in American Indians, Asians, and Hispanics across the world [6,7]. Asian and American Indian women have also been shown to have more aggressive disease, with increased risk of developing GTN compared to other ethnic groups [8,9]. Extremes of maternal age are also correlated with higher rates of CHMs, with an increased incidence among women over 40 years of age and under 20 years of age [10]. Beyond maternal age and ethnicity, prior GTD is the strongest risk factor for GTD, with an incidence of 1.3% [11].

These described risk factors (race/ethnicity and prior GTD) are in accord with an underlying genetic etiology of GTD [1,2,12]. The pathophysiology of hydatidiform moles involves overrepresentation of the paternal genome. The biparental diandric triploid karyotype of partial moles accords with dispermic fertilization, while androgenetic diploid karyotype of most CHMs is consistent with monospermic or dispermic fertilization. Recently, some women with recurrent androgenetic CHMs were shown to have bi-allelic deleterious mutations in *MEI1* (meiotic double-stranded break formation protein 1), *TOP6BL* (type 2 DNA topoisomerase 6 subunit B-like), and *REC114* (meiotic recombination protein REC114), leading to meiotic double-strand break formation and extrusion of all maternal chromosomes [13]. Absence of maternal imprinting of gene expression in hydatidiform moles has also been observed in the rare biparental hydatidiform moles due to *NLRP7* (NLR family pyrin domain containing 7) or *KHDC3L* (KH domain containing 3 like) mutations, suggesting a common endpoint of pathogenesis [12,14,15]. However, for the more common sporadic CHMs, little is known regarding mechanisms responsible for either pathogenesis or progression to GTN.

The few targeted gene expression studies on molar tissue and a recent meta-analysis of these studies showed that the main genes differentially expressed (DE) in molar tissues may be those involved in villous trophoblast differentiation [16]. However, these findings were based on a limited set of molecules, and these studies mostly targeted placenta- or trophoblast-specific transcripts that were known to be differentially expressed during trophoblast differentiation. A more comprehensive approach to identifying genes and pathways involved in the development of molar disease would be a genome-wide gene expression analysis using either microarrays or RNA-Seq, followed by protein-level validation of DE transcripts.

We sought to implement such a high-dimensional and systems biology approach, similar to that used in our recent study on the pathophysiological processes in preeclampsia [17], to gain more in-depth insight into CHM pathogenesis at RNA and protein levels. This high-dimensional, agnostic study is the first to evaluate gene expression levels in CHMs using RNA-Seq followed by protein level validation of selected DE transcripts by immunostaining of tissue microarrays (TMA) and immunoscoring. The aim of our study is to identify genes with expression levels that differ in molar tissue from CHMs in comparison to placental chorionic tissue from uncomplicated pregnancies at similar stages of gestation. More complete understanding of the molecular pathways perturbed in CHMs may inform future efforts to improve procedures for early diagnosis and prognostication.

2. Results

2.1. The Transcriptome of First Trimester Placentas and CHMs

To evaluate absolute gene expression levels, mean expression values were calculated for both groups from RNA-Seq count data by normalizing for housekeeping genes. The highest expression in first trimester placentas was mostly detected for genes with placenta-specific or predominant placental expression [17–19]. Indeed, the 20 most highly-expressed genes (Table 1) included genes previously shown to have predominant placental ($n = 2$) or placenta-specific ($n = 12$) expression and unique placental functions in humans. These encode hormones (*CGA*, chorionic gonadotrophin subunit alpha, *CSH1* and *CSH2*, chorionic somatomammotropin hormone 1 and 2), an estrogen synthesizing enzyme (*CYP19A1*, cytochrome P450 family 19 subfamily A member 1), proteases (*ADAM12*, ADAM metallopeptidase domain 12; *KISS1*, kisspeptin-1; *PAPPA* and *PAPPA2*, pregnancy-associated plasma protein A and A2; *TFPI2*, tissue factor pathway inhibitor 2), and immune proteins (*PSG3* and *PSG4*, placenta-specific glycoprotein 3 and 4).

Table 1. Genes encoding the 20 transcripts most highly expressed in first trimester placentas.

Gene Symbol	Entrez ID	Base Mean	lfcSE	Log$_2$ FC	pFDR	*p*-Value	Control Mean
CGA	1081	164,079.45	0.63	0.07	0.93	0.91	155,778.30
FN1	2335	141,988.22	0.56	0.32	0.64	0.57	135,448.78
TFPI2	7980	125,046.51	0.58	−1.16	0.08	0.05	127,071.15
CSH1	1442	87,798.80	0.67	−4.58	0.00	0.00	93,037.03
EEF1A1	1915	84,023.14	0.15	−1.20	0.00	0.00	87,193.18
PEG10	23089	77,282.40	0.24	−1.74	0.00	0.00	82,299.13
CSH2	1443	61,726.09	0.81	−4.23	0.00	0.00	65,433.13
COL3A1	1281	60,159.32	0.29	−1.30	0.00	0.00	63,035.40
KISS1	3814	65,121.70	0.65	0.13	0.88	0.84	61,457.48
ADAM12	8038	44,435.52	0.35	−0.06	0.90	0.86	43,427.10
CYP19A1	1588	40,376.35	0.53	−1.27	0.03	0.02	41,478.58
SPP1	6696	37,734.38	0.41	−4.02	0.00	0.00	40,796.23
TGM2	7052	39,725.43	0.36	−0.36	0.39	0.31	39,553.03
PSG3	5671	38,487.88	0.51	−1.05	0.07	0.04	39,153.23
PAPPA	5069	35,537.55	0.59	−1.35	0.04	0.02	36,733.78
PSG4	5672	33,870.64	0.59	−3.17	0.00	0.00	36,069.85
COL4A1	1282	36,995.51	0.37	0.33	0.46	0.37	35,055.03
PAPPA2	60676	34,031.77	0.48	0.31	0.60	0.52	32,044.98
FBN2	2201	30,126.49	0.26	−1.57	0.00	0.00	31,930.00
ACTB	60	31,472.69	0.22	0.39	0.11	0.07	29,870.08

Placenta-specific genes are shown in bold blue. False discovery rate, pFDR; fold change, FC; log fold change standard error, lfcSE.

In CHMs, the overall median gene expression levels were ~13% higher than in normal placentas (control placentas: 337.7 vs. CHMs: 382.8). However, placenta-specific transcript levels were 23% lower in CHMs than in normal placentas (placentas: 3,729.6 vs. CHMs: 3,044.3), reflected in a lower number of placenta-specific genes ($n = 8$) among the 20 most highly-expressed transcripts (Table 2). In turn, the 20 most abundant transcripts in CHMs encode proteins with immune, hormone, and oxygen transport functions (*FSTL3*, follistatin-like 3; *HBA2*, hemoglobin-alpha 2; *HBB*, hemoglobin-beta; *IGF2*, insulin-like growth factor 2; *LEP*, leptin; *PAEP*, progestogen-associated endometrial protein).

Table 2. Genes encoding the 20 transcripts most highly expressed in complete hydatidiform moles.

Gene Symbol	Entrez ID	Base Mean	lfcSE	Log$_2$ FC	pFDR	*p*-Value	Control Mean	CHM Mean
HBB *	3043	33,412.13	0.49	7.74	0.00	0.00	1,608.00	373,335.00
CGA	1081	164,079.45	0.63	0.07	0.93	0.91	155,778.30	189,408.50
FN1	2335	141,988.22	0.56	0.32	0.64	0.57	135,448.78	184,274.75
PAEP *	5047	7,444.75	0.96	7.11	0.00	0.00	545.88	95,119.25
KISS1	3814	65,121.70	0.65	0.13	0.88	0.84	61,457.48	78,481.00
GDF15 *	9518	24,551.03	0.58	1.70	0.01	0.00	19,462.58	73,765.25
TFPI2	7980	125,046.51	0.58	−1.16	0.08	0.05	127,071.15	66,657.75
IGF2 *	3481	26,174.25	0.22	1.20	0.00	0.00	23,075.75	60,786.25
HBA2 *	3040	5,986.27	0.56	4.95	0.00	0.00	1,590.28	52,014.25
COL4A1	1282	36,995.51	0.37	0.33	0.46	0.37	35,055.03	49,353.25
ADAM12	8038	44,435.52	0.35	−0.06	0.90	0.86	43,427.10	47,781.00
LEP *	3952	6,372.49	0.65	3.52	0.00	0.00	3,092.85	45,604.50
ACTB	60	31,472.69	0.22	0.39	0.11	0.07	29,870.08	45,004.50
PAPPA2	60676	34,031.77	0.48	0.31	0.60	0.52	32,044.98	44,734.75
EEF1A1 *	1915	84,023.14	0.15	−1.20	0.00	0.00	87,193.18	43,702.00
AHNAK	79026	25,150.93	0.24	0.66	0.01	0.01	23,452.63	43,214.25
CGB5	93659	18,949.99	0.70	1.13	0.16	0.11	16,161.30	40,758.75
FLT1	2321	19,476.41	0.32	0.96	0.01	0.00	17,411.10	38,701.00
FSTL3 *	10272	7,919.03	0.63	2.57	0.00	0.00	5,172.10	36,659.00
CGB3 *	1082	13,840.61	0.66	1.49	0.04	0.02	11,254.20	35,670.25

Placenta-specific genes are shown in bold blue. Differentially expressed genes are shown with asterisks. Complete hydatidiform mole, CHM; false discovery rate, pFDR; fold change, FC; log fold change standard error, lfcSE.

2.2. Differential Gene Expression in CHMs

Among transcripts of 14,022 protein-coding genes analyzed with RNA-Seq, a total of 3,729 (27%) were found to be DE in CHMs in comparison to normal first trimester placental tissues. Of these, 2,667 (72%) were up-regulated while 1,062 (28%) were down-regulated in CHM tissues (Supplementary Table S1, Figure 1). Of note, there were seven genes with placenta-specific expression among the top 20 transcripts most down-regulated but not up-regulated in CHMs (Table 3).

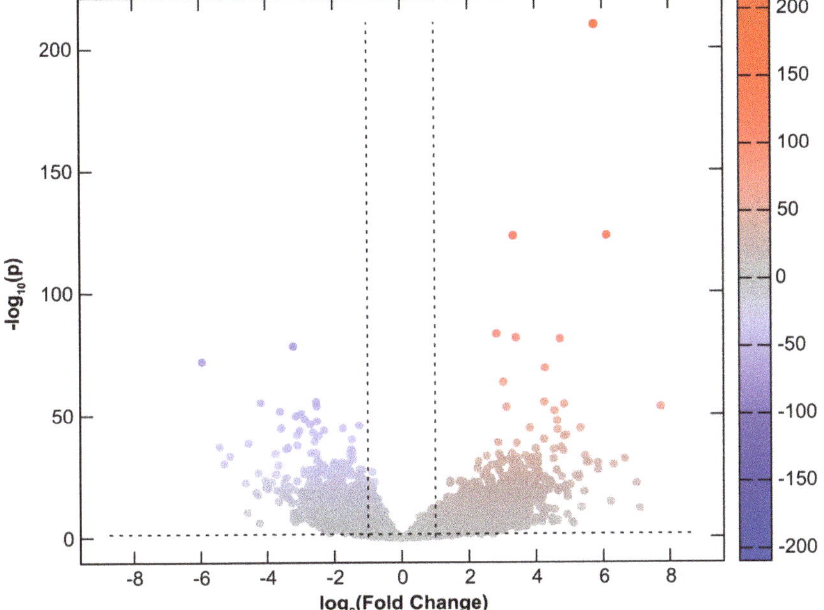

Figure 1. Differential gene expression in complete hydatidiform moles. All 14,022 expressed protein-coding genes are represented in terms of their measured differences in transcript abundance (x-axis) and the significance of the difference (y-axis) on a volcano plot. The significance is represented as negative log (base 10) of the adjusted *p*-value so that more significant differences in expression are plotted higher on the y-axis. Dotted lines represent the thresholds used to select the differentially expressed (DE) genes: <−1 and >1 for the magnitude of differential expression and pFDR <0.05 for statistical significance. According to these criteria, of the 3,729 DE transcripts, 2,667 were up-regulated (depicted with red), while 1062 were down-regulated (depicted with blue) in molar tissues.

Enrichment analysis revealed significant enrichment (OR = 1.88, *p* = 0.0001) of placenta-specific genes (Supplementary Table S2, Figure 2A) among DE genes. Of interest, 50 out of 63 (79%) placenta-specific DE genes, found to be expressed mainly by the trophoblast, were down-regulated. Among functions of products of these genes were growth hormones (*CSHL1*, chorionic somatomammotropin hormone-like 1; *CSH1*, *CSH2*), immune response (*LGALS14*, lectin, galactoside-binding, soluble, 14; *PSG4*), cell attachment (*EGFL6*, EGF-like domain multiple 6; *SMAGP*, small cell adhesion glycoprotein; *SVEP1*, Sushi, Von Willebrand factor type A, EGF and pentraxin domain containing protein 1), coagulation and blood pressure regulation (*AGTR1*, angiotensin II receptor type 1; *F13A1*, coagulation factor XIII A chain), and cell differentiation and developmental processes (*PAGE4*, PAGE family member 4; *PLAC1*, placenta specific 1; *RASA1*, RAS P21 protein activator 1).

Table 3. Genes encoding the 20-20 transcripts with highest and lowest expression in complete hydatidiform moles.

Gene Symbol	Entrez ID	Base Mean	Log$_2$ FC	lfcSE	*p*-Value	pFDR
HBB	3043	33,412.13	7.74	0.49	4.64×10^{-57}	4.06×10^{-54}
PAEP	5047	7,444.75	7.11	0.96	1.70×10^{-13}	2.45×10^{-12}
CP	1356	307.65	7.01	0.69	1.38×10^{-24}	8.39×10^{-23}
C2CD4B	388125	59.15	6.66	0.54	6.42×10^{-35}	1.25×10^{-32}
IRX3	79191	99.72	6.32	0.53	1.90×10^{-32}	2.77×10^{-30}
BEST1	7439	801.46	6.23	0.76	2.40×10^{-16}	5.15×10^{-15}
DSEL	92126	88.53	6.12	0.25	4.36×10^{-128}	3.05×10^{-124}
B4GALNT3	283358	83.11	5.85	0.51	2.80×10^{-30}	3.32×10^{-28}
PGR	5241	62.02	5.85	0.49	3.04×10^{-33}	4.90×10^{-31}
ZNF623	9831	59.42	5.77	0.18	1.30×10^{-214}	1.82×10^{-210}
KIAA1324	57535	79.48	5.58	0.46	1.77×10^{-33}	3.03×10^{-31}
IGFBP7	3490	658.18	5.52	0.46	3.03×10^{-33}	4.90×10^{-31}
RORC	6097	36.06	5.47	0.43	3.04×10^{-36}	6.86×10^{-34}
DEFB1	1672	47.01	5.44	0.58	5.59×10^{-21}	2.16×10^{-19}
PKHD1	5314	74.85	5.36	0.71	5.31×10^{-14}	8.23×10^{-13}
ADCY1	107	53.12	5.33	0.37	6.31×10^{-48}	2.95×10^{-45}
MAPT	4137	23.64	5.30	0.48	4.39×10^{-28}	4.15×10^{-26}
PRUNE2	158471	140.93	5.29	0.49	9.85×10^{-27}	7.89×10^{-25}
GALNT15	117248	23.91	5.18	0.47	8.01×10^{-28}	7.28×10^{-26}
WDR38	401551	74.42	5.12	1.02	5.62×10^{-07}	2.74×10^{-06}
HAPLN1	1404	3,248.02	−3.66	0.31	1.18×10^{-32}	1.80×10^{-30}
CD36	948	2,906.67	−3.71	0.39	6.13×10^{-22}	2.71×10^{-20}
WNT2	7472	1,855.95	−3.72	0.33	8.29×10^{-29}	8.30×10^{-27}
EGFL6	25975	4,141.42	−3.75	0.29	1.50×10^{-37}	3.69×10^{-35}
BMP5	653	1,150.58	−3.86	0.37	3.40×10^{-25}	2.29×10^{-23}
F13A1	2162	7,493.03	−3.98	0.41	1.75×10^{-22}	8.18×10^{-21}
SPP1	6696	37,734.38	−4.02	0.41	6.04×10^{-23}	2.97×10^{-21}
HBG2	3048	4,458.52	−4.05	0.38	4.57×10^{-26}	3.35×10^{-24}
MFSD14A	64645	776.11	−4.19	0.26	5.42×10^{-59}	5.96×10^{-56}
CSH2	1443	61,726.09	−4.23	0.81	1.94×10^{-07}	1.03×10^{-06}
AGTR1	185	664.42	−4.27	0.38	2.85×10^{-29}	3.03×10^{-27}
COX4I2	84701	131.49	−4.28	0.47	1.36×10^{-19}	4.47×10^{-18}
SVEP1	79987	15,505.08	−4.33	0.46	5.20×10^{-21}	2.01×10^{-19}
LYVE1	10894	2,663.78	−4.56	0.34	6.68×10^{-42}	2.23×10^{-39}
CSH1	1442	87,798.80	−4.58	0.67	5.61×10^{-12}	6.40×10^{-11}
AREG	374	710.42	−4.64	0.45	4.88×10^{-25}	3.19×10^{-23}
SPESP1	246777	321.55	−5.11	0.40	1.75×10^{-36}	4.01×10^{-34}
RPS10	6204	742.05	−5.28	0.44	3.32×10^{-33}	5.29×10^{-31}
CSHL1	1444	2,846.50	−5.43	0.41	2.25×10^{-40}	6.85×10^{-38}
ZC3H11A	9877	2,887.90	−5.95	0.32	4.7×10^{-76}	8.29×10^{-73}

Placenta-specific genes are depicted with bold blue. False discovery rate, pFDR; fold change, FC; log fold change standard error, lfcSE.

Since majority of these placenta-specific genes are most highly expressed in the trophoblast, we wished to learn whether their differential expression may reflect alterations in developmental or functional processes of the trophoblast. Therefore, we examined whether genes involved in villous trophoblast differentiation [19] (Figure 2A) are enriched among molar DE genes. However, we found only minimal enrichment (OR = 1.15, p = 0.006).

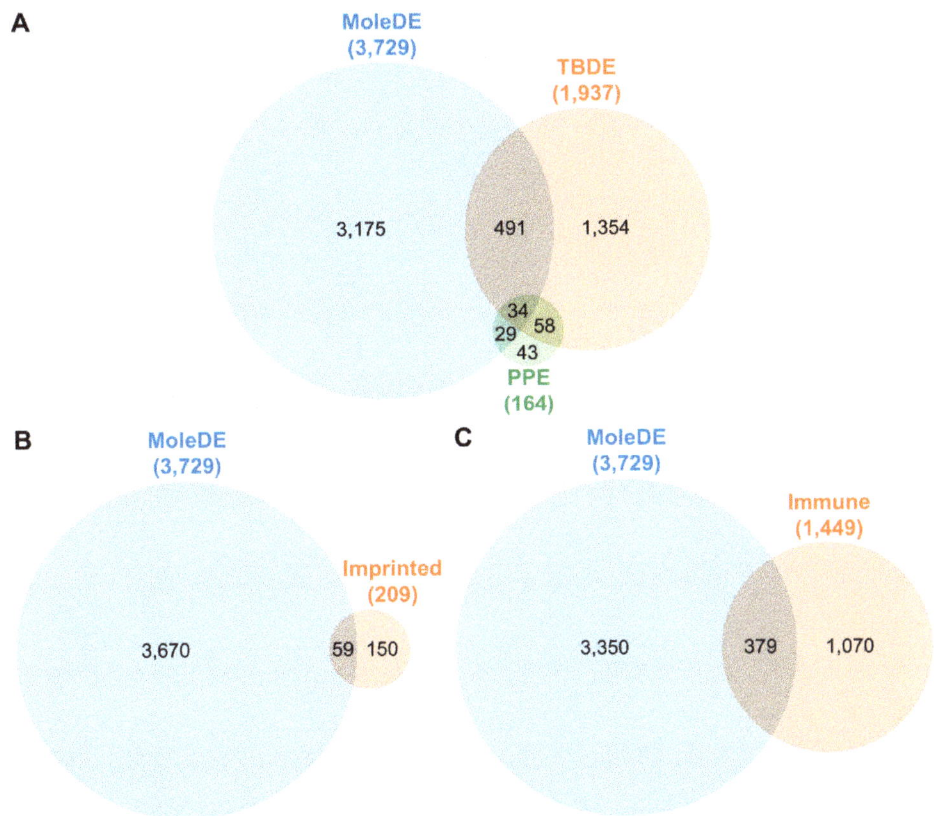

Figure 2. Venn diagrams of gene enrichment in complete hydatidiform moles. (**A**) Overlap between genes differentially expressed (DE) in complete hydatidiform moles (CHMs; MoleDE), and genes previously shown to have specific expression in the placenta (PPE) or during villous trophoblast differentiation (TBDE). (**B**) Overlap between genes DE in CHMs and previously described imprinted genes (Imprinted). (**C**) Overlap between genes DE in CHMs and genes previously shown to be involved in immune-related processes (Immune).

We also analyzed 25 genes that are involved in epigenetic programming (four DNA methyltransferases, three methylcytosine dioxygenases, a cytidine deaminase, and 17 histone demethylases) and found two DNA methylation-related enzymes (*DNMT3A*, DNA methyltransferase 3 alpha; *TET3*, Tet methylcytosine dioxygenase 3) and four histone demethylases (*KDM4C*, lysine demethylase 4C; *KDM4D*; *KDM4E*; *KDM6B*) to be dysregulated in CHMs. *DNMT3A*, which was down-regulated, is required for genome-wide de novo DNA methylation and parental imprinting [20]. *TET3*, which was up-regulated, plays a key role in epigenetic reprogramming of the zygotic paternal DNA [21]. All four histone (lysine) demethylases [22] were up-regulated. In addition, there was an enrichment of genes impacted by parental imprinting (Supplementary Table S3, Figure 2B) among DE genes (OR = 2.38, $p = 1.54 \times 10^{-6}$).

Of note, the DE gene list contained 379 genes involved in immune-related functions (Supplementary Table S4, Figure 2C), of which 278 (73%) were overexpressed in CHMs. Genes contributing to this enrichment (OR = 1.82, $p = 7.34 \times 10^{-18}$) included those encoding several cytokines, chemokines and their receptors (*IL1A*, interleukin 1 alpha; *IL6*, *IL7*, *IL8/CXCL8*, *IL10*, *IL15*, *TGFB1*, transforming growth factor beta 1; *CXCR2*, C-X-C chemokine receptor type 2; *CXCR4*, *IL2RB*, *IL2RG*, *IL7R*, *IL15RA*, *TGFBR1*), integrins (e.g., *ITGA5*, integrin subunit alpha 5; *ITGAE*, *ITGAL*, *ITGAX*, *ITGB7*, integrin subunit beta 7), and galectins (*LGALS4*, *LGALS13*, *LGALS14*).

2.3. Dysregulated Biological Processes and Pathways in CHMs

Use of iPathwayGuide to investigate biological processes and pathways dysregulated in CHMs revealed that among 665 gene ontology (GO) biological processes, the most impacted were "cell adhesion", "biological adhesion", "multicellular organismal process", and "signaling" (Supplementary Table S5). Applying Elim pruning that iteratively removes genes mapped to a significant GO term from more general higher level GO terms, identified "calcium ion binding", "growth factor activity", "extracellular matrix structural constituent", and "G protein-coupled receptor activity" to be the most impacted among 150 dysregulated biological processes (Table 4, Supplementary Table S6).

Table 4. Twenty most impacted Gene Ontology biological processes in complete hydatidiform moles.

GO ID	GO Name	Count DE	Count All	p Elim Pruning
GO:0005509	calcium ion binding	178	426	1.60×10^{-13}
GO:0008083	growth factor activity	50	84	5.50×10^{-11}
GO:0005201	extracellular matrix structural constituent	74	130	3.00×10^{-08}
GO:0004930	G protein-coupled receptor activity	90	188	9.00×10^{-07}
GO:0008201	heparin binding	47	99	2.40×10^{-06}
GO:0005125	cytokine activity	36	70	3.60×10^{-06}
GO:0030020	extracellular matrix structural constituent conferring tensile strength	22	35	4.00×10^{-06}
GO:0004888	transmembrane signaling receptor activity	192	421	4.80×10^{-06}
GO:0042605	peptide antigen binding	13	17	1.70×10^{-05}
GO:0039706	co-receptor binding	8	8	1.90×10^{-05}
GO:0005178	integrin binding	42	91	2.00×10^{-05}
GO:0005044	scavenger receptor activity	24	43	2.60×10^{-05}
GO:0038023	signaling receptor activity	259	568	3.10×10^{-05}
GO:0005249	voltage-gated potassium channel activity	23	27	3.70×10^{-05}
GO:0004252	serine-type endopeptidase activity	30	60	4.60×10^{-05}
GO:0005179	hormone activity	19	34	1.80×10^{-04}
GO:0008331	high voltage-gated calcium channel activity	6	6	2.90×10^{-04}
GO:0020037	heme binding	29	63	3.90×10^{-04}
GO:0030506	ankyrin binding	10	14	4.40×10^{-04}
GO:0003996	acyl-CoA ligase activity	7	8	4.40×10^{-04}

Differentially expressed, DE; Gene Ontology, GO.

We found the most impacted GO molecular functions to be "signaling receptor activity", "molecular transducer activity", and "gated channel activity" among 105 dysregulated molecular functions (Supplementary Table S7). Applying Elim pruning, "regulation of signaling receptor activity", "cell adhesion", "chemical synaptic transmission", and "extracellular matrix organization" were identified as the most impacted among 628 dysregulated molecular functions (Table 5, Supplementary Table S8).

Table 5. Twenty most impacted Gene Ontology molecular functions in complete hydatidiform moles.

GO ID	GO Name	Count DE	Count All	p Elim Pruning
GO:0010469	regulation of signaling receptor activity	143	285	3.90×10^{-14}
GO:0007155	cell adhesion	407	1,014	2.00×10^{-13}
GO:0007268	chemical synaptic transmission	165	397	3.60×10^{-07}
GO:0030198	extracellular matrix organization	122	271	3.80×10^{-07}
GO:0030449	regulation of complement activation	22	33	1.20×10^{-06}
GO:0006958	complement activation, classical pathway	16	22	6.40×10^{-06}
GO:0034765	regulation of ion transmembrane transport	130	295	1.50×10^{-05}
GO:0045669	positive regulation of osteoblast differentiation	29	54	1.50×10^{-05}
GO:0007156	homophilic cell adhesion via plasma membrane adhesion molecules	36	76	5.40×10^{-05}
GO:0007186	G protein-coupled receptor signaling pathway	196	516	6.00×10^{-05}
GO:0007166	cell surface receptor signaling pathway	686	2,116	8.60×10^{-05}
GO:0007601	visual perception	51	123	1.50×10^{-04}
GO:0006957	complement activation, alternative pathway	8	9	1.50×10^{-04}
GO:0006805	xenobiotic metabolic process	32	68	1.60×10^{-04}
GO:0002576	platelet degranulation	44	103	1.80×10^{-04}
GO:0042102	positive regulation of T cell proliferation	32	60	2.20×10^{-04}
GO:0035115	embryonic forelimb morphogenesis	17	29	2.20×10^{-04}
GO:0030501	positive regulation of bone mineralization	19	34	2.20×10^{-04}
GO:0048662	negative regulation of smooth muscle cell proliferation	23	39	2.30×10^{-04}
GO:0035116	embryonic hindlimb morphogenesis	14	22	2.20×10^{-04}

Differentially expressed, DE; Gene Ontology, GO.

The most impacted Kyoto Encyclopedia of Genes and Genomes (KEGG) pathways included "cytokine–cytokine receptor interaction", "cell adhesion molecules", "protein digestion and absorption", and "neuroactive ligand–receptor interaction", all important for placental functions (Table 6, Supplementary Table S9, Figure 3). An unanticipated finding was the most extensive dysregulation of "cytokine–cytokine receptor interaction" pathway (Figure 4) and "cell adhesion" pathway, both required for immune cell influx and activation. In addition, representation of 17 immune-related pathways among 38 dysregulated pathways (Table 6, Supplementary Table S9) reflects a strong immune component of molar pathogenesis.

Table 6. Top 20 most impacted pathways in complete hydatidiform moles.

Pathway Name	*p*-value	pFDR
Cytokine-cytokine receptor interaction	1.38×10^{-07}	2.31×10^{-05}
Cell adhesion molecules (CAMs)	1.43×10^{-07}	2.31×10^{-05}
Protein digestion and absorption	6.20×10^{-07}	6.68×10^{-05}
Neuroactive ligand-receptor interaction	1.32×10^{-06}	1.07×10^{-04}
Complement and coagulation cascades	6.91×10^{-06}	4.46×10^{-04}
Hypertrophic cardiomyopathy (HCM)	1.05×10^{-05}	4.87×10^{-04}
Extracellular matrix (ECM)-receptor interaction	1.06×10^{-05}	4.87×10^{-04}
Autoimmune thyroid disease	2.91×10^{-05}	1.17×10^{-03}
Allograft rejection	7.36×10^{-05}	2.39×10^{-03}
Graft-versus-host disease	8.01×10^{-05}	2.39×10^{-03}
Antigen processing and presentation	8.15×10^{-05}	2.39×10^{-03}
Intestinal immune network for IgA production	9.94×10^{-05}	2.68×10^{-03}
Staphylococcus aureus infection	1.18×10^{-04}	2.94×10^{-03}
Metabolism of xenobiotics by cytochrome P450	1.64×10^{-04}	3.63×10^{-03}
Hematopoietic cell lineage	1.73×10^{-04}	3.63×10^{-03}

Immune-related pathways are shown in bold blue. False discovery rate, pFDR.

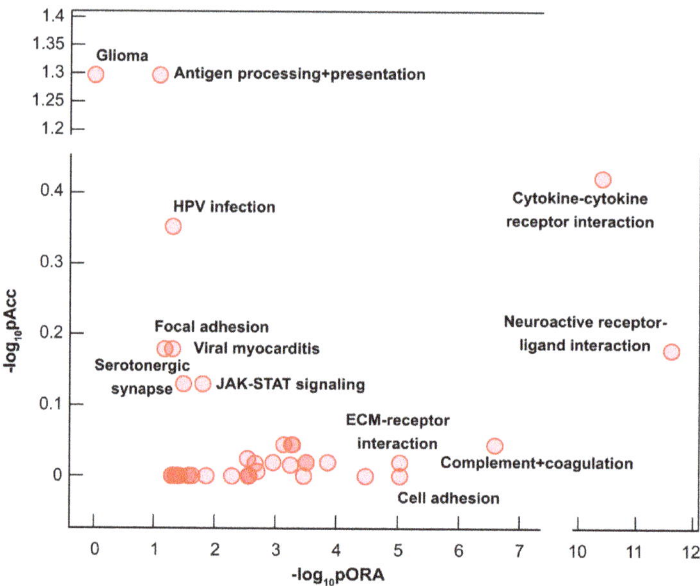

Figure 3. Pathways perturbation vs. over-representation in complete hydatidiform moles. Pathways are plotted according to two types of evidence computed by iPathwayGuide: over-representation on the x-axis (pORA) and the total pathway accumulation on the y-axis (pAcc). For both measures *p*-values are displayed on the negative log (base 10) scale. Extracellular matrix, ECM; human papilloma virus, HPV; Janus kinase, JAK; signal transducer and activator of transcription, STAT.

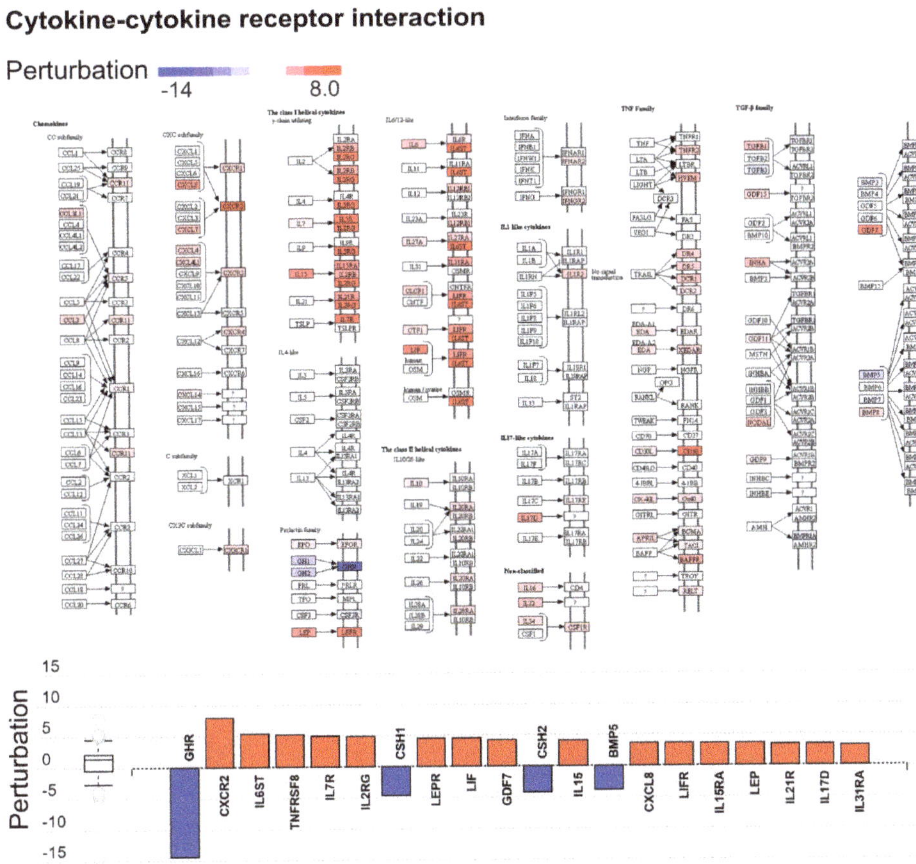

Cytokine-cytokine receptor interaction

Figure 4. Cytokine–cytokine receptor interaction perturbation in complete hydatidiform moles. **Top**: The Kyoto Encyclopedia of Genes and Genomes (KEGG) pathway diagram (KEGG: 04060) is overlaid with the computed perturbation of each gene. Estimates of perturbation account for both for the genes' measured fold change and for the accumulated perturbation propagated from any upstream genes (accumulation). The highest negative perturbation is shown in dark blue, and the highest positive perturbation in dark red. The legend describes values on the gradient. **Bottom**: Gene perturbation bar plot. All genes in the cytokine–cytokine receptor interaction pathway (KEGG: 04060) are ranked according to absolute perturbation values, negative values depicted in blue and positive values in red. The box and whisker plot on the left summarizes the distribution of all gene perturbations in this pathway, the box representing the first quartile, median and third quartile, while circles represent outliers.

2.4. Validation of RNA-Seq Results at the Protein Level

First, we immunostained TMA slides for cyclin-dependent kinase inhibitor p57 (p57) expression to confirm the histopathology diagnosis of CHM at the molecular level. Out of 26 samples with the histopathology diagnosis of CHM, we detected cytotrophoblastic p57 staining in three samples, while 23 (88%) samples were devoid of p57 expression (Supplementary Figure S1), confirming the histopathological diagnosis of CHM in 23 samples. For validation of RNA-level findings, we conducted immunostaining of galectin-14 (gal-14), which is encoded by *LGALS14*, one of the genes most down-regulated in CHMs according to our RNA-Seq study. Average gal-14 immunoscores were

43% lower in CHMs than in gestational age-matched controls (1.47 ± 0.08 and 2.60 ± 0.06, respectively, $p < 0.001$). Additionally, the percentage of low-intensity staining (1+) was higher in molar than in control tissues (58% and 3%, respectively), while the percentage of high-intensity staining (3+) was lower in molar than in control tissues (6% and 61%, respectively), resulting in a significant difference in the distribution of gal-14 immunoscores ($p < 0.001$), consistent with RNA-Seq results for this locus (Figure 5).

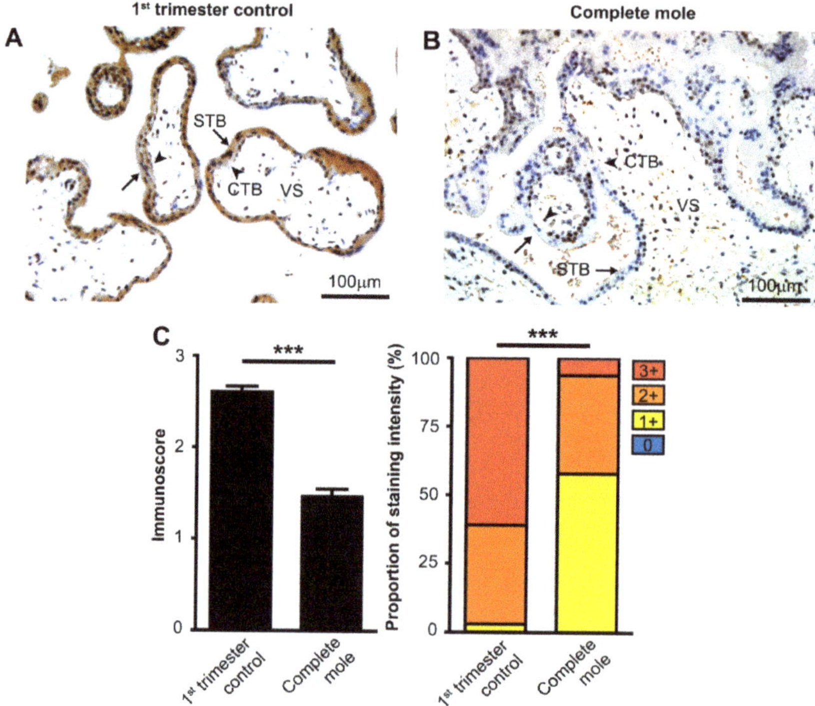

Figure 5. Differential expression of galectin-14 in syncytiotrophoblast in complete hydatidiform moles and first trimester control placentas. Five-μm-thick first trimester placental sections from normal pregnancy (**A**) or from CHMs (**B**) were stained for galectin-14 (gal-14). Chorionic villi exhibited intense syncytiotrophoblast cytoplasmic staining (arrows), while the villus stroma and cytotrophoblasts were negative (arrowheads) in normal placentas, and the syncytiotrophoblast layer had weak staining in CHMs. Representative images, hematoxylin counterstain, 200× magnifications. (**C**) Gal-14 immunoscores (mean ± SEM) and proportion of staining intensities in control placentas ($n = 29$) and CHMs ($n = 23$) are displayed on the left and right graphs, respectively. An unpaired t-test was used to compare mean immunoscores between control and CHM groups. Fisher's exact test was used to test the difference in frequency of gal-14 immunostaining between the two groups. *** $p < 0.001$. Cytotrophoblast, CTB; syncytiotrophoblast, STB; villous stroma, VS.

3. Discussion

3.1. Principal Findings of This Study

High-dimensional transcriptomic analysis identified numerous distinctions between CHM and normal placenta, from which noteworthy patterns can be discerned. (1) the most highly expressed genes in first trimester normal placentas are those previously shown to have placenta-specific or predominantly placental expression; (2) in CHMs, overall gene expression levels are higher, while expression of placenta-specific transcripts are lower than in first trimester normal placentas; (3) the pathogenesis of CHMs involves the dysregulation of 27% of protein-coding genes expressed in both normal placentas and CHMs; (4) most DE genes (72%) in CHMs are up-regulated; (5) placental functions appear to be down-regulated in CHMs, since placenta-specific genes are enriched in DE genes, and most are down-regulated; (6) epigenetic reprogramming of the zygotic DNA and parental imprinting is dysregulated in CHMs; and (7) immune pathways are activated in CHMs as immune-related genes are enriched in DE genes and 17 immune-related pathways are impacted, mostly up-regulated, among 38 dysregulated pathways.

3.2. Placental Gene Expression and Functions are Down-Regulated in CHMs

In accord with previous studies [17,23], here we found genes with placenta-specific or predominant placental expression among the most highly expressed genes—12 out of 20—in first trimester normal placentas. These genes are involved in unique placental functions, including hormones, hormone synthesizing machinery, proteases, and immune proteins, which are pivotal for placental development, growth, signaling, and maternal–fetal immune tolerance. In CHMs, however, in spite of the overall median gene expression level being ~13% higher than in normal first trimester placentas, the overall median placenta-specific transcript level was 23% lower. This is also substantiated by the lower number of placenta-specific genes among the 20 most highly expressed transcripts in CHMs.

When investigating differential gene expression, we found 27% of the expressed protein-coding genes dysregulated in CHMs, of which 72% were up-regulated, underlining the generally higher gene expression levels in CHMs. Placenta-specific genes were enriched among DE genes, and there were much more down-regulated (79%) than up-regulated placenta-specific DE genes. In fact, the 20 most down-regulated genes included seven placenta-specific genes, while the 20 most up-regulated genes were devoid of these. Based on the functions of down-regulated placenta-specific genes, we can infer that placental growth and development, cell attachment, signaling, and immune defense are the most highly impacted processes in CHM. This was also supported by the classical pathway analysis, which revealed among the most impacted GO biological processes "cell adhesion", "signaling", "hormone activity", "growth factor activity", and "extracellular matrix structural constituent", while the most impacted KEGG pathways included "cell adhesion molecules", "neuroactive ligand–receptor interaction", and "ECM–receptor interaction".

3.3. Possible Causes of the Dysregulation of Placental Gene Expression in CHMs

Targeted gene expression studies on molar tissues and a recent meta-analysis of these studies showed that placenta-specific genes involved in villous trophoblast differentiation are also differentially expressed in molar tissues, and concluded that molar pathogenesis is, indeed, rooted in problems with trophoblast differentiation [16]. However, these studies were focusing on just a limited set of placenta-specific transcripts known to be differentially expressed during trophoblast differentiation, thus did not give a comprehensive insight into this important question. We recently discovered that abnormal villous trophoblast differentiation is at the root of the pathogenesis of early-onset preeclampsia, also reflected by profound placental dysregulation of placenta-specific genes [17,19]. Here we applied a similar non-targeted approach to investigate this issue. First, we intersected the list of DE genes in CHMs with the list of DE genes during villous trophoblast differentiation from our parallel study [19] and found only minimal enrichment for trophoblast differentiation genes in dysregulated genes in CHMs, much less than for placenta-specific genes. This suggests that trophoblast differentiation is moderately affected in CHMs, while the expression of placenta-specific genes is more extensively impacted. Among the affected placenta-specific genes we validated the down-regulation of *LGALS14* at the protein level. Since histopathological evidence shows the generally two-layered structure of villous trophoblasts in CHMs as in normal first trimester placentas, in spite of the histological and molecular evidence of locally more proliferative and or immortalized trophoblasts in CHMs [24–31], the histopathology of CHMs also underlines the generally moderate problem with villous trophoblast differentiation in CHMs.

A more compelling explanation would be that there is a problem with placental gene regulation due to the changes in DNA methylation and imprinting [32–34] since CHMs only contain paternal but not maternal genomes [35–37]. To test this hypothesis, we intersected the list of DE genes with the list of imprinted genes obtained from the GeneImprint database, and we found a significant enrichment of imprinted genes, underlining the link between CHM pathogenesis and affected paternal imprinting. In addition, we detected the down-regulation of the de novo DNA methyltransferase *DNMT3A* in CHMs, similar to the down-regulation of *DNMT3A* in absent fetal development, which is also the characteristics of *Dnmt3a* or *Dnmt3b* knockout mice [38]. Moreover, *TET3*, which is enriched specifically in the male pronucleus and responsible for paternal-genome conversion of 5mC into 5hmC [21], was two-fold up-regulated in CHMs. This is in good accordance with the paternal origin of all chromosomes in CHMs. Furthermore, four histone demethylases, which are active in the regulation of chromatin remodeling and gene expression and in influencing cellular differentiation, development, tumorigenesis, and inflammatory diseases [22,39], were found to be up-regulated in CHMs. These findings suggest that the epigenetic reprogramming of the zygotic DNA and parental imprinting are dysregulated in CHMs.

3.4. Immune Functions are Widely Dysregulated in CHMs

The pathogenesis of familial, biparental hydatidiform moles is caused by the inactivating mutations of *NLRP7* and *KHDC3L* [40–42], genes involved in imprinting and inflammation. Inactivating mutations in *NLRP7* inhibit cytokine (IL-1, interleukin-1; TNF, tumor necrosis factor) secretion by interfering with their trafficking, resulting in an in utero environment tolerogenic for the growth of these complete paternal allografts [41]. In this context, it is important that we found a wide dysregulation of immune pathways at the maternal–fetal interface in CHMs, which may also be at the heart of disease pathogenesis. To understand how this would be possible, we need to look first at immune processes in normal pregnancies. In healthy pregnant women, maternal–fetal immune tolerance mechanisms are complex and dynamic since embryo implantation in the first trimester is a pro-inflammatory process at the maternal–fetal interface, while the second trimester brings an anti-inflammatory milieu for the growing fetus in the womb, and the initiation of parturition at the end of pregnancy is characterized by a transition back towards a pro-inflammatory state [43–46]. These dynamic alterations in immune states are driven by the changing placental expression of molecules that regulate maternal immune responses [47–58]. Importantly, the dysregulated expression of immunoregulatory molecules at the maternal–fetal interface and the consequent disturbances in maternal–fetal immune regulation are strongly linked with alterations in the invasiveness of the trophoblast [59,60] and the development of severe pregnancy complications, such as miscarriages [61–65], preterm labor [66–72], or preeclampsia [73–79].

Herein, we detected the enrichment of genes involved in immune-related functions among DE genes in CHMs, and there were much more up-regulated than down-regulated immune-related genes. Of note, we noticed that *IL1A* and *IL33* were down-regulated while *FOXP3* (forkhead box P3), *IL7*, *IL10*, *LIF* (leukemia inhibitory factor), and *TGFB1* were up-regulated, which could theoretically induce a tolerogenic environment, similarly to familial hydatidiform moles. Indeed, previous studies described the infiltration of regulatory T cells, CD3+ T cells, and NK cells in CHMs at the implantation site [80–82]. This is in line with the finding herein on the up-regulation of chemokines and the activation of "cytokine–cytokine receptor interaction" and "cell adhesion molecules" pathways as well as biological processes such as "cell adhesion", "biological adhesion", and "signaling" needed for immune cell influx and immune responses.

On the other hand, we also detected the down-regulation of *LGALS13*, *LGALS14*, the up-regulation of various *HLA-I (A, B, C, E, F, G)* and *HLA-II (DMA, DMB, DPB1, DRA)* genes, complement pathway genes (e.g., *C1S*, *C3*, *C5*, *CR1*) and interleukins (e.g., *IL6*, *IL8/CXCL8*, *IL15*, *IL16*, *IL17D*), and the overall involvement of 17 immune-related pathways among 38 dysregulated pathways. Of note, eight out of 10 (e.g., *CD14, CD36, FCGRT*, neonatal Fc receptor; *LYVE1*, lymphatic vessel endothelial hyaluronan receptor 1) Hofbauer cell marker genes [83] were down-regulated in CHMs in our study in line with the relative lack of Hofbauer cells in CHM villi [84,85]. These findings suggest an enhanced maternal, over fetal, antigen-presenting and pro-inflammatory environment at the implantation site, which is also supported by the activation of the "antigen processing and presentation", "phagosome", "complement and coagulation cascades", "allograft rejection", and "graft-versus-host disease" pathways. This is in line with the fact that CHMs express only paternal antigens and thus represent complete allografts that could stimulate a vigorous alloimmune response from the maternal host [85], including local synthesis of complement C3 and complement activation [86]. Not only paternal alloantigens but also increased necrosis and apoptosis, due to the higher trophoblastic cell proliferation and turnover rate in CHMs [87,88], may trigger the production of several pro-inflammatory cytokines [89] and the activation of the complement system [89,90].

Based on all of these findings, we suggest that there is a complex immune dysregulation at the implantation site in CHMs, which include the parallel induction of anti- and pro-inflammatory processes due to the presentation of excess paternal antigens, aponecrotic trophoblastic debris, and the dysregulation of upstream transcription factors and other gene regulatory mechanisms. These are probably the consequences of abnormal paternal genome dosage and imprinting, resulting in impaired gene expression and placental development. To delineate the exact sequence of events, one would need to collect multiple samples from the implantation site in a time-series from patients and controls, and then to perform the extensive characterization of the localization and expression of RNA and protein signals. However, this type of sample collection is not possible due to ethical and technical reasons in spite of the fact that cellular and molecular characterization of such time-series materials would now be possible using, for example, RNA-Seq after laser capture microdissection or single-cell RNA-Seq as well as mass spectrometry-based tissue imaging. However, these techniques were not available for our current study.

3.5. Strengths and Limitations of the Study

According to our knowledge, this is the first non-hypothesis testing study that utilized various tools of high-dimensional biology to gain in-depth insights into the pathogenesis of CHMs. The strengths of the study are as follows: (1) strict clinical definitions and homogenous patient groups; (2) standardized, quick placental sample collection during surgeries and pregnancy terminations; (3) standardized histopathological examination of molar pregnancies and placentas based on international criteria; (4) high-quality sample processing and state-of-the-art high-dimensional methods for RNA and protein level analysis; (5) parallel expression profiling of various proteins on large tissue sets with tissue microarray and immunostaining followed by semi-quantitative immunoscoring and statistical analysis; and (6) the use of leading bioinformatics tools for RNA-Seq, gene ontology, and pathway analyses.

Limitations of the study are as follows: (1) the relatively modest number of cases due to the strict clinical and histopathological inclusion criteria used for patient enrollment. Paradoxically, this was also one of the most important strengths of our study; (2) data for cases and controls were retrieved from various sources in the RNA-Seq discovery study; (3) TMAs did not allow us the cellular compositional analysis of the investigated tissues due to the small core diameter size; and (4) it is possible that the differences in gene expression levels may originate from differential gene expression of particular cell types, differential cellular composition of CHMs and normal placentas, or the surgical/methodological differences in tissue sampling during elective terminations or CHM surgeries.

3.6. Concluding Remarks

In conclusion, our data shows that placental functions are down-regulated, expression of imprinted genes is altered, and immune pathways are activated in CHMs. Taken together, the results indicate that complex dysregulation of placental developmental and immune processes are pivotal for the pathogenesis of CHMs, providing new biological insight likely to inform molecular translational research addressing multiple stages of the natural history of CHM, such as assays for improved detection and refined diagnosis in early pregnancy [91], risk stratification with respect to malignant potential [92], and differential diagnosis of unexplained pregnancy loss [93].

4. Materials and Methods

4.1. Human Subjects, Clinical Samples, and Definitions

4.1.1. Subjects in the RNA-Seq Discovery Study

All research participants completed a process of informed consent per the University of Southern California (USC) Health Sciences Institutional Review Board (protocols HS-09-00468 [12 November 2009) and HS-11-00095 (15 June 2011)]. Eligible subjects were identified upon presentation to Los Angeles County (LAC) and USC Medical Center. CHMs were identified clinically by a combination of ultrasonographic features ("snowstorm" or "cluster of grapes" appearance) with elevated blood hCG levels. Molar tissues were obtained between 8 2/7 and 14 0/7 weeks of gestational age by therapeutic surgical dilation and curettage. Confirmation of CHM diagnosis was determined by the histopathologic examination of surgical material. One subject who had CHM on the basis of clinical features was subsequently reclassified as having choriocarcinoma after pathological examination and was excluded from the study. Another subject with CHM was excluded due to pooling of several samples from various curettages into one sample, leaving the analyzed number of CHMs at four. Gestational age-matched control chorionic villous tissues were obtained at the Reproductive Options Clinic at LAC + USC Medical Center from surgical elective pregnancy terminations. To increase control group size, we also obtained RNA-Seq data of first trimester control chorionic villous tissues from a published study [23] (Table 7).

Table 7. Demographic characteristics of patients in the RNA-Seq discovery study.

	Patients Providing GTD Tissue Samples (n = 4)	Patients Providing Placenta Samples (n = 40)
Race/ethnicity		
Hispanic	2	6
Caucasian	-	34
Asian	1	-
African-American	-	-
Unknown/other	1	-
Age at procedure	25.7	39.2
Histologic diagnosis		
Complete mole	3	-
Incomplete mole	-	-
Invasive mole	1	-
Sex chromosome content		
XY	-	22
XX	4	18
Gestational age at tissue collection (days)	79.2 (58–98)	84.2 (77–124)

Gestational trophoblastic disease, GTD.

4.1.2. Subjects in the Tissue Microarray Validation Study

GTD tissue samples had been collected during usual care of women who underwent treatment for GTD at the Department of Obstetrics and Gynecology, Keck School of Medicine, USC (Los Angeles, CA, USA). After approvals (protocols HS-14-00808 and HS-15-00720 for IRB approval, Lab agreement: 16-03-03) were obtained from the USC IRB, the institutional pathology database at the LAC Medical Center, Keck Medical Center of USC, and Norris Comprehensive Cancer Center was utilized to identify GTD cases by searching for the keywords "gestational trophoblastic disease", "molar pregnancy", "hydatidiform mole", "complete mole", and "invasive mole" between 1 January 2000 and 11 September

2017. Inclusion criteria included GTD cases from whom archived histopathology specimens were available. GTD cases without salient clinical information were excluded.

For eligible GTD cases, data describing clinical and demographic features, tumor characteristics and markers, treatment course, and survival outcomes were abstracted from the medical record. Clinical and demographic features included patient age at diagnosis of GTD, ethnicity, pregnancy history including interval months from the last menstrual period of index pregnancy, body mass index, and medico-surgical history. Tumor characteristics and markers included the World Health Organization (WHO) score, histology subtype, tumor size, pretreatment beta-hCG, metastatic sites, and number. Treatment course included surgical intervention (dilation and curettage, or hysterectomy) and chemotherapy (type, cycle number, and response). Survival outcomes included progression-free survival (PFS) and overall survival. Progression-free survival was determined as the time interval between treatment initiation for GTD and the date of first recurrence or last follow-up date if there was no recurrence. Overall survival was determined as the time interval between treatment initiation for GTD and the date of death related to GTD or last follow-up date if the patient was alive.

Of 311 cases that were initially identified, 44 CHM cases were chosen for TMA based on tissue availability (Tables 8 and 9). We placed no restriction on the age of gestation and found pregnancies to be dated according to ultrasound scans between 5 and 15 weeks. Patients with multiple pregnancies were excluded, and specimens and data were de-identified for use in research according to procedures approved by the Institutional Review Boards of USC. From the 44 GTD tissues identified between 4 2/7 and 36 1/7 weeks of gestational age, the TMA project included 26 samples obtained from therapeutic surgical dilation and curettage within the first trimester (13 6/7 weeks) and having good tissue quality. The 26 samples were obtained from 23 patients. One patient provided two specimens at the same sampling time, while another patient provided three specimens at different time points.

Table 8. Demographic characteristics of patients in the TMA validation study.

	Patients Providing GTD Tissue Samples (n = 23 *)	Patients Providing Placenta Samples (n = 29)
Race/ethnicity [a]		
Hispanic	18 (78%)	
Non-Hispanic white	2 (9%)	29 (100%)
Asian	0	
African-American	0	
Unknown/other	3 (13%)	
Age (years) at procedure [b]	31.9 ± 9.5	30.4 ± 6.4
Comorbidities	1 (4%)	
Hypertension [a]	1 (4%)	
Diabetes [a]	185,277	
beta-hCG (mIU/mL) [c]	(70,349–387,812)	
Histologic diagnosis [a]		
Complete mole	26 (100%)	0 (0%)
p57 immunostaining confirmation of CHM	23 (88%)	
Gestational age at the procedure (days) [b]	64.4 ± 18.2	62.3 ± 11.8

* A total of 26 samples altogether; one case with two specimens at the same time; another case with three specimens at different time points. [a] Data are presented as number (percentage). [b] Data are presented as mean ± SD. [c] Data are presented as median (IQR). Cyclin-dependent kinase inhibitor p57; human chorionic gonadotropin, hCG; gestational trophoblastic disease, GTD; tissue microarray, TMA.

Control samples of first trimester placental tissue (*n* = 29) matched by gestational age to GTD samples were collected prospectively at the Maternity Private Clinic (Budapest, Hungary) and deposited into the Perinatal Biobank of the Research Centre for Natural Sciences, Hungarian Academy of Sciences (Budapest, Hungary). Informed consent for use of the material in research was obtained from women prior to sample collection, and specimens and data were stored anonymously. The research was approved by the Health Science Board of Hungary (TUKEB 4834-0/2011-1018EKU), and all use of tissue and data in this research conformed to the principles set out in the World Medical Association (WMA) Declaration of Helsinki. Placentas were collected from pregnancies voluntarily terminated by surgical dilation and curettage between 5 and 14 weeks of gestation according to ultrasound scans, excluding multiple pregnancies (Table 8).

Table 9. Criteria for categorizing gestational trophoblastic disease cases and number of samples in each category included in the tissue microarray study.

Grouping	Definition	Sample No. *
1	D&C then cured	15
2a	D&C then persistent GTD, cured by first-line single agent	1
2b	D&C then persistent GTD, cured by upfront hysterectomy	4
2c	D&C then persistent GTD, cured by upfront hysterectomy and adjuvant chemotherapy	0
3a	D&C then persistent GTD, received first-line single agent but failed	3
3b	D&C then persistent GTD, received first-line multi-agent and cured	0
3c	D&C then persistent GTD, received first-line multi-agent and failed	0
3d	D&C then persistent GTD, received upfront hysterectomy but recurred	0
4	Recurrent GTD	3

* A total of 26 samples altogether; one case with two specimens at the same time; another case with three specimens at different time points. Dilatation and curettage, D&C; gestational trophoblastic disease, GTD.

4.2. Experimental Procedures

4.2.1. Sample Collection and Preparation for RNA-Seq

Tissue samples were rinsed with sterile saline and examined under a lightbox for content confirmation and selection of molar tissue or chorionic villi. Samples were then stored in RNA*later* RNA stabilization reagent (approximately 10 µL reagent per 1mg of tissue, Qiagen, Germantown, MD, USA) at 37 °C for a maximum of 24 h until collected for dissection. Samples were then dissected into ~1 g tissue aliquots and stored at −80 °C. Stored samples were then thawed at room temperature and homogenized with electric homogenizer. Total RNA was isolated using the protocol provided with the Qiagen RNeasy Mini Kit. RNA quantity was measured by a Nanodrop ND-8000 analyzer (Thermo Scientific, Waltham, MA, USA). Samples were then stored at −80 °C until used.

4.2.2. RNA Sequencing

Library preparation and paired-end sequencing were performed by the USC Epigenome Core Laboratory. Double-stranded cDNA fragments were synthesized from mRNA, ligated with adapters, and size-selected for library construction according to Illumina protocol [94]. RNA sequencing was carried out on Illumina Genome Analyzer (Illumina, San Diego, CA, USA).

4.2.3. Histopathological Examinations

All slides cut from formalin-fixed, paraffin-embedded (FFPE) molar tissue blocks were retrieved from our USC + LAC histology archives. The slides were reviewed for confirmation of the diagnosis and regions were identified for TMA construction. The blocks from the matching slides were retrieved and submitted for TMA construction.

Samples of tissue from first trimester placentas had been fixed in 10% neutral-buffered formalin and then embedded in paraffin. Five µm sections were cut from tissue blocks, stained with hematoxylin and eosin (H&E), and examined using light microscopy. A perinatal pathologist blinded to patients' clinical information, except for the gestational age, examined the histopathology of placental samples using standard perinatal pathological protocol and previously-published diagnostic criteria [95–97], and identified regions of each tissue block from which to sample cores.

4.2.4. Tissue Microarray Construction

A material transfer agreement between the Hungarian Academy of Sciences and USC (No: 1996/2017) enabled the construction of TMAs containing tissues collected at USC and at Maternity Private Clinic and deposited into the Perinatal Biobank. Four TMAs were constructed, each containing three cylindrical cores two mm in diameter from each sample of first trimester FFPE placental ($n = 29$) or GTD ($n = 26$) tissue. Cores from the same sample were transferred into recipient paraffin blocks adjacent to each other using an automated tissue arrayer (TMA Master II, 3DHISTECH, Budapest, Hungary). Each paraffin block contained three cores from all tissues, with liver included as negative control and third trimester placenta as positive control material.

4.2.5. Immunohistochemistry and Immunoscoring

To validate the histological diagnosis of CHM, we conducted immunostaining for p57, which is expressed from a paternally imprinted gene and is a recognized diagnostic marker in CHM [12,98,99]. To validate RNA level findings at the protein level for one down-regulated gene, five-μm-thick sections were cut from each TMA onto adhesive glass slides (SuperFrost Ultra Plus, TS Labor, Budapest, Hungary), and immunostained for p57 and galectin-14 using antibodies and conditions listed in Table 10. Briefly, sections were dewaxed using xylene and rehydrated in graded alcohol series.

Table 10. Immunohistochemistry conditions.

Primary Antibody (Concentration/Dilution, Distributor)	Secondary Antibody (Dilution, Distributor)	Detection Antibody (Distributor)	Detection System (Distributor)
monoclonal mouse anti-human p57 antibody (1:3000) (code: MA5-11309, Thermo Fisher Scientific, Waltham, MA, USA)	-	Novolink detections system (Leica-Novocastra, Wetzlar, Germany)	Novolink DAB/substrate kit (Leica-Novocastra, Wetzlar, Germany)
recombinant, His$_6$-tagged anti-human gal-14 antibody (0.65 μg/mL) [provided by Prof. R. Romero (PRB, NIH)]	monoclonal mouse anti-His$_6$-tag antibody (1:200) (code: 05-959, Merck-Millipore, Burlington, MA, USA)		

For p57 immunostaining, endogen peroxidases were blocked using 10% H_2O_2 in methanol for 20 min, then antigen retrieval was performed in Tris-EDTA pH 9.0 buffer for 20 min at 96 °C. For galectin-14 immunostaining, endogen peroxidases were blocked using 1% H_2O_2 in methanol for 20 min, then antigen retrieval was performed in Tris-EDTA pH 9.0 buffer for 32 min at 100 °C. In all staining procedure, after 10 min Novolink protein blocking (Leica-Novocastra, Wetzlar, Germany), the sections were incubated at room temperature with antibodies and then with reagents of the Novolink Polymer Detection System (Leica-Novocastra). Between incubation steps, the sections were washed trice for 5 min in Tris-buffered saline (pH 7.4). Finally, sections were counterstained with hematoxylin, and these were mounted (DPX Mountant; Sigma-Aldrich, St. Louis, MO, USA) after dehydration.

Immunostained TMA images were semi-quantitatively scored by two examiners blinded to the clinical information with an immunoreactive score modified from one previously used in our studies [18,65,74,100–102]. Trophoblastic immunostaining intensity was graded as 0 = negative, 1 = weak, 2 = intermediate, and 3 = strong. For each specimen, all villi and trophoblastic tissues in a random field of each of the cores were evaluated by both examiners using Panoramic Viewer 1.15.4 (3DHISTECH Ltd., Budapest, Hungary).

4.3. Data Analysis

4.3.1. Analysis of mRNA Expression (RNA-Seq Data)

For data produced from five samples, FASTQ files were generated by Illumina's pipeline and read quality was assessed by FastQC (http://www.bioinformatics.babraham.ac.uk/projects/fastqc/). Subsequently, reads were submitted to alignment with HISAT2 (v2.1.0) [103]. The mapping was made using default parameters with reference human genome GRCh38 (downloaded from Illumina iGenomes). Aligned BAM files were indexed and sorted with Samtools (v0.1.18) [104] for downstream analysis. Genomic features and read count matrices were obtained using featureCounts (v1.5.2) [105] based on annotation file hg38 (RefSeq track of UCSC Table Browser).

RNA-Seq count data for additional late first trimester control placentas (22 XY and 17 XX) were downloaded from GEO (AccNo: GSE109082) [23]. Differential gene expression analysis was performed using the R package DESeq2 [106,107].

Because data included differing amounts of non-coding RNA leading to high numbers of DE genes presumed to be spurious, we restricted the differential gene expression analysis to 19,690 protein-coding genes based on the ENSEMBL "biotype" annotation. We further excluded genes with zero read count in any sample, resulting in a final analytic set of data on 14,022 genes. As a set of DE transcripts, we selected those with a fold change of ≥ 2 between CHM and control, while holding the false discovery rate (pFDR) to <0.05.

To estimate absolute levels of expression, we normalized RNA-Seq count data by the geometrical mean of measured counts of four housekeeping genes (*TBP*, TATA-box binding protein; *CYC1*, cytochrome C1; *UBC*, ubiquitin C; *TOP1*, topoisomerase (DNA) I), and calculated the mean expression both in CHM and control groups of samples.

4.3.2. Pathway Analysis

To identify significantly impacted pathways, biological processes, and molecular functions, we compared expression between CHM and control groups using Advaita Bio's iPathwayGuide (https://www.advaitabio.com/ipathwayguide). This software analysis tool implements the "impact analysis" approach that takes into consideration the direction and type of all signals on a pathway and the position, role and type of every gene, as described in [108].

4.3.3. Enrichment Analyses

We used Fisher's exact test to test enrichment of the DE transcript set with genes previously demonstrated to have expression patterns of particular relevance: (1) genes with placenta-specific expression (n = 164, Supplementary Table S2) [17]; (2) genes involved in villous trophoblast differentiation (n = 1,937; GEO: GSE130339, the list was provided by authors of an unpublished work [19]); (3) genes imprinted in humans (n = 209; imprinted gene databases, www.geneimprint.com, Supplementary Table S3 and 4 genes related to immune processes (n = 1,449; www.innatedb.com; the final list is the manual curation and combination of Immport, Immunogenetic Related Information Source, and Immunome Database Gene Lists, Supplementary Table S4).

4.3.4. Analysis of TMA Data

For each core, immunostain was scored by each of two examiners using a scale of 0 to 3, 0 representing none and 3 strong immunostaining. Quantitative immunoscores were determined as follows. For each core, immunoscores from examiners were first averaged to represent that core, and resulting values for each of the three cores representing one sample were averaged to represent that sample. Mean immunoscores between control and CHM groups were compared using the unpaired t-test. The Fisher's exact test was performed to test the distribution of gal-14 immunoscores between the control and CHM groups. Results were considered statistically significant at * $p < 0.05$, ** $p < 0.01$, and *** $p < 0.001$.

4.3.5. Data Availability

USC RNA-Seq data was deposited to the Gene Expression Omnibus (GEO) database according to the MIAME guidelines (accession number: GSE138250).

Supplementary Materials: Supplementary materials can be found at http://www.mdpi.com/1422-0067/20/20/4999/s1.

Author Contributions: Conceptualization, M.L.W., L.D.R., V.K.C. and N.G.T.; formal analysis, S.H., P.K., E.T., A.B., A.S., J.M. and N.G.T.; funding acquisition, A.B., A.S., L.D.R. and N.G.T.; investigation, J.R.K., M.L.W., A.V.C., T.K., P.M.-F. and A.B.; methodology, J.R.K., M.L.W., A.S. and N.G.T.; project administration, M.L.W., L.D.R., V.K.C. and N.G.T.; resources, K.M., P.H., Z.P., L.D.R. and N.G.T., software, T.K. and N.G.T.; supervision, M.L.W., L.D.R., V.K.C. and N.G.T.; validation, E.T., A.B., A.S., J.M., L.D.R., V.K.C. and N.G.T.; visualization, A.B., A.S. and N.G.T.; writing—original draft, J.R.K., M.L.W., S.H., P.K., K.M., A.V.C., E.T., T.K., P.H., P.M.-F., A.B., A.S., Z.P., L.D.R., V.K.C. and N.G.T.; writing—review and editing, J.R.K., M.L.W., S.H., P.K., K.M., A.V.C., E.T., T.K., P.H., P.M.-F., A.B., A.S., J.M., Z.P., L.D.R., V.K.C. and N.G.T.

Funding: This research was supported by the Hungarian National Science Fund (K128262 grant to A.S. and K124862 grant to N.G.T.); Hungarian Academy of Sciences (Momentum LP2014-7/2014 grant to N.G.T., Medinprot Protein Science Research Synergy IV and Adhoc Program grants to N.G.T., Premium_2019-436 grant to B.A., Young Research Fellowships to S.H. and E.T.), Hungarian National Research, Development, and Innovation Office (FIEK_16-1-2016-0005 grant to N.G.T.), Hungarian Ministry for National Economy (VEKOP-2.3.3-15-2017-0014 grant to N.G.T.); Gynecologic Oncology Research Fund (USC) to L.D.R..

Acknowledgments: We thank Judit Baunoch, Diana Csala, Zsofia Feher, Nikolett Szenasi (Hungarian Academy of Sciences), Andras Matolcsy, Ilona Kovalszky, Dorottya Csernus-Horvath, Katalin Karaszi, (Semmelweis University), Jolan Csapai, Laszlo Daru, Hajnalka Nyiro, and Erzsebet Szilagyi (Maternity Private Clinic), Brendan Grubbs, Melissa Natavio, and Rachel Stewart (USC) for their support or assistance. The list of genes differentially expressed during trophoblast differentiation was provided by the authors of an unpublished work [19].

Conflicts of Interest: The authors declare no conflicts of interest.

Abbreviations

CHM	complete hydatidiform mole
ECM	extracellular matrix
FFPE	formalin-fixed paraffin-embedded
gal-14	galectin-14
GO	Gene Ontology
GTD	gestational trophoblastic disease
GTN	gestational trophoblastic neoplasia
hCG	human chorionic gonadotropin
IL	interleukin
KEGG	Kyoto Encyclopedia of Genes and Genomes
p57	cyclin-dependent kinase inhibitor p57
pFDR	false discovery rate
TMA	tissue microarray

References

1. Lurain, J.R. Gestational trophoblastic disease I: Epidemiology, pathology, clinical presentation and diagnosis of gestational trophoblastic disease, and management of hydatidiform mole. *Am. J. Obstet. Gynecol.* **2010**, *203*, 531–539. [CrossRef]

2. Garcia-Sayre, J.; Castaneda, A.V.; Roman, L.D.; Matsuo, K. Diagnosis and Management of Gestational Trophoblastic Disease. In *Handbook of Gynecology*; Shoupe, D., Ed.; Springer: New York, NY, USA, 2017; pp. 1–15.

3. Lurain, J.R. Gestational trophoblastic disease II: Classification and management of gestational trophoblastic neoplasia. *Am. J. Obstet. Gynecol.* **2011**, *204*, 11–18. [CrossRef] [PubMed]

4. Seckl, M.J.; Sebire, N.J.; Berkowitz, R.S. Gestational trophoblastic disease. *Lancet* **2010**, *376*, 717–729. [CrossRef]

5. Atrash, H.K.; Hogue, C.J.; Grimes, D.A. Epidemiology of hydatidiform mole during early gestation. *Am. J. Obstet. Gynecol.* **1986**, *154*, 906–909. [CrossRef]

6.	Takeuchi, S. Incidence of gestational trophoblastic disease by regional registration in Japan. *Hum. Reprod.* **1987**, *2*, 729–734. [CrossRef] [PubMed]

7.	Smith, H.O.; Hilgers, R.D.; Bedrick, E.J.; Qualls, C.R.; Wiggins, C.L.; Rayburn, W.F.; Waxman, A.G.; Stephens, N.D.; Cole, L.W.; Swanson, M.; et al. Ethnic differences at risk for gestational trophoblastic disease in New Mexico: A 25-year population-based study. *Am. J. Obstet. Gynecol.* **2003**, *188*, 357–366. [CrossRef] [PubMed]

8.	Smith, H.O.; Qualls, C.R.; Hilgers, R.D.; Verschraegen, C.F.; Rayburn, W.F.; Cole, L.W.; Padilla, L.A.; Key, C.R. Gestational trophoblastic neoplasia in American Indians. *J. Reprod. Med.* **2004**, *49*, 535–544.

9.	Maesta, I.; Berkowitz, R.S.; Goldstein, D.P.; Bernstein, M.R.; Ramirez, L.A.; Horowitz, N.S. Relationship between race and clinical characteristics, extent of disease, and response to chemotherapy in patients with low-risk gestational trophoblastic neoplasia. *Gynecol. Oncol.* **2015**, *138*, 50–54. [CrossRef]

10.	Gockley, A.A.; Joseph, N.T.; Melamed, A.; Sun, S.Y.; Goodwin, B.; Bernstein, M.; Goldstein, D.P.; Berkowitz, R.S.; Horowitz, N.S. Effect of race/ethnicity on clinical presentation and risk of gestational trophoblastic neoplasia in patients with complete and partial molar pregnancy at a tertiary care referral center. *Am. J. Obstet. Gynecol.* **2016**, *215*, 334.e1–334.e6. [CrossRef]

11.	Sand, P.K.; Lurain, J.R.; Brewer, J.I. Repeat gestational trophoblastic disease. *Obstet. Gynecol.* **1984**, *63*, 140–144.

12.	Hui, P.; Buza, N.; Murphy, K.M.; Ronnett, B.M. Hydatidiform Moles: Genetic Basis and Precision Diagnosis. *Annu. Rev. Pathol.* **2017**, *12*, 449–485. [CrossRef] [PubMed]

13.	Nguyen, N.M.P.; Ge, Z.J.; Reddy, R.; Fahiminiya, S.; Sauthier, P.; Bagga, R.; Sahin, F.I.; Mahadevan, S.; Osmond, M.; Breguet, M.; et al. Causative Mutations and Mechanism of Androgenetic Hydatidiform Moles. *Am. J. Hum. Genet.* **2018**, *103*, 740–751. [CrossRef] [PubMed]

14.	Nguyen, N.M.; Slim, R. Genetics and Epigenetics of Recurrent Hydatidiform Moles: Basic Science and Genetic Counselling. *Curr. Obstet. Gynecol. Rep.* **2014**, *3*, 55–64. [CrossRef] [PubMed]

15.	Sanchez-Delgado, M.; Martin-Trujillo, A.; Tayama, C.; Vidal, E.; Esteller, M.; Iglesias-Platas, I.; Deo, N.; Barney, O.; Maclean, K.; Hata, K.; et al. Absence of Maternal Methylation in Biparental Hydatidiform Moles from Women with NLRP7 Maternal-Effect Mutations Reveals Widespread Placenta-Specific Imprinting. *PLoS Genet.* **2015**, *11*, e1005644. [CrossRef] [PubMed]

16.	Desterke, C.; Slim, R.; Candelier, J.J. A bioinformatics transcriptome meta-analysis highlights the importance of trophoblast differentiation in the pathology of hydatidiform moles. *Placenta* **2018**, *65*, 29–36. [CrossRef] [PubMed]

17.	Than, N.G.; Romero, R.; Tarca, A.L.; Kekesi, K.A.; Xu, Y.; Xu, Z.; Juhasz, K.; Bhatti, G.; Leavitt, R.J.; Gelencser, Z.; et al. Integrated Systems Biology Approach Identifies Novel Maternal and Placental Pathways of Preeclampsia. *Front. Immunol.* **2018**, *9*, 1661. [CrossRef] [PubMed]

18.	Karaszi, K.; Szabo, S.; Juhasz, K.; Kiraly, P.; Kocsis-Deak, B.; Hargitai, B.; Krenacs, T.; Hupuczi, P.; Erez, O.; Papp, Z.; et al. Increased placental expression of Placental Protein 5 (PP5) / Tissue Factor Pathway Inhibitor-2 (TFPI-2) in women with preeclampsia and HELLP syndrome: Relevance to impaired trophoblast invasion? *Placenta* **2019**, *76*, 30–39. [CrossRef] [PubMed]

19.	Szilagyi, A.; Romero, R.; Xu, Y.; Kiraly, P.; Palhalmi, J.; Gyorffy, B.A.; Juhasz, K.; Hupuczi, P.; Kekesi, K.A.; Meinhardt, G.; et al. Placenta-specific genes and their regulation during villous trophoblast differentiation. 2019; Submitted.

20.	Jia, D.; Jurkowska, R.Z.; Zhang, X.; Jeltsch, A.; Cheng, X. Structure of Dnmt3a bound to Dnmt3L suggests a model for de novo DNA methylation. *Nature* **2007**, *449*, 248–251. [CrossRef]

21.	Gu, T.P.; Guo, F.; Yang, H.; Wu, H.P.; Xu, G.F.; Liu, W.; Xie, Z.G.; Shi, L.; He, X.; Jin, S.G.; et al. The role of Tet3 DNA dioxygenase in epigenetic reprogramming by oocytes. *Nature* **2011**, *477*, 606–610. [CrossRef]

22.	Nottke, A.; Colaiacovo, M.P.; Shi, Y. Developmental roles of the histone lysine demethylases. *Development* **2009**, *136*, 879–889. [CrossRef] [PubMed]

23.	Gonzalez, T.L.; Sun, T.; Koeppel, A.F.; Lee, B.; Wang, E.T.; Farber, C.R.; Rich, S.S.; Sundheimer, L.W.; Buttle, R.A.; Chen, Y.I.; et al. Sex differences in the late first trimester human placenta transcriptome. *Biol. Sex Differ.* **2018**, *9*, 4. [CrossRef] [PubMed]

24.	Nishi, H.; Yahata, N.; Ohyashiki, K.; Isaka, K.; Shiraishi, K.; Ohyashiki, J.H.; Toyama, K.; Takayama, M. Comparison of telomerase activity in normal chorionic villi to trophoblastic diseases. *Int. J. Oncol.* **1998**, *12*, 81–85. [CrossRef] [PubMed]

25. Nishi, H.; Ohyashiki, K.; Fujito, A.; Yahata, N.; Ohyashiki, J.H.; Isaka, K.; Takayama, M. Expression of telomerase subunits and localization of telomerase activation in hydatidiform mole. *Placenta* **1999**, *20*, 317–323. [CrossRef]

26. Sebire, N.J.; Lindsay, I. Current issues in the histopathology of gestational trophoblastic tumors. *Fetal Pediatr. Pathol.* **2010**, *29*, 30–44. [CrossRef] [PubMed]

27. Knofler, M.; Pollheimer, J. Human placental trophoblast invasion and differentiation: A particular focus on Wnt signaling. *Front. Genet.* **2013**, *4*, 190. [CrossRef] [PubMed]

28. Buza, N.; Hui, P. Immunohistochemistry and other ancillary techniques in the diagnosis of gestational trophoblastic diseases. *Semin. Diagn. Pathol.* **2014**, *31*, 223–232. [CrossRef]

29. Petts, G.; Fisher, R.A.; Short, D.; Lindsay, I.; Seckl, M.J.; Sebire, N.J. Histopathological and immunohistochemical features of early hydatidiform mole in relation to subsequent development of persistent gestational trophoblastic disease. *J. Reprod. Med.* **2014**, *59*, 213–220.

30. Fock, V.; Plessl, K.; Fuchs, R.; Dekan, S.; Milla, S.K.; Haider, S.; Fiala, C.; Knofler, M.; Pollheimer, J. Trophoblast subtype-specific EGFR/ERBB4 expression correlates with cell cycle progression and hyperplasia in complete hydatidiform moles. *Hum. Reprod.* **2015**, *30*, 789–799. [CrossRef]

31. Velicky, P.; Meinhardt, G.; Plessl, K.; Vondra, S.; Weiss, T.; Haslinger, P.; Lendl, T.; Aumayr, K.; Mairhofer, M.; Zhu, X.; et al. Genome amplification and cellular senescence are hallmarks of human placenta development. *PLoS Genet.* **2018**, *14*, e1007698. [CrossRef]

32. Lim, D.H.; Maher, E.R. Human imprinting syndromes. *Epigenomics* **2009**, *1*, 347–369. [CrossRef] [PubMed]

33. Ferguson-Smith, A.C. Genomic imprinting: The emergence of an epigenetic paradigm. *Nat. Rev. Genet.* **2011**, *12*, 565–575. [CrossRef] [PubMed]

34. Tomizawa, S.; Sasaki, H. Genomic imprinting and its relevance to congenital disease, infertility, molar pregnancy and induced pluripotent stem cell. *J. Hum. Genet.* **2012**, *57*, 84–91. [CrossRef] [PubMed]

35. Fan, J.B.; Surti, U.; Taillon-Miller, P.; Hsie, L.; Kennedy, G.C.; Hoffner, L.; Ryder, T.; Mutch, D.G.; Kwok, P.Y. Paternal origins of complete hydatidiform moles proven by whole genome single-nucleotide polymorphism haplotyping. *Genomics* **2002**, *79*, 58–62. [CrossRef] [PubMed]

36. Hauzman, E.E.; Papp, Z. Conception without the development of a human being. *J. Perinat. Med.* **2008**, *36*, 175–177. [CrossRef]

37. Carey, L.; Nash, B.M.; Wright, D.C. Molecular genetic studies of complete hydatidiform moles. *Transl. Pediatr.* **2015**, *4*, 181–188. [PubMed]

38. Gu, H.; Gao, J.; Guo, W.; Zhou, Y.; Kong, Q. The expression of DNA methyltransferases3A is specifically downregulated in chorionic villi of early embryo growth arrest cases. *Mol. Med. Rep.* **2017**, *16*, 591–596. [CrossRef] [PubMed]

39. Dimitrova, E.; Turberfield, A.H.; Klose, R.J. Histone demethylases in chromatin biology and beyond. *EMBO Rep.* **2015**, *16*, 1620–1639. [CrossRef]

40. Slim, R.; Mehio, A. The genetics of hydatidiform moles: New lights on an ancient disease. *Clin. Genet.* **2007**, *71*, 25–34. [CrossRef]

41. Messaed, C.; Akoury, E.; Djuric, U.; Zeng, J.; Saleh, M.; Gilbert, L.; Seoud, M.; Qureshi, S.; Slim, R. NLRP7, a nucleotide oligomerization domain-like receptor protein, is required for normal cytokine secretion and co-localizes with Golgi and the microtubule-organizing center. *J. Biol. Chem.* **2011**, *286*, 43313–43323. [CrossRef]

42. Kalogiannidis, I.; Kalinderi, K.; Kalinderis, M.; Miliaras, D.; Tarlatzis, B.; Athanasiadis, A. Recurrent complete hydatidiform mole: Where we are, is there a safe gestational horizon? Opinion and mini-review. *J. Assist. Reprod. Genet.* **2018**, *35*, 967–973. [CrossRef] [PubMed]

43. Mor, G.; Cardenas, I.; Abrahams, V.; Guller, S. Inflammation and pregnancy: The role of the immune system at the implantation site. *Ann. N. Y. Acad. Sci.* **2011**, *1221*, 80–87. [CrossRef] [PubMed]

44. Racicot, K.; Kwon, J.Y.; Aldo, P.; Silasi, M.; Mor, G. Understanding the complexity of the immune system during pregnancy. *Am. J. Reprod. Immunol.* **2014**, *72*, 107–116. [CrossRef] [PubMed]

45. Aghaeepour, N.; Ganio, E.A.; McIlwain, D.; Tsai, A.S.; Tingle, M.; Van Gassen, S.; Gaudilliere, D.K.; Baca, Q.; McNeil, L.; Okada, R.; et al. An immune clock of human pregnancy. *Sci. Immunol.* **2017**, *2*. [CrossRef] [PubMed]

46. Fulop, V.; Vermes, G.; Demeter, J. The relationship between inflammatory and immunological processes during pregnancy. Practical aspects. *Orv. Hetil.* **2019**, *160*, 1247–1259. [PubMed]

47. Rusterholz, C.; Hahn, S.; Holzgreve, W. Role of placentally produced inflammatory and regulatory cytokines in pregnancy and the etiology of preeclampsia. *Semin. Immunopathol.* **2007**, *29*, 151–162. [CrossRef] [PubMed]

48. Than, N.G.; Romero, R.; Erez, O.; Weckle, A.; Tarca, A.L.; Hotra, J.; Abbas, A.; Han, Y.M.; Kim, S.S.; Kusanovic, J.P.; et al. Emergence of hormonal and redox regulation of galectin-1 in placental mammals: Implication in maternal-fetal immune tolerance. *Proc. Natl. Acad. Sci. USA* **2008**, *105*, 15819–15824. [CrossRef] [PubMed]

49. Than, N.G.; Romero, R.; Goodman, M.; Weckle, A.; Xing, J.; Dong, Z.; Xu, Y.; Tarquini, F.; Szilagyi, A.; Gal, P.; et al. A primate subfamily of galectins expressed at the maternal-fetal interface that promote immune cell death. *Proc. Natl. Acad. Sci. USA* **2009**, *106*, 9731–9736. [CrossRef] [PubMed]

50. Hunt, J.S.; Pace, J.L.; Gill, R.M. Immunoregulatory molecules in human placentas: Potential for diverse roles in pregnancy. *Int. J. Dev. Biol.* **2010**, *54*, 457–467. [CrossRef]

51. Than, N.G.; Balogh, A.; Romero, R.; Karpati, E.; Erez, O.; Szilagyi, A.; Kovalszky, I.; Sammar, M.; Gizurarson, S.; Matko, J.; et al. Placental protein 13 (PP13)—A placental immunoregulatory galectin protecting pregnancy. *Front. Immunol.* **2014**, *5*, 348. [CrossRef]

52. Cheng, S.B.; Sharma, S. Interleukin-10: A pleiotropic regulator in pregnancy. *Am. J. Reprod. Immunol.* **2015**, *73*, 487–500. [CrossRef] [PubMed]

53. Than, N.G.; Romero, R.; Balogh, A.; Karpati, E.; Mastrolia, S.A.; Staretz-Chacham, O.; Hahn, S.; Erez, O.; Papp, Z.; Kim, C.J. Galectins: Double-edged swords in the cross-roads of pregnancy complications and female reproductive tract inflammation and neoplasia. *J. Pathol. Transl. Med.* **2015**, *49*, 181–208. [CrossRef] [PubMed]

54. Zhang, Y.H.; Tian, M.; Tang, M.X.; Liu, Z.Z.; Liao, A.H. Recent insight into the role of the PD-1/PD-L1 pathway in feto-maternal tolerance and pregnancy. *Am. J. Reprod. Immunol.* **2015**, *74*, 201–208. [CrossRef] [PubMed]

55. Ferreira, L.M.R.; Meissner, T.B.; Tilburgs, T.; Strominger, J.L. HLA-G: At the interface of maternal-fetal tolerance. *Trends Immunol.* **2017**, *38*, 272–286. [CrossRef] [PubMed]

56. Le Bouteiller, P.; Bensussan, A. Up-and-down immunity of pregnancy in humans. *F1000Res* **2017**, *6*, 1216. [CrossRef]

57. Schumacher, A. Human chorionic gonadotropin as a pivotal endocrine immune regulator initiating and preserving fetal tolerance. *Int. J. Mol. Sci.* **2017**, *18*, 2166. [CrossRef]

58. Robertson, S.A.; Care, A.S.; Moldenhauer, L.M. Regulatory T cells in embryo implantation and the immune response to pregnancy. *J. Clin. Investig.* **2018**, *128*, 4224–4235. [CrossRef]

59. Takahashi, H.; Takizawa, T.; Matsubara, S.; Ohkuchi, A.; Kuwata, T.; Usui, R.; Matsumoto, H.; Sato, Y.; Fujiwara, H.; Okamoto, A.; et al. Extravillous trophoblast cell invasion is promoted by the CD44-hyaluronic acid interaction. *Placenta* **2014**, *35*, 163–170. [CrossRef]

60. Mori, A.; Nishi, H.; Sasaki, T.; Nagamitsu, Y.; Kawaguchi, R.; Okamoto, A.; Kuroda, M.; Isaka, K. HLA-G expression is regulated by miR-365 in trophoblasts under hypoxic conditions. *Placenta* **2016**, *45*, 37–41. [CrossRef]

61. Raghupathy, R.; Al-Mutawa, E.; Al-Azemi, M.; Makhseed, M.; Azizieh, F.; Szekeres-Bartho, J. Progesterone-induced blocking factor (PIBF) modulates cytokine production by lymphocytes from women with recurrent miscarriage or preterm delivery. *J. Reprod. Immunol.* **2009**, *80*, 91–99. [CrossRef]

62. Prins, J.R.; Gomez-Lopez, N.; Robertson, S.A. Interleukin-6 in pregnancy and gestational disorders. *J. Reprod. Immunol.* **2012**, *95*, 1–14. [CrossRef] [PubMed]

63. Whitten, A.E.; Romero, R.; Korzeniewski, S.J.; Tarca, A.L.; Schwartz, A.G.; Yeo, L.; Dong, Z.; Hassan, S.S.; Chaiworapongsa, T. Evidence of an imbalance of angiogenic/antiangiogenic factors in massive perivillous fibrin deposition (maternal floor infarction): A placental lesion associated with recurrent miscarriage and fetal death. *Am. J. Obstet. Gynecol.* **2013**, *208*, 310.e1–310.e11. [CrossRef] [PubMed]

64. Shemesh, A.; Tirosh, D.; Sheiner, E.; Tirosh, N.B.; Brusilovsky, M.; Segev, R.; Rosental, B.; Porgador, A. First trimester pregnancy loss and the expression of alternatively spliced NKp30 isoforms in maternal blood and placental tissue. *Front. Immunol.* **2015**, *6*, 189. [CrossRef] [PubMed]

65. Balogh, A.; Toth, E.; Romero, R.; Parej, K.; Csala, D.; Szenasi, N.L.; Hajdu, I.; Juhasz, K.; Kovacs, A.F.; Meiri, H.; et al. Placental Galectins Are Key Players in Regulating the Maternal Adaptive Immune Response. *Front. Immunol.* **2019**, *10*, 1240. [CrossRef] [PubMed]

66. Romero, R.; Mazor, M.; Brandt, F.; Sepulveda, W.; Avila, C.; Cotton, D.B.; Dinarello, C.A. Interleukin-1 alpha and interleukin-1 beta in preterm and term human parturition. *Am. J. Reprod. Immunol.* **1992**, *27*, 117–123. [CrossRef] [PubMed]

67. Romero, R.; Mazor, M.; Sepulveda, W.; Avila, C.; Copeland, D.; Williams, J. Tumor necrosis factor in preterm and term labor. *Am. J. Obstet. Gynecol.* **1992**, *166*, 1576–1587. [CrossRef]

68. Cherouny, P.H.; Pankuch, G.A.; Romero, R.; Botti, J.J.; Kuhn, D.C.; Demers, L.M.; Appelbaum, P.C. Neutrophil attractant/activating peptide-1/interleukin-8: Association with histologic chorioamnionitis, preterm delivery, and bioactive amniotic fluid leukoattractants. *Am. J. Obstet. Gynecol.* **1993**, *169*, 1299–1303. [CrossRef]

69. Gotsch, F.; Romero, R.; Kusanovic, J.P.; Erez, O.; Espinoza, J.; Kim, C.J.; Vaisbuch, E.; Than, N.G.; Mazaki-Tovi, S.; Chaiworapongsa, T.; et al. The anti-inflammatory limb of the immune response in preterm labor, intra-amniotic infection/inflammation, and spontaneous parturition at term: A role for interleukin-10. *J. Matern. Fetal Neonatal Med.* **2008**, *21*, 529–547. [CrossRef] [PubMed]

70. Gomez-Lopez, N.; Laresgoiti-Servitje, E.; Olson, D.M.; Estrada-Gutierrez, G.; Vadillo-Ortega, F. The role of chemokines in term and premature rupture of the fetal membranes: A review. *Biol. Reprod.* **2010**, *82*, 809–814. [CrossRef] [PubMed]

71. Romero, R.; Dey, S.K.; Fisher, S.J. Preterm labor: One syndrome, many causes. *Science* **2014**, *345*, 760–765. [CrossRef] [PubMed]

72. Romero, R.; Grivel, J.C.; Tarca, A.L.; Chaemsaithong, P.; Xu, Z.; Fitzgerald, W.; Hassan, S.S.; Chaiworapongsa, T.; Margolis, L. Evidence of perturbations of the cytokine network in preterm labor. *Am. J. Obstet. Gynecol.* **2015**, *213*, 836.e1–836.e18. [CrossRef] [PubMed]

73. Than, N.G.; Erez, O.; Wildman, D.E.; Tarca, A.L.; Edwin, S.S.; Abbas, A.; Hotra, J.; Kusanovic, J.P.; Gotsch, F.; Hassan, S.S.; et al. Severe preeclampsia is characterized by increased placental expression of galectin-1. *J. Matern. Fetal Neonatal Med.* **2008**, *21*, 429–442. [CrossRef] [PubMed]

74. Than, N.G.; Abdul Rahman, O.; Magenheim, R.; Nagy, B.; Fule, T.; Hargitai, B.; Sammar, M.; Hupuczi, P.; Tarca, A.L.; Szabo, G.; et al. Placental protein 13 (galectin-13) has decreased placental expression but increased shedding and maternal serum concentrations in patients presenting with preterm pre-eclampsia and HELLP syndrome. *Virchows Arch.* **2008**, *453*, 387–400. [CrossRef] [PubMed]

75. Than, N.G.; Romero, R.; Meiri, H.; Erez, O.; Xu, Y.; Tarquini, F.; Barna, L.; Szilagyi, A.; Ackerman, R.; Sammar, M.; et al. PP13, maternal ABO blood groups and the risk assessment of pregnancy complications. *PLoS ONE* **2011**, *6*, e21564. [CrossRef] [PubMed]

76. Hsu, P.; Nanan, R.K. Innate and adaptive immune interactions at the fetal-maternal interface in healthy human pregnancy and pre-eclampsia. *Front. Immunol.* **2014**, *5*, 125. [CrossRef]

77. Stoikou, M.; Grimolizzi, F.; Giaglis, S.; Schafer, G.; van Breda, S.V.; Hoesli, I.M.; Lapaire, O.; Huhn, E.A.; Hasler, P.; Rossi, S.W.; et al. Gestational diabetes mellitus Is associated with altered neutrophil activity. *Front. Immunol.* **2017**, *8*, 702. [CrossRef]

78. Geldenhuys, J.; Rossouw, T.M.; Lombaard, H.A.; Ehlers, M.M.; Kock, M.M. Disruption in the regulation of immune responses in the placental subtype of preeclampsia. *Front. Immunol.* **2018**, *9*, 1659. [CrossRef]

79. Tsuda, S.; Zhang, X.; Hamana, H.; Shima, T.; Ushijima, A.; Tsuda, K.; Muraguchi, A.; Kishi, H.; Saito, S. Clonally expanded decidual effector regulatory T cells increase in late gestation of normal pregnancy, but not in preeclampsia, in humans. *Front. Immunol.* **2018**, *9*, 1934. [CrossRef]

80. Nagymanyoki, Z.; Callahan, M.J.; Parast, M.M.; Fulop, V.; Mok, S.C.; Berkowitz, R.S. Immune cell profiling in normal pregnancy, partial and complete molar pregnancy. *Gynecol. Oncol.* **2007**, *107*, 292–297. [CrossRef]

81. Hussein, M.R.; Abd-Elwahed, A.R.; Abodeif, E.S.; Abdulwahed, S.R. Decidual immune cell infiltrate in hydatidiform mole. *Cancer Investig.* **2009**, *27*, 60–66. [CrossRef]

82. Sundara, Y.T.; Jordanova, E.S.; Hernowo, B.S.; Gandamihardja, S.; Fleuren, G.J. Decidual infiltration of FoxP3(+) regulatory T cells, CD3(+) T cells, CD56(+) decidual natural killer cells and Ki-67 trophoblast cells in hydatidiform mole compared to normal and ectopic pregnancies. *Mol. Med. Rep.* **2012**, *5*, 275–281. [CrossRef] [PubMed]

83. Suryawanshi, H.; Morozov, P.; Straus, A.; Sahasrabudhe, N.; Max, K.E.A.; Garzia, A.; Kustagi, M.; Tuschl, T.; Williams, Z. A single-cell survey of the human first-trimester placenta and decidua. *Sci. Adv.* **2018**, *4*, eaau4788. [CrossRef] [PubMed]

84. Driscoll, S.G. Trophoblastic growths: Morphologic aspects and taxonomy. *J. Reprod. Med.* **1981**, *26*, 181–191. [PubMed]

85. Berkowitz, R.S.; Faris, H.M.; Hill, J.A.; Anderson, D.J. Localization of leukocytes and cytokines in chorionic villi of normal placentas and complete hydatidiform moles. *Gynecol. Oncol.* **1990**, *37*, 396–400. [CrossRef]
86. Sacks, S.H.; Chowdhury, P.; Zhou, W. Role of the complement system in rejection. *Curr. Opin. Immunol.* **2003**, *15*, 487–492. [CrossRef]
87. Qiao, S.; Nagasaka, T.; Harada, T.; Nakashima, N. p53, Bax and Bcl-2 expression, and apoptosis in gestational trophoblast of complete hydatidiform mole. *Placenta* **1998**, *19*, 361–369. [CrossRef]
88. Wargasetia, T.L.; Shahib, N.; Martaadisoebrata, D.; Dhianawaty, D.; Hernowo, B. Characterization of apoptosis and autophagy through Bcl-2 and Beclin-1 immunoexpression in gestational trophoblastic disease. *Iran. J. Reprod. Med.* **2015**, *13*, 413–420.
89. Davidovich, P.; Kearney, C.J.; Martin, S.J. Inflammatory outcomes of apoptosis, necrosis and necroptosis. *Biol. Chem.* **2014**, *395*, 1163–1171. [CrossRef]
90. Fishelson, Z.; Attali, G.; Mevorach, D. Complement and apoptosis. *Mol. Immunol.* **2001**, *38*, 207–219. [CrossRef]
91. Ronnett, B.M. Hydatidiform Moles: Ancillary Techniques to Refine Diagnosis. *Arch. Pathol. Lab. Med.* **2018**, *142*, 1485–1502. [CrossRef]
92. Ning, F.; Hou, H.; Morse, A.N.; Lash, G.E. Understanding and management of gestational trophoblastic disease. *F1000Res* **2019**, *8*, 1–8. [CrossRef] [PubMed]
93. Colley, E.; Hamilton, S.; Smith, P.; Morgan, N.V.; Coomarasamy, A.; Allen, S. Potential genetic causes of miscarriage in euploid pregnancies: A systematic review. *Hum. Reprod. Update* **2019**, *25*, 452–472. [CrossRef] [PubMed]
94. Meyer, M.; Kircher, M. Illumina sequencing library preparation for highly multiplexed target capture and sequencing. *Cold Spring Harb. Protoc.* **2010**, *2010*, pdb prot5448. [CrossRef]
95. Hargitai, B.; Marton, T.; Cox, P.M. Best practice no 178. Examination of the human placenta. *J. Clin. Pathol.* **2004**, *57*, 785–792. [CrossRef] [PubMed]
96. Redline, R.W. Placental pathology: A systematic approach with clinical correlations. *Placenta* **2008**, *29* Suppl. A, 86–91. [CrossRef]
97. Khong, T.Y.; Mooney, E.E.; Ariel, I.; Balmus, N.C.; Boyd, T.K.; Brundler, M.A.; Derricott, H.; Evans, M.J.; Faye-Petersen, O.M.; Gillan, J.E.; et al. Sampling and Definitions of Placental Lesions: Amsterdam Placental Workshop Group Consensus Statement. *Arch. Pathol. Lab. Med.* **2016**, *140*, 698–713. [CrossRef]
98. Soma, H.; Osawa, H.; Oguro, T.; Yoshihama, I.; Fujita, K.; Mineo, S.; Kudo, M.; Tanaka, K.; Akita, M.; Urabe, S.; et al. P57kip2 immunohistochemical expression and ultrastructural findings of gestational trophoblastic disease and related disorders. *Med. Mol. Morphol.* **2007**, *40*, 95–102. [CrossRef]
99. Madi, J.M.; Braga, A.; Paganella, M.P.; Litvin, I.E.; Wendland, E.M. Accuracy of p57(KIP)(2) compared with genotyping to diagnose complete hydatidiform mole: A systematic review and meta-analysis. *BJOG* **2018**, *125*, 1226–1233. [CrossRef]
100. Szabo, S.; Xu, Y.; Romero, R.; Fule, T.; Karaszi, K.; Bhatti, G.; Varkonyi, T.; Varkonyi, I.; Krenacs, T.; Dong, Z.; et al. Changes of placental syndecan-1 expression in preeclampsia and HELLP syndrome. *Virchows Arch.* **2013**, *463*, 445–458. [CrossRef]
101. Szabo, S.; Mody, M.; Romero, R.; Xu, Y.; Karaszi, K.; Mihalik, N.; Xu, Z.; Bhatti, G.; Fule, T.; Hupuczi, P.; et al. Activation of villous trophoblastic p38 and ERK1/2 signaling pathways in preterm preeclampsia and HELLP syndrome. *Pathol. Oncol. Res.* **2015**, *21*, 659–668. [CrossRef]
102. Szenasi, N.L.; Toth, E.; Balogh, A.; Juhasz, K.; Karaszi, K.; Ozohanics, O.; Gelencser, Z.; Kiraly, P.; Hargitai, B.; Drahos, L.; et al. Proteomic identification of membrane-associated placental protein 4 (MP4) as perlecan and characterization of its placental expression in normal and pathologic pregnancies. *PeerJ* **2019**, *7*, e6982. [CrossRef] [PubMed]
103. Kim, D.; Langmead, B.; Salzberg, S.L. HISAT: A fast spliced aligner with low memory requirements. *Nat. Methods* **2015**, *12*, 357–360. [CrossRef] [PubMed]
104. Li, H.; Handsaker, B.; Wysoker, A.; Fennell, T.; Ruan, J.; Homer, N.; Marth, G.; Abecasis, G.; Durbin, R.; Genome Project Data Processing Subgroup. The Sequence Alignment/Map format and SAMtools. *Bioinformatics* **2009**, *25*, 2078–2079. [CrossRef] [PubMed]
105. Liao, Y.; Smyth, G.K.; Shi, W. featureCounts: An efficient general purpose program for assigning sequence reads to genomic features. *Bioinformatics* **2014**, *30*, 923–930. [CrossRef] [PubMed]

106. Anders, S.; Huber, W. Differential expression analysis for sequence count data. *Genome Biol.* **2010**, *11*, R106. [CrossRef]
107. Love, M.I.; Huber, W.; Anders, S. Moderated estimation of fold change and dispersion for RNA-seq data with DESeq2. *Genome Biol.* **2014**, *15*, 550. [CrossRef]
108. Draghici, S.; Khatri, P.; Tarca, A.L.; Amin, K.; Done, A.; Voichita, C.; Georgescu, C.; Romero, R. A systems biology approach for pathway level analysis. *Genome Res.* **2007**, *17*, 1537–1545. [CrossRef]

Article

Toll-Like Receptor Mediated Activation of Natural Autoantibody Producing B Cell Subpopulations in an Autoimmune Disease Model

Szabina Erdő-Bonyár [1,2], Judit Rapp [1], Tünde Minier [2], Gábor Ráth [3], József Najbauer [1], László Czirják [2], Péter Németh [1], Timea Berki [1,*] and Diána Simon [1]

[1]   Department of Immunology and Biotechnology, Clinical Center, University of Pécs Medical School, H-7624 Pécs, Hungary; erdo-bonyar.szabina@pte.hu (S.E.-B.); rapp.judit@pte.hu (J.R.); najbauer.jozsef@pte.hu (J.N.); nemeth.peter@pte.hu (P.N.); simon.diana@pte.hu (D.S.)

[2]   Department of Rheumatology and Immunology, Clinical Center, University of Pécs Medical School, H-7632 Pécs, Hungary; minier.tunde@pte.hu (T.M.); czirjak.laszlo@pte.hu (L.C.)

[3]   Department of Pediatrics, Clinical Center, University of Pécs Medical School, H-7623 Pécs, Hungary; rath.gabor@pte.hu

*   Correspondence: berki.timea@pte.hu; Tel.: +36-72-536-291; Fax: +36-72-536-289

Received: 15 October 2019; Accepted: 3 December 2019; Published: 6 December 2019

Abstract: Altered expression and function of the Toll-like receptor (TLR) homologue CD180 molecule in B cells have been associated with autoimmune disorders. In this study, we report decreased expression of CD180 at protein and mRNA levels in peripheral blood B cells of diffuse cutaneous systemic sclerosis (dcSSc) patients. To analyze the effect of CD180 stimulation, together with CpG (TLR9 ligand) treatment, on the phenotype defined by CD19/CD27/IgD/CD24/CD38 staining, and function (CD69 and CD180 expression, cytokine and antibody secretion) of B cell subpopulations, we used tonsillar B cells. After stimulation, we found reduced expression of CD180 protein and mRNA in total B cells, and CD180 protein in B cell subpopulations. The frequency of CD180$^+$ cells was the highest in the CD19$^+$CD27$^+$IgD$^+$ non-switched (NS) B cell subset, and they showed the strongest activation after anti-CD180 stimulation. Furthermore, B cell activation via CD180 induced IL-6 and natural autoantibody secretion. Treatment with the combination of anti-CD180 antibody and CpG resulted in increased IL-6 and IL-10 secretion and natural autoantibody production of B cells. Our results support the role of CD180 in the induction of natural autoantibody production, possibly by NS B cells, and suggest an imbalance between the pathologic and natural autoantibody production in SSc patients.

Keywords: B cells; non-switched B cells; systemic sclerosis; dcSSc; TLR; CD180; RP105; CpG; IL-6; IL-10; natural autoantibodies; IgM; citrate synthase; DNA topoisomerase I

1. Introduction

The production of scleroderma-specific autoantibodies and secretion of pro-inflammatory and pro-fibrotic cytokines by B cells is a well-described result that reflects immune dysregulation affecting B cells in systemic sclerosis (SSc) [1–4]. However, a large number of autoantibodies directed against well-conserved functional structures of the cell (e.g., nucleosome, DNA, nuclear and mitochondrial proteins, and receptors), the so-called natural autoantibodies that serve a protective function, can also be detected in healthy subjects and are dysregulated in patients with systemic autoimmune diseases [4,5].

Autoantibody production is a widely investigated function of B cells in SSc, but less attention has been devoted to their activation by innate immune receptors, including Toll-like receptors (TLRs), that are involved in recognizing pathogen- and damage-associated molecular patterns. CD180, or RP105 (radioprotective 105 kDa), is a TLR-like membrane protein that lacks an intracellular Toll-IL-1R (TIR) signaling domain [6]. CD180 was originally defined as a B cell surface molecule mediating polyclonal B cell activation, proliferation, and immunoglobulin production [7,8]. It was later described as a TLR homologue also expressed by monocytes and dendritic cells (antigen presenting cells), and the expression of CD180 correlated with TLR4 expression. CD180 and its helper molecule, MD-1, interact directly with the TLR4 signaling complex, inhibiting its ability to bind microbial ligands; thus, it serves as a negative regulator of TLR4 responses of antigen-presenting cells [6,9].

Differential expression and functions of CD180 on B cells have been associated with immune-mediated pathologies, including infection, chronic inflammation, and autoimmune disorders [6]. The severity of the disease in systemic lupus erythematosus (SLE) patients correlated with the amount of CD180-negative B cells in the peripheral blood [10,11]. CD180-negative peripheral blood B cells were also increased in patients with Sjögren's syndrome; furthermore, these cells extensively infiltrated the salivary glands [12].

As the natural ligand of CD180 remains unknown, effects of crosslinking CD180 with monoclonal anti-CD180 antibody has been investigated. Anti-CD180 antibody activates over 85% of human and mouse B cells in vitro and induces robust immunoglobulin production [8]. The stimulation with anti-CD180 antibody synergizes with TLR9 ligands [8,13]. When CpG and anti-CD180 were used simultaneously, the proliferation of peripheral blood B cells was enhanced, and IgG and IgM production increased [13]. Simultaneous treatment with anti-CD180 antibody and LPS or CpG resulted in increased cytokine production of murine B cells [14].

In this study, we investigated the expression of CD180 at protein and mRNA levels in peripheral blood B cells of early diffuse cutaneous SSc (dcSSc) patients, and compared to healthy control (HC) B cells. We found that CD180 expression of dcSSc B cells was significantly lower than in HC B cells. To further investigate the role of CD180 in B cell activation, we used tonsillar B lymphocytes as a model. To investigate the CD180-mediated activation of B cell subsets, we used anti-CD180 antibody to ligate the receptor, and combined with treatment with CpG, a TLR9 ligand. Expression of CD180 in B cell subsets, and molecules of B cell activation, cytokine, and autoantibody production were analyzed. The frequency of CD180$^+$ cells was the highest in non-switched memory (NS) B cells, which showed the strongest activation (CD69$^+$) upon anti-CD180 stimulation. This activation was not influenced by the addition of CpG. Stimulation with anti-CD180 antibody alone, and in combination with CpG resulted in downregulation of CD180 protein and mRNA expression in total B cells, and decreased CD180 protein expression in B cell subsets. Activation via CD180 induced the elevation of IL-6 production and the anti-DNA topoisomerase I (anti-topo I) IgM natural autoantibody secretion, which were enhanced by the addition of CpG. Furthermore, treatment with the combination of anti-CD180 antibody and CpG resulted in increased IL-10 secretion and anti-citrate synthase IgM natural autoantibody production of B cells. Our results support the role of CD180 in the induction of natural autoantibody production, possibly by NS B cells, which are diminished in SSc patients, resulting in an imbalance between the pathologic and natural autoantibody production.

2. Results

2.1. CD180 Expression Is Decreased in dcSSc B Cells

Since it was described in other autoimmune diseases that differential expression of CD180 (RP105) might have a pathological role in B cell activation and autoantibody production [10–12,15], first we determined the CD180 expression of monocytes, T cells, and B cells. We compared the level (mean fluorescence intensity, MFI) of CD180 in PBMC samples of early, untreated dcSSc patients and HCs with flow cytometry. We found that the MFI of CD180 labeling was the highest in B cells, followed by monocytes, and T cells showed the lowest expression. The expression of CD180 in monocytes and T cells was similar in patients and HCs, while its level was significantly lower in B cells of dcSSc patients than in HC B cells (Figure 1A,B). Next, we examined the mRNA expression of the CD180 gene in purified B cells of early, untreated dcSSc patients and compared to HCs. We found highly downregulated expression of CD180 mRNA in dcSSc patients (Figure 1C).

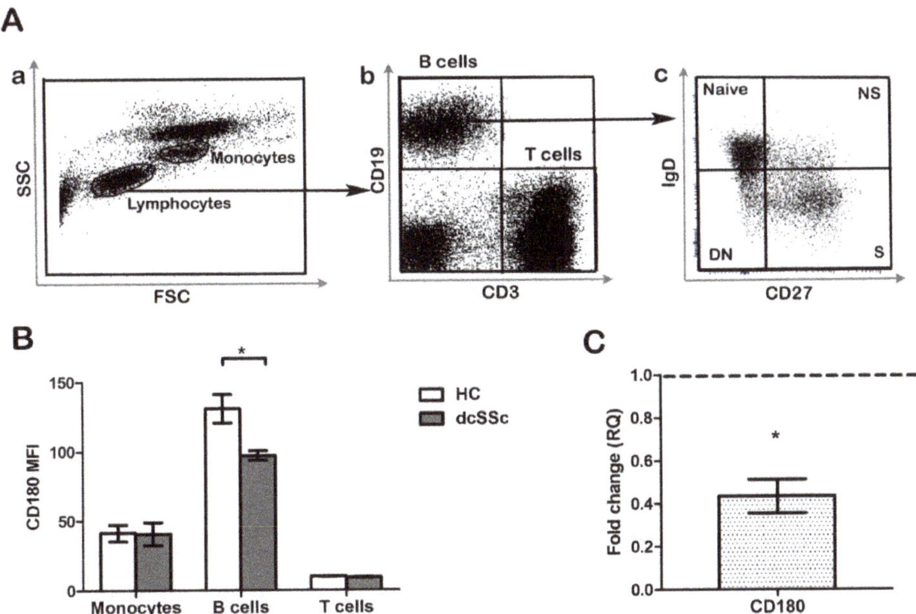

Figure 1. Analysis of CD180 expression in diffuse cutaneous systemic sclerosis (dcSSc) B cells. (**A**) Representative flow cytometry plots of peripheral blood leukocytes (**a**), B and T cells (**b**), and CD19$^+$ B cells stained with CD27 and IgD defining the following four subsets: CD27$^+$IgD$^+$ non-switched memory (NS) B cells, CD27$^+$IgD$^-$ switched memory (S) B cells, CD27$^-$IgD$^+$ naive B cells, and CD27$^-$IgD$^-$ double negative (DN) B cells (**c**). (**B**) Flow cytometric analysis of CD180 expression in peripheral blood B cells, T cells, and monocytes of early untreated dcSSc patients compared to healthy controls (HCs). (**C**) CD180 mRNA expression in B cells of early untreated dcSSc patients compared to HCs. Gene expression was normalized to HCs and the horizontal line (value 1) represents the expression of control samples. Changes in gene expression are shown as relative quantification (RQ) values. Data are shown as mean ± standard error of the mean (SEM), $n = 4$ HC and $n = 4$ dcSSc, * $p < 0.05$.

2.2. TLR Ligation Results in Reduced CD180 mRNA and Protein Expression of B Cells

The CD180-negative B cells were described as highly activated cells in SLE [11], and stimulation via CD180 is known to activate B cells [6]. Furthermore, TLR ligands were reported to downregulate the mRNA expression of CD180 molecule [16], thus we hypothesized that the decreased CD180 expression of dcSSc B cells could be a result of activation through TLRs. To investigate whether TLR stimulation leads to diminished expression of CD180 molecules in B cells, we stimulated tonsillar B cells with anti-CD180 antibody. We measured the expression of CD180 at protein and mRNA levels, and found that following anti-CD180 ligation, the MFI and mRNA levels of CD180 significantly decreased (Figure 2A,B). To study the influence of other TLR ligands on the activation via CD180, we co-treated the B cells with CpG, and found that the expression of CD180 was similar to anti-CD180-stimulated cells both at protein (Figure 2A) and mRNA (Figure 2B) levels. Treatment with CpG alone did not result in changes of CD180 MFI (Figure 2A) or CD180 mRNA (Figure 2B) levels in B cells.

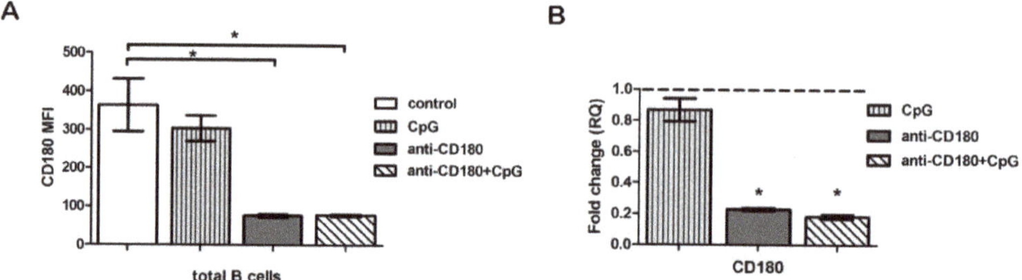

Figure 2. Effect of Toll-like receptor (TLR) stimulation on CD180 protein and mRNA expression. (**A**) CD180 expression of unstimulated (control), CpG, anti-CD180 antibody-stimulated, and anti-CD180 + CpG-treated (24 h) tonsillar B cells (mean fluorescence intensity, MFI). (**B**) CD180 mRNA expression in tonsillar B cells following CpG, anti-CD180, and anti-CD180 + CpG stimulation (24 h). Changes in gene expression are shown as RQ values, normalized to unstimulated controls. The horizontal line (value 1) represents the CD180 mRNA of unstimulated control samples. Data are shown as mean ± SEM, $n = 4$, * $p < 0.05$.

2.3. The Frequency of CD180$^+$ Cells Is the Highest in the Non-Switched Memory B Cell Subset

To assess phenotypical and functional alterations of B cells upon anti-CD180 stimulation, first we investigated the expression of CD180 in B cell subsets, defined by CD27 and IgD labeling (Figure 1A). Using tonsillar B cells, we analyzed the following subpopulations: CD27$^+$IgD$^+$ non-switched memory (NS) B cells, CD27$^+$IgD$^-$ switched memory (S) B cells, CD27$^-$IgD$^+$ naive B cells (N), and CD27$^-$IgD$^-$ double negative (DN) B cells. We found that the percentage of CD180$^+$ cells was significantly higher in NS B cells compared to all other subsets, namely, naive, S, and DN B cells (Figure 3A,B). Next, we measured the changes in the percentage of CD180$^+$ B cells in the NS, S, naive, and DN B cell subpopulations upon anti-CD180 stimulation, and found that the frequency of CD180$^+$ cells was significantly decreased in all four B cell subsets (Figure 3B). Addition of CpG to the anti-CD180 antibody-treated B cells did not result in further changes in the ratio of CD180$^+$ B cell subpopulations (Figure 3B). Treatment with CpG alone did not reduce the percentage of CD180$^+$ cells in the investigated B cell subsets (Figure 3B). The overall pattern of the changes in CD180 MFI in the investigated B cell subsets was similar to that found in the frequency of CD180$^+$ cells, but the CD180 MFI in unstimulated B cells was the highest in naive B cells (Figure 3C). We also investigated the expression of CD180 in regulatory B cells (Bregs). There is still no consensus on the phenotype of Bregs, multiple subsets with many similarities in phenotype and effector functions have been described [17]. In humans, both CD19$^+$CD24^{high}CD38^{high} [18] and CD19$^+$CD24^{high}CD27$^+$ [19] Bregs have been defined. Based on these findings, we analyzed the CD180 expression of Breg subsets with these phenotypes using flow cytometry. We found that the percentage of CD180+ cells and the MFI of CD180 labeling was significantly decreased after anti-CD180 antibody treatment, and also after combined treatment with anti-CD180 antibody and CpG (Figure 3D,E).

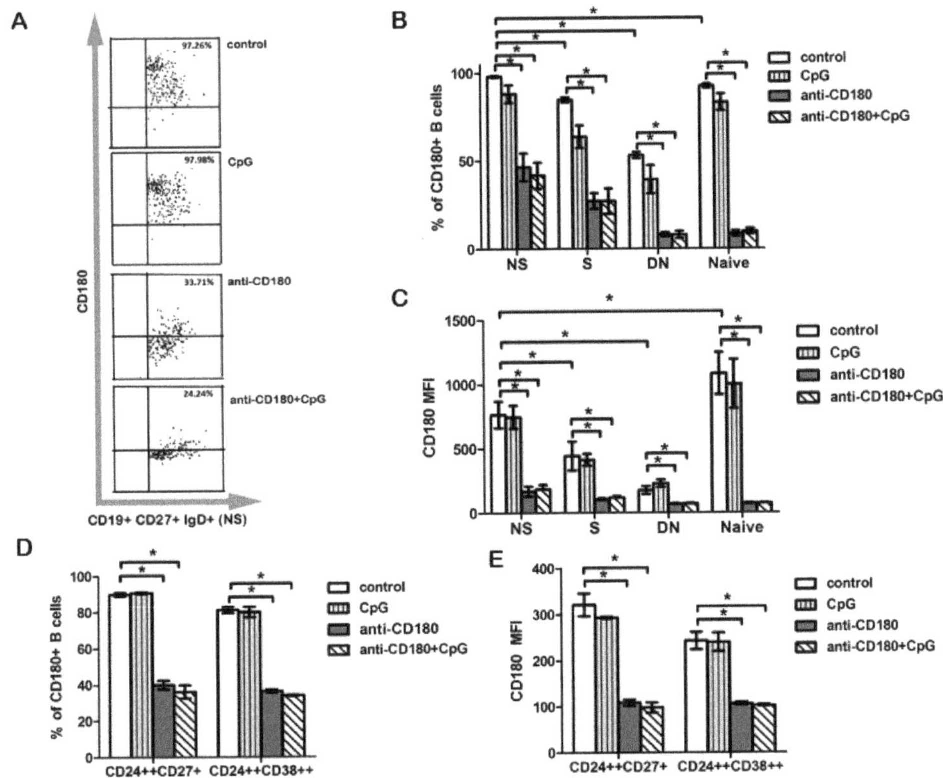

Figure 3. Effect of TLR stimulation on CD180 expression in B cell subpopulations. Flow cytometric analysis of the distribution of CD180$^+$ B cells and CD180 MFI in CD27$^+$IgD$^+$ non-switched memory (NS), CD27$^+$IgD$^-$ switched memory (S), CD27$^-$IgD$^+$ naive, and CD27$^-$IgD$^-$ double negative (DN) B cell subsets (**A–C**), and in CD24^{high}CD38^{high} and CD24^{high}CD27$^+$ Breg subsets (**D,E**). Representative dot plots show the changes of CD180 positivity in control and treated NS B cells (**A**). Changes of CD180$^+$ cell ratios (**B**) and CD180 MFI (**C**) in different B cell subsets defined by CD27 and IgD staining after CpG, anti-CD180 antibody stimulation or anti-CD180 + CpG treatment. Changes of CD180$^+$ cell ratios (**D**) and CD180 MFI (**E**) in Breg subsets defined by CD24, CD27, and CD38 staining after CpG, anti-CD180 antibody stimulation or anti-CD180 + CpG treatment. Data are presented as means ± SEM, $n = 4$, * $p < 0.05$.

2.4. Anti-CD180 Stimulation Resulted in Activation of B Cell Subsets

We also examined the activation of the B cell subsets defined by CD27 and IgD labeling after 24 h of anti-CD180, CpG, and anti-CD180 + CpG stimulation by detecting CD69, an early activation marker. Upon anti-CD180 stimulation, the frequency of CD69$^+$ cells was increased in all investigated subsets compared to CpG stimulation, showing the highest ratio in NS B cells (Figure 4A,B). This result is consistent with our observation that NS B cells show the highest ratio of CD180$^+$ cells. CpG, when used together with anti-CD180 antibody stimulation, did not alter the percentage of CD69$^+$ cells in NS, S, or DN B cells compared to anti-CD180 stimulation alone. However, a significant decrease in the frequency of CD69$^+$ naive B cells was observed, compared to anti-CD180 ligation alone (Figure 4B). The overall pattern of the changes in CD69 MFI was similar to that found in the frequency of CD69$^+$ cells, and after anti-CD180 antibody stimulation, the CD69 MFI was the highest in NS B cells. Yet, CD69 MFI was not significantly different between CpG- and anti-CD180 + CpG-stimulated DN B cells, while combined anti-CD180 and CpG treatment compared to anti-CD180 antibody stimulation alone significantly reduced the MFI of CD69 in DN B cells (Figure 4C).

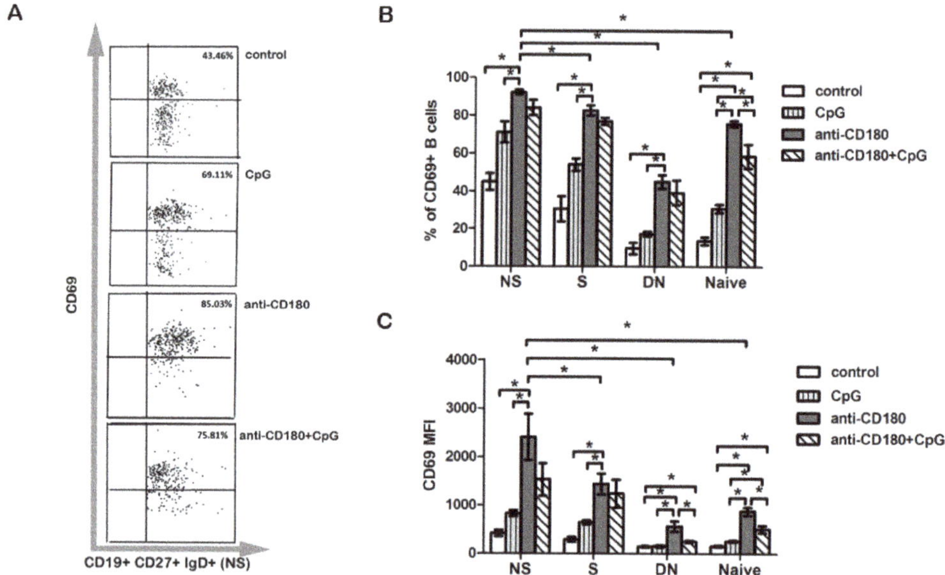

Figure 4. Expression of CD69 activation marker on B cell subpopulations. Representative dot plots show the changes of CD69 positivity in control and treated NS B cells (**A**). The percentage of CD69$^+$ (**B**) B cells and CD69 MFI (**C**) in CD27$^+$IgD$^+$ non-switched memory (NS), CD27$^+$IgD$^-$ switched memory (S), CD27$^-$IgD$^+$ naive, and CD27$^-$IgD$^-$ double negative (DN) B cell subsets defined by CD27 and IgD in anti-CD180 antibody, CpG-stimulated, and in anti-CD180 and CpG co-treated tonsillar B cells. Data are presented as means ± SEM, $n = 4$, * $p < 0.05$.

2.5. TLR Stimulation Differentially Affects IL-6 and IL-10 Production of B Cells

Recently, it was described that prominent IL-6 production by activated B cells promotes spontaneous germinal center formation and plasma cell differentiation, further supporting the role of this cytokine in the pathogenesis of systemic autoimmune diseases [20]. To test whether anti-CD180 stimulation influences B cell–derived IL-6 production, or induces the production of Breg cytokine IL-10, we measured the concentration of these cytokines in the supernatant of separated B cells after 24 h of stimulation. Anti-CD180 stimulation alone significantly increased the concentration of IL-6 in the supernatant of tonsillar B cells compared to both unstimulated and CpG-stimulated cells, while it had no significant effect on the production of IL-10. CpG on its own significantly elevated the concentration of IL-6 in the supernatant, but did not increase the production of IL-10. However, supplementing anti-CD180 antibody with CpG significantly augmented the production of both IL-6 and IL-10 compared to unstimulated, CpG-, and anti-CD180-stimulated B cells (Figure 5A,B).

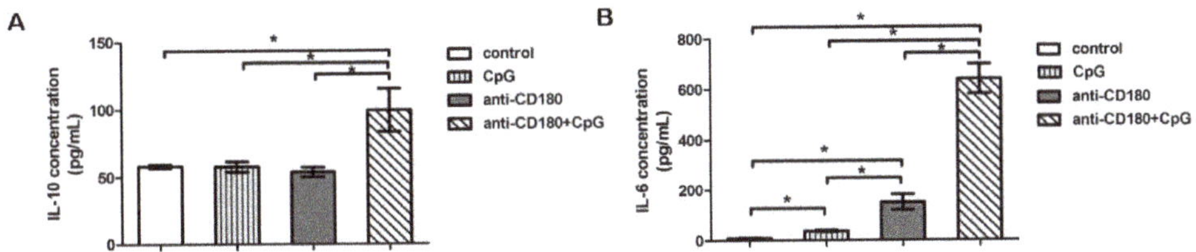

Figure 5. TLR induced cytokine production of total B cells. Detection of IL-6 (**A**) and IL-10 cytokines (**B**) in the supernatant of tonsillar B cells stimulated with CpG, anti-CD180 antibody, or anti-CD180 + CpG, or left unstimulated (control), as measured by ELISA. Data are presented as means ± SEM, $n = 4$, * $p < 0.05$.

2.6. CD180 Stimulation Induces Natural Autoantibody Production

Since CD180 stimulation resulted in the most pronounced activation of NS memory B cells, resembling the B1 B cell population, which includes cells responsible for natural IgM autoantibody production [21], we investigated the effect of CD180 ligation on the production of natural autoantibodies by tonsillar B cells. In our previous studies, we detected anti-citrate synthase (CS) IgM antibodies in HCs and patients with autoimmune diseases [22,23]. Here we measured the anti-CS IgM autoantibody in the supernatant of anti-CD180-stimulated and control tonsillar B cells after 7 days of culture, but did not find any differences between the control and anti-CD180-stimulated samples. Treatment with CpG alone resulted in a significant increase in the level of anti-CS IgM in the supernatant. Combined treatment with anti-CD180 antibody and CpG significantly increased the production of anti-CS IgM autoantibody compared to unstimulated, CpG-, and anti-CD180-stimulated B cells (Figure 6A).

Previously, we reported that different epitopes of topoisomerase I (Scl-70) induce not only pathologic autoantibodies in dcSSc patients, but the antigen has epitopes (including F4) that induce natural autoantibody production in healthy individuals as well [24]. Therefore, we also measured the level of anti-topoisomerase I F4 fragment (anti-topo I) antibody in the supernatant of CD180-stimulated B cells. Anti-CD180 itself significantly raised the level of anti-topo I IgM antibodies in the supernatant of tonsillar B cells similarly to CpG, and the level was further elevated with addition of CpG (Figure 6B). Further, we measured the production of pathologic autoantibodies in the supernatant of anti-CD180- and anti-CD180 + CpG-stimulated tonsillar B cells. Antinuclear antibodies (ANA), anti-dsDNA, and anti-nucleosome antibodies were not detectable in the supernatant of B cells under the investigated conditions.

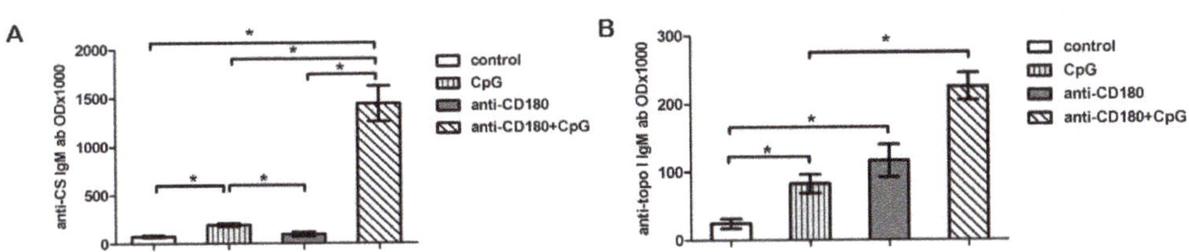

Figure 6. Induction of natural autoantibody production by TLR stimulation. Anti-citrate synthase IgM and (**A**) anti-DNA topoisomerase I IgM (**B**) production of B cells stimulated with CpG, anti-CD180 antibody, or anti-CD180 + CpG, or left unstimulated (control). Data are presented as means ± SEM, $n = 3$, * $p < 0.05$.

3. Discussion

In SSc, data on the role of innate immune molecules in the dysregulated B cell functions are scarce. The TLR homologue CD180 molecule activates the majority of B cells, resulting in phenotypic and functional alterations [8,14,25]. CD180-negative B cells, macrophages, and dendritic cells were shown to be elevated in SLE and in lupus-prone mice [15,26]. Here, we demonstrated that the expression of CD180 is decreased in early dcSSc peripheral blood B cells compared to HCs, while the percentage of CD180-negative monocytes was unaltered, emphasizing the potential role of CD180 in B cell dysfunction in SSc. CD180-negative B cells in SLE were described as highly activated cells [15], and CD180 can be internalized after ligation by anti-CD180 antibody [14]; thus, B cell activation via CD180 might be a possible explanation for the decreased expression of CD180 in SSc B cells.

We were the first to investigate the distribution of CD180 molecules in tonsillar B cell subpopulations defined by CD27 and IgD labeling, and found that the frequency of CD180$^+$ cells was the highest among NS cells. The percentage of CD180$^+$ cells was significantly decreased in all subsets upon anti-CD180 stimulation, suggesting that ligation of CD180 triggered its internalization. Interestingly, we also found that anti-CD180 stimulation downregulated the mRNA expression of CD180, strengthening the possibility of autoregulation of CD180 expression by B cells. Signaling via TLRs can interfere with CD180 signals, since ligands of TLR7 and TLR9 have already been shown to downregulate CD180 expression in human peripheral blood B cells [16]. Nevertheless, according to our results, co-treatment with anti-CD180 and CpG reduced the expression of CD180 to a similar level as anti-CD180 did on its own, while CpG treatment alone had no significant effect on the expression of CD180. All investigated B cell subsets were activated by anti-CD180 and anti-CD180 and CpG co-treatment, as indicated by CD69 expression.

Anti-CD180 antibody is known to activate marginal zone (MZ) B cells and induces overexpression of CD69 in mice [14]. NS B cells in human peripheral blood represent MZ-derived B cells [21], and according to our results, ratio of CD69$^+$ cells was the highest among NS B cells upon anti-CD180 stimulation. Moreover, the combination of anti-CD180 stimulation with CpG showed no additional effect on the activation of NS B cells.

We have already described that the percentage of NS B cells is lower in SSc than in HC [27]. NS B cells resemble B1 B cells with innate-like features [21], suggesting that they produce natural autoantibodies. Natural IgM autoantibodies are polyreactive and serve as scavengers of damaged molecules and cells, and therefore have been implicated in the control of inflammation and autoimmune diseases [28]. The therapeutic effects of intravenous immunoglobulins (IVIg) could partly be due to the appropriate pool of natural autoantibodies [29], and IVIg seems to be beneficial in SSc [30]; therefore, we also examined the effect of anti-CD180 and anti-CD180 and CpG co-treatment on the production of natural and pathologic autoantibodies of tonsillar B cells. Pathologic autoantibodies were undetectable in the supernatant of cells, while anti-CD180 and CpG co-treatment significantly enhanced the production of both anti-CS and anti-topoisomerase I IgM antibodies, suggesting the synergistic beneficial effect of anti-CD180 antibody and TLR9 ligand on natural antibody production of B cells. Interestingly, anti-CD180 stimulation increased the level of anti-topoisomerase I IgM, while it had no influence on anti-CS IgM production. However, CpG stimulation increased the level of anti-topoisomerase I IgM and anti-CS IgM as well. CS molecule is not the target of a disease-specific pathologic antibody, while natural anti-topoisomerase I autoantibodies recognize distinct epitopes of topoisomerase I, which is the target antigen of an SSc-specific pathologic autoantibody (anti-Scl-70) [24]. The onset of autoimmune diseases may correlate with a switch from production of self-reactive low-affinity IgM to high-affinity IgG isotype autoantibodies by B cells. B cell activation via CD180 may have a role in regulating the level of natural IgM antibodies directed against the target antigens of pathologic antibodies.

The serum concentration of IL-6 is elevated both in SSc [31] and SLE patients [32]; furthermore, IL-6 production by B cells drives autoimmune germinal center formation in a mouse model of SLE, promoting the disease [20]. We found that ani-CD180 stimulation enhanced the IL-6 production by tonsillar B cells, which was further augmented by the addition of CpG. IL-6 was shown to have a pronounced effect on plasma cell survival [33], and if combined with IL-2 or IL-10, IL-6 enhanced the generation of early plasma cells [34]. Moreover, in SLE patients, endogenous IL-6 produced by B cells bound to IL-6R of B cells and drove them into terminal differentiation [35]. Consequently, B cell activation via CD180 alone or together with the TLR9 ligand may help plasma cell differentiation and antibody production. Bregs are decreased and functionally impaired in SSc patients [36], thus we also investigated the effect of anti-CD180 treatment on the IL-10 production of B cells. We found that CpG alone did not induce the production of IL-10, which is in agreement with the findings of Gantner et al. [37]. The concentration of IL-10 was increased only when anti-CD180 antibody was combined with CpG, suggesting their synergisitc effect on Bregs. However, CD180 expression was significantly decreased in CD19$^+$CD24^{high}CD38^{high} and CD19$^+$CD24^{high}CD27$^+$ Breg subsets after anti-CD180 antibody treatment, and also after combined treatment with anti-CD180 antibody and CpG, suggesting that anti-CD180 antibody stimulation alone may influence Breg functions other than the production of IL-10.

Our findings suggest that the TLR homologue CD180 molecule may be involved in B cell dysfunction in early dcSSc. We also described the effects of anti-CD180 activation on natural autoantibody and cytokine production of tonsillar B cells and conclude that our data raise the possibility that anti-CD180 stimulation might shift the type of antibodies produced by B cells towards the natural autoantibodies.

4. Materials and Methods

4.1. Patients

Diffuse cutaneous systemic sclerosis patients (dcSSc) enrolled for the gene expression and flow cytometric studies all fulfilled the 2013 ACR/EULAR SSc classification criteria. Disease duration was 1.75 ± 0.96 years, age at enrollment was 39 ± 22.73 years. None of the patients were on current or previous immunosuppressive therapy. The controls were age- and sex-matched healthy individuals (HC). All participants gave their informed consent to the study following approval by the Hungarian National Ethics Committee (ETT TUKEB 47861-6/2018/EKU) and Regional Clinical Research Ethics Committee (7724-PTE 2019).

4.2. Mononuclear Cell Isolation, B Cell Separation

Peripheral blood mononuclear cells (PBMCs) were isolated by Ficoll-Paque Plus (GE Healthcare, Chicago, IL, USA) density gradient centrifugation of peripheral blood samples (SSc $n = 4$, HC $n = 4$). Negative selection of B cells was performed using the MACS B cell isolation kit II (Miltenyi Biotech, Bergisch Gladbach, Germany), according to the manufacturer's protocol. B cell purity was >95%.

Tonsil tissue was collected from asymptomatic children who underwent routine tonsillectomy at the Department of Pediatrics, University of Pécs. After surgical removal, the tonsils were transported immediately to the laboratory and they were prepared the same day.

Tonsils were manually homogenized, and the cell suspension was filtered through a 70 μm sterile cell strainer, followed by isolation of mononuclear cells and separation of B cells, as described above.

4.3. Cell Stimulation

For RNA isolation, B cell subset analysis and cytokine measurements, 3×10^5 B cells were stimulated with LEAF Purified anti-human CD180 (RP105) antibody (Clone: MHR73-11) (BioLegend, San Diego, CA, USA) at 1 μg/mL (hereafter referred to as anti-CD180) or anti-CD180 in combination with 1 μg/mL CpG (Hu/Ms CpG-B DNA, ODN2006, Hycult Biotech, Wayne, NJ, USA) or with CpG alone or were left unstimulated for 24 h at 37 °C. To assess natural and pathologic antibody production, 4×10^5 B cells were stimulated with anti-CD180, or CpG or anti-CD180 in combination with CpG, or left unstimulated for 7 days at 37 °C.

4.4. RNA Isolation, cDNA Synthesis, and qPCR

Total RNA was extracted from isolated B cells using the NucleoSpin RNA XS kit (Macherey-Nagel Inc, Bethlehem, PA, USA). Following cDNA generation (High Capacity cDNA Reverse Transcription Kit,

Thermo Fisher Scientific, Waltham, MA, USA), the CD180 mRNA expression was analyzed individually in dcSSc patients ($n = 4$) and HCs ($n = 4$) using Applied Biosystems TaqMan Gene Expression Assays (Thermo Fisher Scientific, Waltham, MA, USA). To determine the CD180 mRNA expression of tonsillar B cells ($n = 4$), the SensiFAST SYBR Lo-ROX Kit (Bioline, London, UK) was used. Amplifications were performed using an Applied Biosystems 7500 RT-PCR System (Thermo Fisher Scientific, Waltham, MA, USA). Gene expression was analyzed with 7500 Software v2.0.6 (Thermo Fisher Scientific, Waltham, MA, USA) and normalized to GAPDH (a "housekeeping" gene) as reference. Fold changes (RQ) were calculated according to the 2-ddC$_T$ method.

4.5. Flow Cytometric Analysis

To measure the expression of CD180 of B cells, PBMCs from dcSSc patients ($n = 4$) and HCs ($n = 4$) were labeled with the combination of anti-human CD19-AmCyan (SJ25C1, Becton Dickinson, Franklin Lakes, NJ, USA) and anti-human CD180-PE (G28-8, Becton Dickinson, Franklin Lakes, NJ, USA) antibodies. Multiparametric flow cytometry was performed on tonsillar B cells ($n = 4$) with antibodies specific for CD19, CD27, IgD, CD24, CD38, CD180, and CD69. Purity of B cells was assessed using anti-human CD19-AmCyan antibody. To distinguish between memory B cell subsets and to evaluate the expression of CD180 and the activation marker CD69, four-color analysis was conducted using the combination of anti-human CD27-PE/Cy7 (M-T271, BioLegend, San Diego, CA, USA), anti-human IgD-PerCP (IA6-2, BioLegend, San Diego, CA, USA), anti-CD180-PE (G28-8, Becton Dickinson, Franklin Lakes, NJ, USA), and anti-human CD69-APC/Cy7 (FN50, BioLegend, San Diego, CA, USA), while Breg subsets were defined by anti-human CD27-PE/Cy7, anti-human CD24-BV421 (ML5, BioLegend, San Diego, CA, USA), and anti-human CD38-APC/Cy7 (HIT2, BioLegend, San Diego, CA, USA) antibodies following the manufacturer's instructions. Briefly, PBMCs or separated B cells were incubated with the appropriate antibodies for 30 min on ice, washed twice in phosphate-buffered saline (PBS), and fixed with FACSFix (0.5% PFA in PBS). Fluorescence of the labeled cells was recorded using a FACS Canto II flow cytometer (Becton Dickinson, Franklin Lakes, NJ, USA) and analyzed with FCS Express 6 software (De Novo Software, Pasadena, CA, USA).

4.6. IL-10 and IL-6 ELISA

Supernatant of anti-CD180, CpG, and CpG + anti-CD180 stimulated, and unstimulated B cells were collected ($n = 4$) and stored at −80 °C until being measured. IL-10 and IL-6 production was quantified using Human IL-10 and IL-6 DuoSet ELISA kits (Bio-Techne, Minneapolis, MN, USA) according to the manufacturer's protocols. The reaction was developed with TMB and measured at 450 nm using an iEMS MF microphotometer (Thermo Labsystem, Beverly, MA, USA).

4.7. Measurement of Natural and Pathologic Autoantibodies

The supernatant obtained from tonsillar B cells treated with anti-CD180, CpG, or CpG + anti-CD180, or left unstimulated were collected ($n = 3$) and stored at −80 °C until being measured. The levels of anti-citrate synthase (anti-CS) IgM autoantibodies and anti-topoisomerase I (fragment F4) IgM autoantibodies were determined with in-house ELISAs, as previously described [23,24]. The amounts of autoantibodies against dsDNA, nucleosome, and antinuclear antibodies (ANA) in the supernatant of tonsillar B cells were measured using commercial ELISA kits (ORGENTEC Diagnostika GmbH, Mainz, Germany). The reaction was developed with TMB and measured at 450 nm using an iEMS MF microphotometer (Thermo Labsystem, Beverly, MA, USA).

4.8. Statistical Analysis

Statistical evaluation was performed with SPSS v. 25.0 statistics package (IBM, Armonk, NY, USA) using Student's *t*-tests and ANOVA where *p*-values < 0.05 were considered significant.

Author Contributions: Conceptualization: D.S. and T.B.; methodology: D.S., J.R., and T.B.; software: D.S. and T.B.; validation: D.S. and T.B.; formal analysis: D.S. and T.B.; investigation: S.E.-B. and D.S.; resources: T.B., P.N., and L.C.; data curation: S.E.-B., D.S., T.M., and G.R.; writing—original draft preparation: D.S., S.E.-B., and T.B.; writing—review and editing: D.S., T.B., and J.N.; visualization: S.E.-B., D.S., and T.B.; supervision: D.S. and T.B.; project administration: D.S. and T.B.; funding acquisition: T.B., L.C., and P.N.

Funding: This work was supported by the Hungarian Scientific Research Fund (OTKA) K-75912, K-112939 to L.C. and K-105962 to T.B. The work was also supported by the European Union, co-financed by the European Social Fund as part of the project "PEPSYS—Complexity of peptide-signalization and its role in systemic diseases" of GINOP 2.3.2-15-2016-00050 and EFOP 3.6.1-16-2016-00004 grants.

Conflicts of Interest: The authors declare no conflicts of interest. The funders had no role in the design of the study; in the collection, analyses, or interpretation of data; in the writing of the manuscript; or in the decision to publish the results.

Abbreviations

ANA	Antinuclear antibodies
Breg	Regulatory B cell
CpG	5′—C—phosphate—G—3′ dinucleotide
CS	Citrate synthase
dcSSc	Diffuse cutaneous systemic sclerosis
DN B	Double negative B cell
ELISA	Enzyme-linked immunosorbent assay
FSC	Forward scatter
GAPDH	Glyceraldehyde 3-phosphate dehydrogenase
HC	Healthy control
Hu	Human
LEAF	Low endotoxin, azide-free
MFI	Mean fluorescence intensity
Ms	Mouse
MZ	Marginal zone
NS B	Non-switched B cell
PBMC	Peripheral blood mononuclear cells
PFA	Paraformaldehyde
qPCR	Quantitative polymerase chain reaction
RQ	Relative quantification
S B	Switched B cell
SLE	Systemic lupus erythematosus
SSc	Systemic sclerosis
SSC	Side scatter
TLR	Toll-like receptor
TMB	3,3′,5,5′-Tetramethylbenzidine
Topo I	Topoisomerase I

References

1. Varga, J.; Trojanowska, M.; Kuwana, M. Pathogenesis of systemic sclerosis: Recent insights of molecular and cellular mechanisms and therapeutic opportunities. *J. Scleroderma Relat. Disord.* **2017**, *2*, 137–152. [CrossRef]
2. Cabral-Marques, O.; Riemekasten, G. Functional autoantibodies directed against cell surface receptors in systemic sclerosis. *J. Scleroderma Relat. Disord.* **2017**, *2*, 160–168. [CrossRef]
3. Johnson, M.E.; Grassetti, A.V.; Taroni, J.N.; Lyons, S.M.; Schweppe, D.; Gordon, J.K.; Spiera, R.F.; Lafyatis, R.; Anderson, P.J.; Gerber, S.A.; et al. Stress granules and RNA processing bodies are novel autoantibody targets in systemic sclerosis. *Arthritis Res.* **2016**, *18*, 27. [CrossRef] [PubMed]
4. Senécal, J.; Hoa, S.; Yang, R.; Koenig, M. Pathogenic roles of autoantibodies in systemic sclerosis: Current understandings in pathogenesis. *J. Scleroderma Relat. Disord.* **2019**. [CrossRef]
5. Mannoor, K.; Xu, Y.; Chen, C. Natural autoantibodies and associated B cells in immunity and autoimmunity. *Autoimmunity* **2013**, *46*, 138–147. [CrossRef]

6. Schultz, T.E.; Blumenthal, A. The RP105/MD-1 complex: Molecular signaling mechanisms and pathophysiological implications. *J. Leukoc. Biol.* **2017**, *101*, 183–192. [CrossRef]

7. Valentine, M.A.; Clark, E.A.; Shu, G.L.; Norris, N.A.; Ledbetter, J.A. Antibody to a novel 95-kDa surface glycoprotein on human B cells induces calcium mobilization and B cell activation. *J. Immunol.* **1988**, *140*, 4071–4078.

8. Chaplin, J.W.; Kasahara, S.; Clark, E.A.; Ledbetter, J.A. Anti-CD180 (RP105) activates B cells to rapidly produce polyclonal Ig via a T cell and MyD88-independent pathway. *J. Immunol.* **2011**, *187*, 4199–4209. [CrossRef]

9. Divanovic, S.; Trompette, A.; Atabani, S.F.; Madan, R.; Golenbock, D.T.; Visintin, A.; Finberg, R.W.; Tarakhovsky, A.; Vogel, S.N.; Belkaid, Y.; et al. Negative regulation of Toll-like receptor 4 signaling by the Toll-like receptor homolog RP105. *Nat. Immunol.* **2005**, *6*, 571–578. [CrossRef]

10. Fujita, K.; Akasaka, Y.; Kuwabara, T.; Wang, B.; Tanaka, K.; Kamata, I.; Yokoo, T.; Kinoshita, T.; Iuchi, A.; Akishima-Fukasawa, Y.; et al. Pathogenesis of lupus-like nephritis through autoimmune antibody produced by CD180-negative B lymphocytes in NZBWF1 mouse. *Immunol. Lett.* **2012**, *144*, 1–6. [CrossRef]

11. Koarada, S.; Tada, Y.; Ushiyama, O.; Morito, F.; Suzuki, N.; Ohta, A.; Miyake, K.; Kimoto, M.; Nagasawa, K. B cells lacking RP105, a novel B cell antigen, in systemic lupus erythematosus. *Arthritis Rheum.* **1999**, *42*, 2593–2600. [CrossRef]

12. Kikuchi, Y.; Koarada, S.; Nakamura, S.; Yonemitsu, N.; Tada, Y.; Haruta, Y.; Morito, F.; Ohta, A.; Miyake, K.; Horiuchi, T.; et al. Increase of RP105-lacking activated B cells in the peripheral blood and salivary glands in patients with Sjögren's syndrome. *Clin. Exp. Rheumatol.* **2008**, *26*, 5–12. [PubMed]

13. Eriksen, A.B.; Indrevær, R.L.; Holm, K.L.; Landskron, J.; Blomhoff, H.K. TLR9-signaling is required for turning retinoic acid into a potent stimulator of RP105 (CD180)-mediated proliferation and IgG synthesis in human memory B cells. *Cell Immunol.* **2012**, *279*, 87–95. [CrossRef] [PubMed]

14. Chaplin, J.W.; Chappell, C.P.; Clark, E.A. Targeting antigens to CD180 rapidly induces antigen-specific IgG, affinity maturation, and immunological memory. *J. Exp. Med.* **2013**, *210*, 2135–2146. [CrossRef] [PubMed]

15. Koarada, S.; Tada, Y. RP105-negative B cells in systemic lupus erythematosus. *Clin. Dev. Immunol.* **2012**, *2012*, 259186. [CrossRef] [PubMed]

16. You, M.; Dong, G.; Li, F.; Ma, F.; Ren, J.; Xu, Y.; Yue, H.; Tang, R.; Ren, D.; Hou, Y. Ligation of CD180 inhibits IFN-α signaling in a Lyn-PI3K-BTK-dependent manner in B cells. *Cell Mol. Immunol.* **2017**, *14*, 192–202. [CrossRef]

17. Rosser, E.C.; Mauri, C. Regulatory B cells: Origin, phenotype, and function. *Immunity* **2015**, *42*, 607–612. [CrossRef]

18. Blair, P.A.; Noreña, L.Y.; Flores-Borja, F.; Rawlings, D.J.; Isenberg, D.A.; Ehrenstein, M.R.; Mauri, C. CD19+CD24hiCD38hi B cells exhibit regulatory capacity in healthy individuals but are functionally impaired in systemic Lupus Erythematosus patients. *Immunity* **2010**, *32*, 129–140. [CrossRef]

19. Iwata, Y.; Matsushita, T.; Horikawa, M.; Dilillo, D.J.; Yanaba, K.; Venturi, G.M.; Szabolcs, P.M.; Bernstein, S.H.; Magro, C.M.; Williams, A.D.; et al. Characterization of a rare IL-10-competent B-cell subset in humans that parallels mouse regulatory B10 cells. *Blood* **2011**, *117*, 530–541. [CrossRef]

20. Arkatkar, T.; Du, S.W.; Jacobs, H.M.; Dam, E.M.; Hou, B.; Buckner, J.H.; Rawlings, D.J.; Jackson, S.W. B cell-derived IL-6 initiates spontaneous germinal center formation during systemic autoimmunity. *J. Exp. Med.* **2017**, *214*, 3207–3217. [CrossRef]

21. Weller, S.; Braun, M.C.; Tan, B.K.; Rosenwald, A.; Cordier, C.; Conley, M.E.; Plebani, A.; Kumararatne, D.S.; Bonnet, D.; Tournilhac, O.; et al. Human blood IgM "memory" B cells are circulating splenic marginal zone B cells harboring a prediversified immunoglobulin repertoire. *Blood* **2004**, *104*, 3647–3654. [CrossRef] [PubMed]

22. Czömpöly, T.; Olasz, K.; Nyárády, Z.; Simon, D.; Bovári, J.; Németh, P. Detailed analyses of antibodies recognizing mitochondrial antigens suggest similar or identical mechanism for production of natural antibodies and natural autoantibodies. *Autoimmun Rev.* **2008**, *7*, 463–467. [CrossRef] [PubMed]

23. Czömpöly, T.; Olasz, K.; Simon, D.; Nyárády, Z.; Pálinkás, L.; Czirják, L.; Berki, T.; Németh, P. A possible new bridge between innate and adaptive immunity: Are the anti-mitochondrial citrate synthase autoantibodies components of the natural antibody network? *Mol. Immunol.* **2006**, *43*, 1761–1768. [CrossRef] [PubMed]

24. Simon, D.; Czömpöly, T.; Berki, T.; Minier, T.; Peti, A.; Tóth, E.; Czirják, L.; Németh, P. Naturally occurring and disease-associated auto-antibodies against topoisomerase I: A fine epitope mapping study in systemic sclerosis and systemic lupus erythematosus. *Int. Immunol.* **2009**, *21*, 415–422. [CrossRef] [PubMed]

25. Roe, K.; Shu, G.L.; Draves, K.E.; Giordano, D.; Pepper, M.; Clark, E.A. Targeting Antigens to CD180 but Not CD40 Programs Immature and Mature B Cell Subsets to Become Efficient APCs. *J. Immunol.* **2019**, *203*, 1715–1729. [CrossRef]

26. Yang, Y.; Wang, C.; Cheng, P.; Zhang, X.; Li, X.; Hu, Y.; Xu, F.; Hong, F.; Dong, G.; Xiong, H. CD180 Ligation Inhibits TLR7- and TLR9-Mediated Activation of Macrophages and Dendritic Cells Through the Lyn-SHP-1/2 Axis in Murine Lupus. *Front. Immunol.* **2018**, *9*, 2643. [CrossRef] [PubMed]

27. Simon, D.; Balogh, P.; Bognár, A.; Kellermayer, Z.; Engelmann, P.; Németh, P.; Farkas, N.; Minier, T.; Lóránd, V.; Czirják, L.; et al. Reduced non-switched memory B cell subsets cause imbalance in B cell repertoire in systemic sclerosis. *Clin. Exp. Rheumatol.* **2016**, *34*, 30–36. [PubMed]

28. Ehrenstein, M.R.; Notley, C.A. The importance of natural IgM: Scavenger, protector and regulator. *Nat. Rev. Immunol.* **2010**, *10*, 778–786. [CrossRef]

29. Vani, J.; Elluru, S.; Negi, V.S.; Lacroix-Desmazes, S.; Kazatchkine, M.D.; Bayry, J.; Bayary, J.; Kaveri, S.V. Role of natural antibodies in immune homeostasis: IVIg perspective. *Autoimmun. Rev.* **2008**, *7*, 440–444. [CrossRef]

30. Baleva, M.; Nikolov, K. The role of intravenous immunoglobulin preparations in the treatment of systemic sclerosis. *Int. J. Rheumatol.* **2011**, *2011*, 829751. [CrossRef]

31. Sato, S.; Hasegawa, M.; Takehara, K. Serum levels of interleukin-6 and interleukin-10 correlate with total skin thickness score in patients with systemic sclerosis. *J. Derm. Sci.* **2001**, *27*, 140–146. [CrossRef]

32. Chun, H.Y.; Chung, J.W.; Kim, H.A.; Yun, J.M.; Jeon, J.Y.; Ye, Y.M.; Kim, S.H.; Park, H.S.; Suh, C.H. Cytokine IL-6 and IL-10 as biomarkers in systemic lupus erythematosus. *J. Clin. Immunol.* **2007**, *27*, 461–466. [CrossRef] [PubMed]

33. Cassese, G.; Arce, S.; Hauser, A.E.; Lehnert, K.; Moewes, B.; Mostarac, M.; Muehlinghaus, G.; Szyska, M.; Radbruch, A.; Manz, R.A. Plasma cell survival is mediated by synergistic effects of cytokines and adhesion-dependent signals. *J. Immunol.* **2003**, *171*, 1684–1690. [CrossRef] [PubMed]

34. Jego, G.; Bataille, R.; Pellat-Deceunynck, C. Interleukin-6 is a growth factor for nonmalignant human plasmablasts. *Blood* **2001**, *97*, 1817–1822. [CrossRef]

35. Kitani, A.; Hara, M.; Hirose, T.; Harigai, M.; Suzuki, K.; Kawakami, M.; Kawaguchi, Y.; Hidaka, T.; Kawagoe, M.; Nakamura, H. Autostimulatory effects of IL-6 on excessive B cell differentiation in patients with systemic lupus erythematosus: Analysis of IL-6 production and IL-6R expression. *Clin. Exp. Immunol.* **1992**, *88*, 75–83. [CrossRef]

36. Mavropoulos, A.; Simopoulou, T.; Varna, A.; Liaskos, C.; Katsiari, C.G.; Bogdanos, D.P.; Sakkas, L.I. Breg Cells Are Numerically Decreased and Functionally Impaired in Patients With Systemic Sclerosis. *Arthritis Rheumatol.* **2016**, *68*, 494–504. [CrossRef]

37. Gantner, F.; Hermann, P.; Nakashima, K.; Matsukawa, S.; Sakai, K.; Bacon, K.B. CD40-dependent and -independent activation of human tonsil B cells by CpG oligodeoxynucleotides. *Eur. J. Immunol.* **2003**, *33*, 1576–1585. [CrossRef]

International Journal of
Molecular Sciences

Article

Altered Cell Surface N-Glycosylation of Resting and Activated T Cells in Systemic Lupus Erythematosus

Enikő Szabó [1], Ákos Hornung [2], Éva Monostori [1], Márta Bocskai [2], Ágnes Czibula [1,*] and László Kovács [2,*]

[1] Institute of Genetics, Biological Research Centre of the Hungarian Academy of Sciences 6726 Szeged, Hungary

[2] Department of Rheumatology and Immunology, Faculty of Medicine, University of Szeged, 6725 Szeged, Hungary

* Correspondence: czibula.agnes@brc.hu (Á.C.); kovacs.laszlo@med.u-szeged.hu (L.K.)

Received: 13 August 2019; Accepted: 5 September 2019; Published: 10 September 2019

Abstract: Altered cell surface glycosylation in congenital and acquired diseases has been shown to affect cell differentiation and cellular responses to external signals. Hence, it may have an important role in immune regulation; however, T cell surface glycosylation has not been studied in systemic lupus erythematosus (SLE), a prototype of autoimmune diseases. Analysis of the glycosylation of T cells from patients suffering from SLE was performed by lectin-binding assay, flow cytometry, and quantitative real-time PCR. The results showed that resting SLE T cells presented an activated-like phenotype in terms of their glycosylation pattern. Additionally, activated SLE T cells bound significantly less galectin-1 (Gal-1), an important immunoregulatory lectin, while other lectins bound similarly to the controls. Differential lectin binding, specifically Gal-1, to SLE T cells was explained by the increased gene expression ratio of sialyltransferases and neuraminidase 1 (*NEU1*), particularly by elevated ST6 beta-galactosamide alpha-2,6-sialyltranferase 1 (*ST6GAL1*)/*NEU1* and ST3 beta-galactoside alpha-2,3-sialyltransferase 6 (*ST3GAL6*)/*NEU1* ratios. These findings indicated an increased terminal sialylation. Indeed, neuraminidase treatment of cells resulted in the increase of Gal-1 binding. Altered T cell surface glycosylation may predispose the cells to resistance to the immunoregulatory effects of Gal-1, and may thus contribute to the pathomechanism of SLE.

Keywords: systemic lupus erythematosus; T cells; glycosylation; sialylation; lectin binding; glycosylation enzymes; galectin 1

---

1. Introduction

Numerous congenital and acquired diseases show altered cell surface glycosylation, including several types of cancer and autoimmune syndromes [1,2]. Altered oligosaccharide structures have been identified in tumors and have proven to be diagnostic markers of malignant phenotypes [1,3,4]. Protein glycosylation has become an integral part of research in autoimmunity, as defective glycan structures have been described on serum immunoglobulins [5] and the different glycans at certain residues on IgG subclasses affected the effector function of autoantibodies [6,7]. Surface glycosylation of immune cells has also been studied, and glycoconjugates have been proven to play a role in many fields of cellular physiology, such as migration and signal transduction [8]. T cell functions can also be modulated by interaction between cell surface glycoproteins and endogenous lectins, including galectins [9].

Glycoconjugates are created in the endoplasmic reticulum, the Golgi apparatus, and on the cell surface by enzymes, including mannosidases, glycosyltransferases, sialyltransferases, and neuraminidases (NEU) [10]. The concerted action of these enzymes produces the specific sugar 'code' presented on the cell surface, which then regulates further signaling and adhesion properties of a particular cell type. The expression of enzymes participating in glycosylation can determine the sensitivity of the cells to numerous extracellular signals.

Galectin 1 (Gal-1) is a member of the β-galactoside binding mammalian lectin family with specific affinity to terminal N-acetyllactosamine motifs on multi-antennary cell surface glycans [11]. One of the major effects of Gal-1 in immunoregulation is the induction of apoptosis of the activated T cell subpopulations Th1 and Th17, whereas Th2 and Treg cell functions are promoted by Gal-1 [12,13]. This selectivity is caused by the differences in surface glycosylation of various T cell subtypes [9]. Gal-1-triggered cell death has been extensively studied in vitro, and its mechanism has been described [14–18]. Lactosamine sequences, required for Gal-1 binding, are synthesized by specific glycosyltransferases, such as beta-N-acetylglucosaminyltransferases and beta-galactosyltransferases. The expression of such enzymes controls T cell susceptibility to Gal-1-driven apoptosis [19].

We have recently demonstrated that activated T cells from patients with active systemic lupus erythematosus (SLE) are resistant to the apoptotic effect of Gal-1 [20], and we suggested that this finding is relevant to the immunoregulatory dysfunction observed in SLE. As a potential cause of this resistance may be an impaired binding of Gal-1 to the T cell surface, we set out to examine cell surface glycosylation and the expression of glycosylation enzymes in SLE T cells in comparison with healthy control T cells. The glycosylation pattern of resting SLE T cells resembled the activated phenotype of T cells. Activated SLE T cells bound significantly less galectin 1 (Gal-1) than the controls, while other lectins bound similarly. To understand the distinct lectin binding, specifically Gal-1, to SLE T cells, we found that the terminal sialylation increased in the autoimmune cells, and accordingly, neuraminidase treatment resulted in a remarkable increase in Gal-1 binding.

2. Results

The N-glycome diversity of T cell surface glycans was analyzed by the binding of lectins derived from plants (concanavalin-A (ConA), *Lens culinaris* agglutinin (LCA), wheat germ agglutinin (WGA), *Phaseolus vulgaris* leukoagglutinin (PHA-L), and *Sambucus nigra* agglutinin (SNA)) or of a human lectin, Gal-1, with known sugar binding specificity (Figure S1 and Table 1). Lectin binding to resting and phytohaemagglutinin (PHA)-activated T cells obtained from SLE patients and healthy controls was measured. Analysis of resting T cells from SLE patients and control individuals revealed that resting SLE T cells bound significantly more ConA, LCA, and WGA than healthy T cells (Figure 1A). ConA couples with mannoses present in early high-mannose glycans and mannoses in complex sugars [21,22], while LCA has high affinity to fucosylated core mannoses present in bi-antennary complex N-glycans and does not bind to tri- and tetra-antennary N-glycans [23]. WGA binds to N-acetyl glucosamines present in hybrid-type sugar chains (early and complex sugars) or to sialic acid, which can be terminally attached to complex multi-antennary glycans, and its affinity to the sialylated version of tri- or tetra-antennary glycan-containing glycoproteins was shown to be higher than to the desialylated form [24,25].

Table 1. Names, abbreviations, and binding specificities of lectins.

Lectins	Abbreviation	Specificity	Reference
Concanavalin A	ConA	mannose, glucose (low affinity)	[21,22]
Lens culinaris agglutinin	LCA	core-fucosylated bi-antennary N-glycan	[22,23]
Wheat germ agglutinin	WGA	GlcNAc, sialic acid	[24,25]
Phaseolus vulgaris leucoagglutinin	PHA-L	β-1,6-branched tri- and tetra-antennary N-glycans	[26]
Sambucus nigra agglutinin	SNA	α-2,6-linked sialic acid	[27]
Galectin-1	Gal-1	LAcNAc	[28]

Abbreviation: GlcNAc: N-acetylgucosamine; LacNAc: N-acetyllactoseamine.

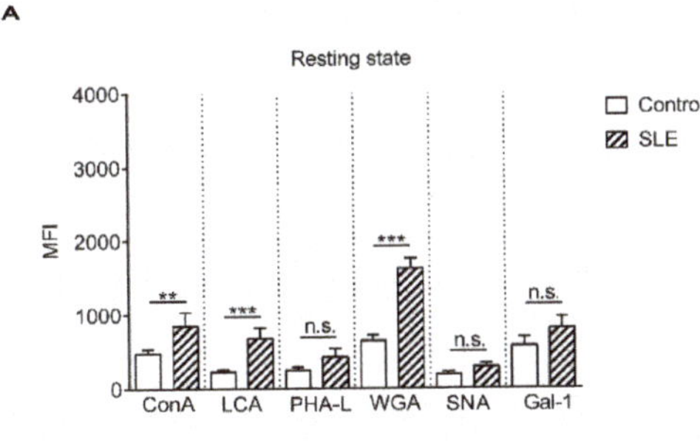

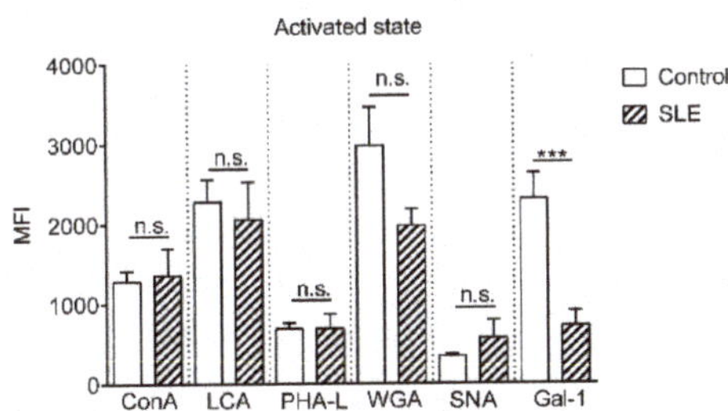

Figure 1. Lectin binding properties of resting and activated T cells from healthy donors and from systemic lupus erythematosus (SLE) patients. Peripheral blood T cells were obtained from healthy controls and SLE patients. The cells were left unstimulated (resting state, **A**) or were activated with 1 µg/mL phytohaemagglutinin L (PHA L) for 72 h (activated state, **B**). Cells were stained with viability dye, fixed then labeled with anti-CD3-PE-Cy5 antibody, followed by fluorescein isothiocyanate (FITC)-conjugated lectin. The samples were evaluated with flow cytometry. Binding of FITC-conjugated lectins is shown as mean (±SEM) of the median fluorescence intensity (MFI) values of flow cytometry histograms of resting (**A**) or activated (**B**) CD3-positive live T cells. Lectin names are listed in Table 1. MFI: mean fluorescence intensity, ConA: concanavalin-A, LCA: *Lens culinaris* agglutinin, WGA: wheat germ agglutinin, PHA-L: *Phaseolus vulgaris* leukoagglutinin, SNA: *Sambucus nigra* agglutinin, Gal-1: galectin 1. Statistical analysis was performed using an unpaired Student *t*-test. ** *p* < *0.01*; *** *p* < *0.001*; n. s.: not significant. SLE: *n* = 18, and healthy controls: *n* = 19.

Comparing healthy and autoimmune-activated T cells, we found that activated SLE T cells bound lectins in levels similar to control cells with the exception of Gal-1. SLE cells bound significantly less Gal-1 than control cells, indicating that terminal N-acetyllactosamine side chains, the Gal-1 ligands, were less accessible on these cells (Figure 1B). The changes in the pattern of lectin bindings did not occur preferentially on either CD4+ or CD4- (CD8+) cells, as these were similar in the control as well as in SLE activated T cells (Figure S2).

Glycosylation of proteins is regulated by multiple factors in the Golgi apparatus, such as sub-Golgi localization of glycosylation enzymes, transporters, pH, endoplasmic reticulum stress, or substrate availability (reviewed in [29]). However, a major element is the expression and function of glycosylation enzymes [30,31]. Therefore, expression levels of the genes involved in N-linked glycosylation (Figure S1 and Table 2) were examined by qPCR analysis of activated T cells. Gene expression of alpha mannosidases (*MAN1A1, MAN1A2, MAN2A1* and *MAN2A2*) in activated SLE T cells did not differ from the controls (Figure 2A). Analysis of beta-N-acetylglucosaminyltransferases (*MGAT1–5*) presented a slight but significant difference in the cases of *MGAT4A* and *MGAT4B* (Figure 2B).

Table 2. Symbols and full names of glycosylation enzyme genes.

Enzyme Genes	Gene Symbol	Full Gene Name
Mannosidases	*MAN1A1*	Mannosidase alpha class 1A member 1
	MAN1A2	Mannosidase alpha class 1A member 2
	MAN2A1	Mannosidase alpha class 2A member 1
	MAN2A2	Mannosidase alpha class 2A member 2
N-Acetylglucosaminyltransferase	*MGAT1*	Mannosyl (alpha-1,3-)-glycoprotein beta-1,2-N-acetylglucosaminyl-transferase
	MGAT4A	Mannosyl (alpha-1,3-)-glycoprotein beta-1,4-N-acetylglucosaminyl-transferase isozyme A
	MGAT4B	Mannosyl (alpha-1,3-)-glycoprotein beta-1,4-N-acetylglucosaminyl-transferase isozyme B
	MGAT5	Mannosyl (alpha-1,6-)-glycoprotein beta-1,6-N-Acetyl-glucosaminyltransferase
Sialyltransferases	*ST3GAL3*	ST3 beta-galactosidealpha-2,3-sialyltransferase 3
	ST3GAL4	ST3 beta-galactosidealpha-2,3-sialyltransferase 4
	ST3GAL6	ST3 beta-galactosidealpha-2,3-sialyltransferase 6
	ST6GAL1	ST6 beta-galactosamidealpha-2,6-sialyltranferase 1
Neuraminidases	*NEU1*	Neuraminidase 1

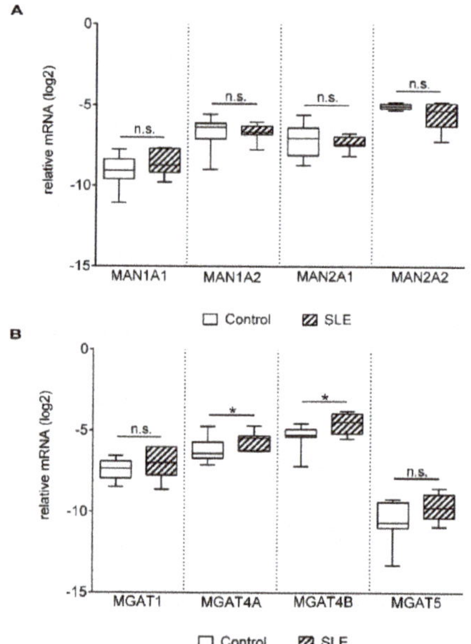

Figure 2. Gene expression of mannosidases (MANs) (**A**) and N-acetyl glucosaminyltransferases (MGATs) (**B**) in activated T cells. Total RNA was extracted from activated T cells and mRNA expression levels were analyzed by qPCR. Results of the relative expression were normalized to the expression levels of the *RPL27* housekeeping gene (log$_2$ transformation, ΔCt). Gene names and primer sequences are listed in Table 2 and Table 4, respectively. Upper and lower quartiles and whiskers of boxes extend to the minimum and maximum values, and the band inside the box is the median. Statistical analysis was performed using an unpaired Student's *t*-test, where * $p < 0.05$; SLE: $n = 18$, and healthy controls: $n = 19$.

Poly-N-acetyllactosamine chains on N glycans can be capped with the attachment of α-2,6 sialic acid by ST6 beta-galactosamidealpha-2,6-sialyltranferase 1 (*ST6GAL1*) and α-2,3 sialic acid by *ST3GAL3*, *ST3GAL4*, and *ST3GAL6* [32],and cleaved by neuraminidases. In the control and patient groups, *ST6GAL1*, *ST3GAL3*, *ST3GAL4*, and neuraminidase 1 (*NEU1*) gene expression levels were similar, whereas the mRNA level of *ST3GAL6* was significantly elevated in SLE T cells (Figure 3A).

During T cell activation, the gene expression of *NEU1* is strongly upregulated, while *NEU3* expression remains constant [33]; hence, only *NEU1* was analyzed.

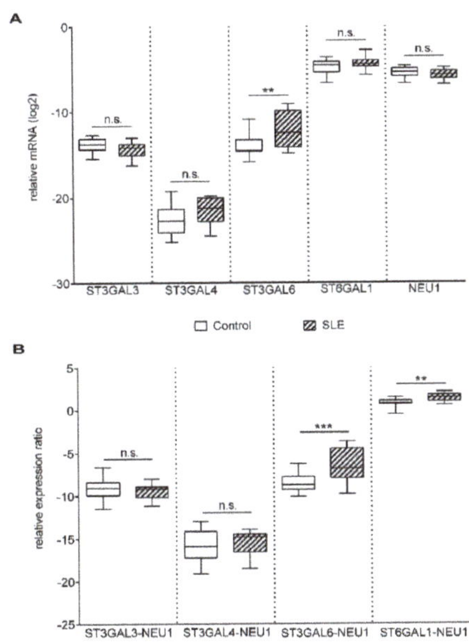

Figure 3. Gene expression of sialyltransferases (ST) and neuraminidase-1 (*NEU1*) in activated T cells. Total RNA was extracted from activated T cells and mRNA expression levels were analyzed by qPCR. (**A**) Results of the relative expression were normalized to the expression levels of *RPL27* housekeeping gene ($\log_2$ transformation, ΔCt). Gene names and primer sequences are listed in Table 2 and Table 4, respectively. (**B**) ST/*NEU1* mRNA expression ratios. The sialyltransferase-neuraminidase mRNA expression ratios of individual persons were calculated as follows: ΔCt ST/$\Delta CtNEU1$. Upper and lower quartiles and whiskers of boxes extend to the minimum and maximum values, and the band inside the box is the median. Statistical analysis was performed using an unpaired Student *t*-test, where ** $p< 0.01$; *** $p< 0.001$. SLE: $n = 18$, and healthy controls: $n = 19$.

Concerted action of sialyltransferases and neuraminidases determine the sialylation pattern. The large variations between the Gal-1 binding of the control and SLE T cells (Figure 1B) indicated an alteration in the sialylation of SLE surface glycans. As this is determined by the net effect of enzymes that sialylate (sialyltransferase) and desialylate (neuraminidase) the glycans, gene expression ratios of the opposing acting enzymes were calculated. A significantly higher *ST3GAL6/NEU1* and *ST6GAL1/NEU1* mRNA ratio was observed in SLE compared to control T cells (Figure 3B), indicating higher sialylation of SLE T cells. Other sialyltransferase/neuraminidase mRNA ratios, such as *ST3GAL3/NEU1* and*ST3GAL4/NEU1*, remained similar in the control and SLE groups (Figure 3B). These results indicated that reduced Gal-1 binding to SLE T cells may be a result of a more densely sialylated glycan profile. Indeed, cleaving sialic acid from the surface of SLE activated T cells by α2-3,6,8 Neu (specific to α2-3,6,8 linked sialic acid) resulting in the elevation of Gal-1 binding (Figure 4). The increase of Gal-1 binding to SLE cells was similar to that of control T cells (data not shown).

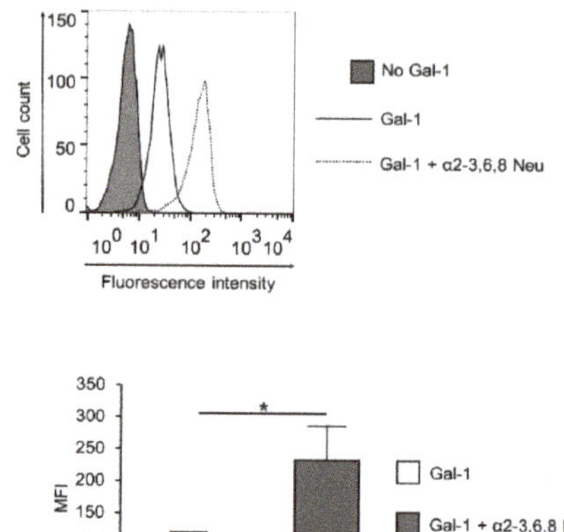

Figure 4. Effect of neuraminidase treatment on Gal-1 binding of SLE T cells. Activated SLE T cells were treated with α2-3,6,8 neuraminidase (Gal-1 + α2-3,6,8 Neu; dotted line) or left untreated (Gal-1; empty, continuous line), and then Gal-1 binding was investigated by cytofluorimetry, as described in the materials and methods section. The grey shadowed histogram shows the negative control: no Gal-1, +streptavidin—FITC. The upper image shows a representative profile of Gal-1 binding, the lower graph shows means (±SEM) of the mean fluorescence intensity (MFI) values of activated SLE T cells. Statistical analysis was performed using a two-tailed paired *t*-test. * $p < 0.05$, $n = 3$.

3. Discussion

Selected steps of mammalian N linked glycosylation and lectin binding to specific sugar side chains are summarized in Figure S1. It must be noted that binding of the used lectins was more degenerated than what is shown in the simplified Figure S1; however, it may help in a better apprehension of this work.

Remarkable differences were detected in ConA, LCA, and WGA binding between resting SLE and control T cells, since SLE T cells bound significantly higher amounts of these lectins. As LCA and WGA recognize matured sugar side chains and ConA couples both unmatured (early) and matured (complex) glycans, these results indicated that resting SLE T cells present a glycan structure similar to their activated phenotype. Detection of other activation markers, such as heightened CD40L expression [34], CD44 expression [35–37], exhibition of constant membrane raft polarization and increased GM1 content [38], measured by others, also suggest similarity to an activated state. On the other hand, this activated phenotype did not manifest in terms of CD25 expression on resting SLE T cells, as CD25 levels were similar on resting and activated T cells, suggesting that the activated phenotype is limited to several, but not all activation markers (Figure S3).

Stimulation of control T cells with PHA-L resulted in an elevation of binding of all used lectins, except the terminal α-2,6 sialic acid binding SNA (data not shown), indicating a generally increased complexity of glycosylation pattern upon activation. These results were in accordance with previous findings, arguing that N-glycan abundance, branching, glycan chain elongation, and hereby complexity enhanced [39,40], and terminal α-2,6 sialic acid residues declined [41–43] on freshly activated T cells. However, the increase in glycan complexity after activation was hardly seen in SLE T cells, as an increase in lectin binding upon activation was rather low or was absent in SLE, a phenomenon that is also explained by the activated phenotype of SLE T cells, even without treatment with activating agents. Furthermore, the decreased Gal-1 binding of SLE T cells was observed, not only in resting state, but also persisting after activation. No difference was found between CD4+ and CD4- (CD8) cells in terms of binding of any of the lectins (Figure S2). Some previous findings indicate that the proportion of effector memory (C-C Motif Chemokine Receptor 7{CCR7}-CD27+) and terminally differentiated effector memory (CCR7-CD27-) cells increase in SLE, and may correlate with disease activity or damage [44,45]. It would be interesting to compare the cell surface glycosylation patterns of naive and various subtypes of memory T cells; however, it was outside the scope of our present study. It is also to be noted that SLE memory effector T cells are crippled in response to antigen stimulation, as they respond to stimulation with apoptosis instead of proliferation [45], which might be a consequence of the altered glycosylation.

Variation in the binding of lectins tested was only confined to Gal-1, whose binding is determined by the presence of asialylated terminal N-acetyl lactose residues [46]. To clarify the background of this specific variability of Gal-1 binding, the expression of enzymes involved in creating the glycosylation pattern was determined. Analysis of the mRNA expression of glycosylation enzymes was chosen, since previous data indicated that glycosylation was primarily regulated at the transcriptional level of the appropriate enzymes [31]. Expressions of glycosyltransferases, alpha mannosidases, beta-N-acetylglucosaminetransferases, sialtransferases, and a neuraminidase, *NEU1*, were similar in control and SLE T cells. These findings are in accordance with the results of lectin assays, as all lectins tested bound similarly to the activated control and autoimmune T cells, with the exception ofGal-1. The difference in Gal-1 binding between control and patient activated T cells might therefore result from the distinct sialylation of the Gal-1 binding glycoconjugates. This presumption seemed to be supported by the finding that ratios of the expression of sialyltransferases (*ST3GAL6* and *ST6GAL1*) and neuraminidase shifted towards the sialtransferases, indicating a more intensified sialylation of SLE T cell glycans, including Gal-1 binding structures. Sialylation plays an important role in masking terminal carbohydrate chains, hence regulating lectin binding and signal transduction processes. Sialyltransferases attach, while neuraminidases remove sialic acid residues of terminal carbohydrate groups. Consequently, the accessibility of lectin binding sites is specifically regulated by the concerted action of sialyltransferases and neuraminidases. This point of view was supported with the finding, that ablation of sialic acid from surface glycoconjugates of living activated SLE T cells by neuraminidase treatment resulted in an increase in Gal-1 binding.

An important issue is how glycosylation affects autoimmune T cell activation, cell–cell interactions and autoantibody production. The available literature data are limited, and further detailed investigation is required. However, several studies suggest that pathological glycosylation results in disturbed T cell receptor (TCR)-major histocompatibility complex (MHC) interactions [38], cell adhesion [40], necrotic cell death- and glycan-specific autoantibody production [47], and deviant antigen presentation [48,49]. How closely the pathomechanism of SLE is associated with the surface glycosylation pattern and its abnormalities remains to be elucidated. Nevertheless, the present findings seem to corroborate our hypothesis regarding the resistance of activated SLE T cells to the apoptotic effects of Gal-1. As we described earlier [20], activated SLE T cells showed a reduced response to Gal-1 due to their defective expression of intracellular Gal-1. The present work suggests that altered glycosylation and, hence, the decreased binding of extracellular Gal-1 to SLE T cells can be another cause of the resistance to Gal-1-mediated immunomodulation, serving a putative novel pathogenic mechanism in SLE. It has to also be clarified whether removal of sialic acid from cell surface glycoconjugates results in the restoration of Gal-1-induced apoptotic sensitivity. Nevertheless, it has become clear from this work that analyzing the glycosylation process, especially the expression ratio of sialyltransferases and the neuraminidases and the binding of Gal-1 to the cell surface, may emerge as a novel approach to connecting disease phenotypes with functional pathways within T cells (differential-diagnosis of diseases or patient subset analysis within particular multisystem autoimmune diseases).

Since SLE is an autoimmune disease with multiple alterations on genetic, protein, signaling, and glycosylation levels, it is difficult to determine the primary cause of the disorder. It is likely that all the more-or-less relational changes result in the final manifestation of SLE. As glycosylation affects cell migration, adhesion, and signal transduction, its changes must be an important factor contributing to the pathomechanism of the disease. This view is supported by the finding that the binding of Gal-1 to SLE T cells, an anti-inflammatory human lectin, decreases because of the different sialylation of SLE T cells from that of healthy T cells, and thus it must be another reason that SLE T cells are more resistant to Gal-1-induced apoptosis [20].

The glycosylation phenotype resembles an activated state of SLE T cells. Though the fundamental causes behind the development of SLE are unclear, it is known that alpha-mannosidase II knock out mice develop an SLE-like disease [50]. This enzyme removes early mannose from maturing glycoconjugates, therefore, it is crucial to the final formation of healthy complex N-glycan structures observed in mammals. Its deficiency leads to immature, mannose-rich glycan chains being upregulated because of a disruption in their stepwise disassembly before the complex chain can be built in their place [51]. This effect is similar to our findings indicating a higher-than-normal distribution of mannose-rich glycans on SLE T cells before activation. It is known that mannose-rich chains are much more common on many strains of fungi and are easily recognized as non-self structures leading to auto-immune reactions [50].

Altered glycosylation in SLE is not confined to T cells, as the glycan profile of IgG is a primary predictor of the inflammatory capability of the molecule. Asialylated, agalactosylated glycan chains are stronger activators of complement than sialylated chains, and as such, pro-inflammatory responses are upregulated by the IgG molecule if it contains less terminal syalic acid units [5].

Altogether, the above data indicate that alteration in glycosylation in SLE is likely to be a primary phenomenon, and it contributes to the pathomechanism of the disease.

To our knowledge, the current work provides the first evidence for altered T cell surface glycosylation in SLE. Our major findings are that resting SLE T cells show an activated phenotype from the glycosylation point of view and lectin binding of activated SLE T cells is similar to the controls, with exception of the significantly lower Gal-1 binding. Furthermore, this is a consequence of a shift toward terminal sialylation of glycan structures due to an increased *ST6GAL1/NEU1* and/or *ST3GAL6/NEU1* ratio. Indeed, desialylation of the surface glycans on SLE T cells results in a remarkable increase in Gal-1 binding.

4. Materials and Methods

4.1. Ethical Statement

The study was designed in accordance with the guidelines of the Declaration of Helsinki and was approved by the Human Investigation Review Board, University of Szeged reference, No. 2833/2011 on 21 February 2011.

4.2. Patients

Patients with SLE ($n = 18$) and healthy controls ($n = 19$) were examined, except in one experiment where $n = 3$ (Figure 3). All patients met the 2012 SLICC classification criteria for SLE [52,53] and had active disease, as reflected by relevant disease activity indices. Eligible patients had an SLE Disease Activity-Index-2000 (SLEDAI-2K) ≥ 6 [54], did not have a co-existent inflammatory condition (overlapping autoimmune disease or infection), and did not have diabetes mellitus. Treatment with potent immunosuppressive drugs (mycophenolate mofetil, cyclophosphamide, rituximab) or corticosteroid at a dose >5 mg prednisolone equivalent was also an exclusion criterion. Controls were healthy individuals without any inflammatory disease or diabetes mellitus.

Demographics and the relevant disease activity data are presented in Table 3.

Table 3. Demographics and disease activity parameters. Numbers before and in brackets indicate mean and range, respectively. SLEDAI-2K: SLE Disease Activity-Index-2000, anti-dsDNA: antibody to double-stranded DNA.

Subject Characteristics	Age	Female/Male	Disease Activity Parameter
SLE	42 (23–54)	17/1	
SLEDAI-2K			14 (6–30)
anti-dsDNA (IU/mL)			88 (2–220)
Control	54 (31–75)	17/2	

4.3. Cells

Peripheral blood mononuclear cells (PBMC) were isolated from SLE patients and healthy donors using Ficoll (GE Healthcare, Amersham, UK) gradient centrifugation. A part of the PBMCs was stimulated with 1 µg/mL phytohaemagglutinin-L (PHA-L, Sigma-Aldrich, St. Louis, MO, USA) and the cells were cultured for 72 h in a humidified incubator with 5% CO_2 at 37 °C in Roswell Park Memorial Institute (RPMI)-1640 medium (Gibco, Life Technologies, Paisley, UK) supplemented with 10% fetal bovine serum (FBS) (Gibco, Life Technologies, Paisley, UK), 2 mM L-glutamine (Gibco, Life Technologies, Paisley, UK) and penicillin-streptomycin (Sigma-Aldrich, St. Louis, MO, USA), henceforward referred to as activated T cells. Activated T cell cultures were > 90% pure, as controlled with flow cytometry using anti-human CD3 antibody (BioLegend, San Diego, CA, USA) (data not shown).

4.4. Lectin Binding Assay

Resting or activated T cells were washed twice with cold phosphate buffered saline (PBS) and incubated at 4 °C for 30 min with eFluor 660 fixable viability dye (eBioscience, Thermo Fisher Scientific, Waltham, MA, USA), then fixed with 4% paraformaldehyde for 4 min at room temperature. After washing the samples twice in PBS supplemented with 1% FBS and 0.1% sodium-azide (fluorescence-activated cell sorting (FACS) buffer), the cells were incubated at 4 °C for 20 min with fluorescein-labeled lectins or unlabeled Gal-1 (lectins used are listed in Table 1). The fluorescein-labeled plant lectin kit (Vector Laboratories, Burlingame, CA, USA) was used according to the manufacturer's instructions. Recombinant galectin-1 was produced and characterized in our laboratory, as previously described [15]. Gal-1-binding was detected as follows: after washing with FACS buffer, biotinylated mouse monoclonal antibody to Gal-1 (2C1/6) was added and incubated at 4 °C for 45 min. Samples were washed with cold FACS buffer before adding fluorescein isothiocyanate(FITC)-labeled streptavidin and incubating the cells at 4 °C for 20 min. Finally, the samples were washed twice in FACS buffer. Samples were analyzed with a FACSCalibur system (BD Biosciences, Franklin Lakes, NJ, USA) and data were evaluated using FlowJo V10 software (BD Biosciences, Franklin Lakes, NJ, USA). The lectin binding was evaluated on resting T cells within PBMCs or activated T cells by gating CD3+ cells with PE/Cy5-conjugated anti-human CD3 antibody (BioLegend, San Diego, CA, USA).

4.5. Neuraminidase Treatment

Activated T cells (2×10^6/sample) were treated with α2-3,6,8 neuraminidase (α2-3,6,8 Neu, New England BioLabs, Ipswich, MA, USA) according to the manufacturer's instructions or left untreated.

Briefly, the cells were washed with PBS then incubated with 200 U of α2-3,6,8Neu in a total volume of 40 µL glycobuffer (provided by the manufacturer) for 15 min at 37 °C. After washing the samples with cold PBS, Gal-1 binding, viability staining, fixation, and flow cytometry analysis were done, as described above.

4.6. Quantitative Real-Time PCR (qPCR)

The qPCR assays were performed according to the MIQE (Minimum Information for Publication of Quantitative Real-Time PCR Experiments) guidelines [55]. The names of genes are listed in Table 2. Total RNA was extracted from activated T cells ($1–3 \times 10^6$ cells) using PerfectPure RNA Cultured Cell kit (5 PRIME, Gaithersburg, MD, USA) according to the manufacturer's instructions with on-column DNase digestion. The amount and quality of RNA were measured using NanoDrop-1000 spectrophotometer (Thermo Fisher Scientific, Waltham, MA, USA). For cDNA synthesis, 2 µg of total RNA/reaction was reverse transcribed using RevertAid H Minus First Strand cDNA Synthesis Kit (Thermo Fisher Scientific, Waltham, MA, USA) in the presence of 1.66 µM of oligo (dT) 18 and random hexamer primers, 0.5 mMdNTP, 10 U RiboLock RNase Inhibitor, and 200 U RevertAid H Minus Reverse Transcriptase for 60 min at 42 °C, then heated for 10 min at 70 °C. Quantitative PCR amplifications were carried out with appropriate negative controls at least in duplicate. The reaction volume was 20 µL and composed of AccuPower 2 × Greenstar qPCR Master Mix (Bioneer, Alameda, CA, USA), 300–300 nM primers, and 40-fold diluted cDNA. For PCR amplifications, RotoGene3000 instrument (Corbett Research, Sydney, Australia) was used. The qPCR program included: initial 95 °C for 15 min, followed by repeated 45 cycles (95 °C for 15 s, 60–62 °C for 20 s, 72 °C for 20 s), and melting temperature analysis increasing the temperature from 55 °C to 98 °C at 0. 5 °C/step with 8 sec stops between each step. Quantitative real-time PCR data were analyzed using the Rotor Gene software (v6.1 build 93). Relative mRNA levels, normalized to *RPL27*, were presented as $\log_2$ transformation of relative gene expression, namely subtraction of Ct values ($\Delta Ct = CtRPL27 − CtGOI$). The mRNA expression ratios were calculated using the following formula: $\Delta Ct \times$ gene $−\Delta Ct$ Y gene. Primer sequences were partly taken from [56–59], or designed using the Universal Probe Library Assay Design program (Roche Applied Science, Basel, Switzerland). Primer sequences used in the study are shown in Table 4.

Table 4. Primer sequences used in the study for the mRNA amplification of glycosylation enzymes.

Name	Forward Primer	Reverse Primer
RPL27	5′-CGCAAAGCTGTCATCGTG-3′	5′-GTCACTTTGCGGGGGTAG-3′
MAN1A1	5′-TTGGGCATTGCTGAATATGA-3′	5′-CAGAATACTGCTGCCTCCAGA-3′
MAN1A2	5′-GGAGGCCTACTTGCAGCATA-3′	5′-GAGTTTCTCAGCCAATTGCAC-3′
MAN2A1	5′-CCTGGAAATGTCCAAAGCA-3′	5′-GCGGAAATCATCTCCTAGTGG-3′
MAN2A2	5′-TCCACCTGCTCAACCTACG-3′	5′-TGTAAGATGAGTGCGGTCTCC-3′
MGAT1	5′-CGGAGCAGGCCAAGTTC-3′	5′-CCTTGCCCGCAGTCCTA-3′
MGAT4A	5′-CATAGCGGCAACCAAGAAC-3′	5′-TGCTTATTTCCAAACCTTCACTC-3′
MGAT4B	5′-CACTCTGCACTCGCTCATCT-3′	5′-CACTGCCGAAGTGTACTGTGA-3′
MGAT5	5′-GCTCATCTGCGAGCCTTCT-3′	5′-TTGGCAGGTCACCTTGTACTT-3′
ST3GAL3	5′-TATGCTTCAGCCTTGATG-3′	5′-TTGGTGACTGACAAGATGG-3′
ST3GAL4	5′-ATGTTGGCTCTGGTCCTG-3′	5′-AGGAAGATGGGCTGATCC-3′
ST3GAL6	5′-TCTATTGGGTGGCACCTGTGGAAA-3	5′-TGATGAAACCTCAGCAGAGAGGCA-3′
ST6GAL1	5′-TGGGACCCATCTGTATACCACT-3′	5′-ATTGGGGTGCAGCTTACGAT-3′
NEU1	5′-CCTGGATATTGGCACTGAA-3′	5′-CATCGCTGAGGAGACAGAAG-3′

4.7. Statistical Analysis

Statistical analysis was performed using GraphPad Prism Version 7.01. In all statistical analyses, an unpaired, two-tailed *t*-test was used to compare data obtained in T cells of healthy controls and SLE patients. There was only one exception (Figure 4B), where paired two-tailed Student's *t*-test was chosen for the comparison of values from T cells of the same patient with or without neuraminidase treatment. The significant differences were indicated as follows: * $p < 0.05$, ** $p < 0.01$, *** $p < 0.001$.

Supplementary Materials: Supplementary materials can be found at http://www.mdpi.com/1422-0067/20/18/4455/s1.

Author Contributions: Conceptualization: É.M., Á.C., L.K.; methodology: É.M., Á.C., E.S., Á.H., M.B.; software: E.S., Á.H.; validation: Á.C., É.M.; formal analysis: E.S., Á.H., Á.C.; investigation: M.B., L.K.; resources: É.M., Á.C., L.K.; data curation: Á.H., Á.C.; writing—original draft preparation: E.S., Á.H.; writing—review and editing: Á.C., É.M., L.K.; visualization: E.S.; supervision: É.M., L.K.; project administration: Á.C.; funding acquisition: É.M., L.K.

Funding: This research was funded by GINOP 2.2.1-15-2016-00007 and by a research grant of the Faculty of Medicine, University of Szeged No IV-3606-222/b-10/2017.

Acknowledgments: The authors would like to thank Andrea Gercsó for excellent technical assistance, and Edit Kotogány for flow cytometric analysis. We thank Ágnes Zvara for discussions about gene expression data analysis and László Puskás for the access to the Rotogene3000 instrument.

Conflicts of Interest: The authors declare no conflict of interest.

Abbreviations

SLE	Systemic lupus erythematosus
Gal-1	Galectin-1
NEU	Neuraminidase
ConA	Concanavalin-A
LCA	*Lens culinaris* agglutinin
WGA	Wheat germ agglutinin
PHA	phytohaemagglutinin
PHA-L	*Phaseolus vulgaris* leukoagglutinin
SNA	*Sambucus nigra* agglutinin
MFI	Median fluorescence intensity
MGAT1–5-	Beta-N acetylglucosaminyltransferases
MAN	Mannosidase
MGAT	N-Acetyl glucosaminyltransferase
ST	Sialyltransferase
SLEDAI-2K	SLE disease activity index-2000
Anti-dsDNA	Antibody to double-stranded DNA
PBMC	Peripheral blood mononuclear cell
PBS	Phosphate buffered saline
FACS	Fluorescence-activated cell sorting

References

1. Chachadi, V.B.; Cheng, H.; Klinkebiel, D.; Christman, J.K.; Cheng, P.W. 5-Aza-2′-deoxycytidine increases sialyl Lewis X on MUC1 by stimulating β-galactoside:α2,3-sialyltransferase 6 gene. *Int. J. Biochem. Cell Biol.* **2011**, *43*, 586–593. [CrossRef] [PubMed]
2. Stowell, S.R.; Ju, T.; Cummings, R.D. Protein glycosylation in cancer. *Annu. Rev. Pathol.* **2015**, *10*, 473–510. [CrossRef] [PubMed]
3. Axford, J.S. Glycosylation and rheumatic disease. *Biochim. Biophys. Acta-Mol. Basis Dis.* **1999**, *1455*, 219–229. [CrossRef]
4. Mackiewicz, A.; Mackiewicz, K. Glycoforms of serum alpha 1-acid glycoprotein as markers of inflammation and cancer. *Glycoconj. J.* **1995**, *12*, 241–247. [CrossRef]
5. Sell, S. Progress in pathology cancer-associated carbohydrates identified by monoclonal antibodies. *Hum. Pathol.* **1990**, *21*, 1003–1019. [CrossRef]
6. Gudelj, I.; Lauc, G.; Pezer, M. Immunoglobulin G glycosylation in aging and diseases. *Cell. Immunol.* **2018**, *333*, 65–79. [CrossRef] [PubMed]
7. Arnold, J.N.; Wormald, M.R.; Sim, R.B.; Rudd, P.M.; Dwek, R.A. The impact of glycosylation on the biological function and structure of human immunoglobulins. *Annu. Rev. Immunol.* **2007**, *25*, 21–50. [CrossRef]
8. Maverakis, E.; Kim, K.; Shimoda, M.; Gershwin, M.E.; Patel, F.; Wilken, R.; Raychaudhuri, S.; Ruhaak, L.R.; Lebrilla, C.B. Glycans in the immune system and the altered glycan theory of autoimmunity: A critical review. *J. Autoimmun.* **2015**, *57*, 1–13. [CrossRef]

9. Hauser, M.A.; Kindinger, I.; Laufer, J.M.; Späte, A.-K.; Bucher, D.; Vanes, S.L.; Krueger, W.A.; Wittmann, V.; Legler, D.F. Distinct CCR7 glycosylation pattern shapes receptor signaling and endocytosis to modulate chemotactic responses. *J. Leukoc. Biol.* **2016**, *99*, 993–1007. [CrossRef]

10. Toscano, M.A.; Bianco, G.A.; Ilarregui, J.M.; Croci, D.O.; Correale, J.; Hernandez, J.D.; Zwirner, N.W.; Poirier, F.; Riley, E.M.; Baum, L.G.; et al. Differential glycosylation of TH1, TH2 and TH-17 effector cells selectively regulates susceptibility to cell death. *Nat. Immunol.* **2007**, *8*, 825–834. [CrossRef]

11. Bieberich, E. Synthesis, Processing, and Function of N-glycans in N-glycoproteins. *Adv. Neurobiol.* **2014**, *9*, 47–70. [PubMed]

12. Camby, I.; Le Mercier, M.; Lefranc, F.; Kiss, R. Galectin-1: A small protein with major functions. *Glycobiology* **2006**, *16*, 137R–157R. [CrossRef] [PubMed]

13. Garin, M.I.; Chu, C.-C.; Golshayan, D.; Cernuda-Morollon, E.; Wait, R.; Lechler, R.I. Galectin-1: A key effector of regulation mediated by CD4+CD25+ T cells. *Blood* **2007**, *109*, 2058–2065. [CrossRef] [PubMed]

14. Motran, C.C.; Molinder, K.M.; Liu, S.D.; Poirier, F.; Miceli, M.C. Galectin-1 functions as a {Th}2 cytokine that selectively induces {Th}1 apoptosis and promotes {Th}2 function. *Eur. J. Immunol.* **2008**, *38*, 3015–3027. [CrossRef] [PubMed]

15. Ion, G.; Fajka-Boja, R.; Tóth, G.K.; Caron, M.; Monostori, É. Role of p56lck and ZAP70-mediated tyrosine phosphorylation in galectin-1-induced cell death. *Cell Death Differ.* **2005**, *12*, 1145–1147. [CrossRef]

16. Ion, G.; Fajka-Boja, R.; Kovács, F.; Szebeni, G.; Gombos, I.; Czibula, Á.; Matkó, J.; Monostori, É. Acid sphingomyelinase mediated release of ceramide is essential to trigger the mitochondrial pathway of apoptosis by galectin-1. *Cell Signal.* **2006**, *18*, 1887–1896. [CrossRef] [PubMed]

17. Kovács-Sólyom, F.; Blaskó, A.; Fajka-Boja, R.; Katona, R.L.; Végh, L.; Novák, J.; Szebeni, G.J.; Krenács, L.; Uher, F.; Tubak, V.; et al. Mechanism of tumor cell-induced T-cell apoptosis mediated by galectin-1. *Immunol. Lett.* **2010**, *127*, 108–118. [CrossRef]

18. Blaskó, A.; Fajka-Boja, R.; Ion, G.; Monostori, É. How does it act when soluble? Critical evaluation of mechanism of galectin-1 induced T-cell apoptosis. *Acta Biol. Hung.* **2011**, *62*, 106–111. [CrossRef]

19. Novák, J.; Kriston-Pál, É.; Czibula, Á.; Deák, M.; Kovács, L.; Monostori, É.; Fajka-Boja, R. GM1 controlled lateral segregation of tyrosine kinase Lck predispose T-cells to cell-derived galectin-1-induced apoptosis. *Mol. Immunol.* **2014**, *57*, 302–309. [CrossRef]

20. Cabrera, P.V.; Amano, M.; Mitoma, J.; Chan, J.; Said, J.; Fukuda, M.; Baum, L.G. Haploinsufficiency of C2GnT-I glycosyltransferase renders T lymphoma cells resistant to cell death. *Blood* **2006**, *108*, 2399–2406. [CrossRef]

21. Deák, M.; Hornung, Á.; Novák, J.; Demydenko, D.; Szabó, E.; Czibula, Á.; Fajka-Boja, R.; Kriston-Pál, É.; Monostori, É.; Kovács, L. Novel role for galectin-1 in T-cells under physiological and pathological conditions. *Immunobiology* **2015**, *220*, 483–489. [CrossRef] [PubMed]

22. Cummings, R.D.; Etzler, M.E. *Essentials of Glycobiology*; Varki, A., Cummings, R.D., Esko, J.D., Eds.; Cold Spring Harbor Laboratory Press: Cold Spring Harbor, NY, USA, 2009.

23. Maupin, K.A.; Liden, D.; Haab, B.B. The fine specificity of mannose-binding and galactose-binding lectins revealed using outlier motif analysis of glycan array data. *Glycobiology* **2012**, *22*, 160–169. [CrossRef] [PubMed]

24. Tateno, H.; Nakamura-Tsuruta, S.; Hirabayashi, J. Comparative analysis of core-fucose-binding lectins from Lens culinaris and Pisumsativum using frontal affinity chromatography. *Glycobiology* **2009**, *19*, 527–536. [CrossRef] [PubMed]

25. Peters, B.P.; Goldstein, I.J.; Flashner, M.; Ebisu, S. Interaction of Wheat Germ Agglutinin with Sialic Acid. *Biochemistry* **1979**, *18*, 5505–5511. [CrossRef] [PubMed]

26. Gladman, D.D.; Ibanez, D.; Urowitz, M.B. Systemic lupus erythematosus disease activity index 2000. *J. Rheumatol.* **2002**, *29*, 288–291. [PubMed]

27. Schwarz, R.E.; Wojciechowicz, D.C.; Park, P.Y.; Paty, P.B. Phytohemagglutinin-L (PHA-L) lectin surface binding of N-linked β1-6 carbohydrate and its relationship to activated mutant ras in human pancreatic cancer cell-lines. *Cancer Lett.* **1996**, *107*, 285–291. [CrossRef]

28. Fischer, E.; Brossmer, R. Sialic acid-binding lectins: Submolecular specificity and interaction with sialoglycoproteins and tumour cells. *Glycoconj. J.* **1995**, *12*, 707–713. [CrossRef] [PubMed]

29. Itakura, Y.; Nakamura-Tsuruta, S.; Kominami, J.; Tateno, H.; Hirabayashi, J. Sugar-binding profiles of chitin-binding lectins from the hevein family: A comprehensive study. *Int. J. Mol. Sci.* **2017**, *18*, 1160. [CrossRef] [PubMed]

30. Pothukuchi, P.; Agliarulo, I.; Russo, D.; Rizzo, R.; Russo, F.; Parashuraman, S. Translation of genome to glycome: Role of the Golgi apparatus. *FEBS Lett.* **2019**. [CrossRef]

31. Comelli, E.M.; Head, S.R.; Gilmartin, T.; Whisenant, T.; Haslam, S.M.; North, S.J.; Wong, N.K.; Kudo, T.; Narimatsu, H.; Esko, J.D.; et al. A focused microarray approach to functional glycomics: Transcriptional regulation of the glycome. *Glycobiology* **2006**, *16*, 117–131. [CrossRef]

32. Nairn, A.V.; York, W.S.; Harris, K.; Hall, E.M.; Pierce, J.M.; Moremen, K.W. Regulation of glycan structures in animal tissues. *J. Biol. Chem.* **2008**, *283*, 17298–17313. [CrossRef] [PubMed]

33. Altheide, T.K.; Hayakawa, T.; Mikkelsen, T.S.; Diaz, S.; Varki, N.; Varki, A. System-wide genomic and biochemical comparisons of sialic acid biology among primates and rodents. *J. Biol. Chem.* **2006**, *281*, 25689–25702. [CrossRef] [PubMed]

34. Nan, X.; Carubelli, I.; Stamatos, N.M. Sialidase expression in activated human T lymphocytes influences production of IFN-γ. *J. Leukoc. Biol.* **2007**, *81*, 284–296. [CrossRef] [PubMed]

35. Katsiari, C.G.; Liossis, S.N.C.; Dimopoulos, A.M.; Charalambopoulos, D.V.; Mavrikakis, M.; Sfikakis, P.P. CD40L overexpression on T cells and monocytes from patients with systemic lupus erythematosus is resistant to calcineurin inhibition. *Lupus* **2002**, *11*, 370–378. [CrossRef] [PubMed]

36. Lesley, J. CD44 structure and function. *Front. Biosci.* **2016**, *3*, 616–630. [CrossRef]

37. Guan, H.; Nagarkatti, P.S.; Nagarkatti, M. Role of CD44 in the differentiation of Th1 and Th2 cells: CD44-deficiency enhances the development of Th2 effectors in response to sheep RBC and chicken ovalbumin. *J. Immunol.* **2009**, *183*, 172–180. [CrossRef] [PubMed]

38. Li, Y.; Harada, T.; Juang, Y.-T.; Kyttaris, V.C.; Wang, Y.; Zidanic, M.; Tung, K.; Tsokos, G.C. Phosphorylated ERM Is Responsible for Increased T Cell Polarization, Adhesion, and Migration in Patients with Systemic Lupus Erythematosus. *J. Immunol.* **2007**, *178*, 1938–1947. [CrossRef]

39. Jury, E.C.; Flores-Borja, F.; Kabouridis, P.S. Lipid rafts in T cell signalling and disease. *Semin. Cell Dev. Biol.* **2007**, *18*, 608–615. [CrossRef]

40. Shental-Bechor, D.; Levy, Y. Effect of glycosylation on protein folding: A close look at thermodynamic stabilization. *Proc. Natl. Acad. Sci. USA* **2008**, *105*, 8256–8261. [CrossRef]

41. Polley, A.; Orłowski, A.; Danne, R.; Gurtovenko, A.A.; Bernardino de la Serna, J.; Eggeling, C.; Davis, S.J.; Róg, T.; Vattulainen, I. Glycosylation and Lipids Working in Concert Direct CD2 Ectodomain Orientation and Presentation. *J. Phys. Chem. Lett.* **2017**, *8*, 1060–1066. [CrossRef]

42. Chen, H.-L.; Li, C.F.; Grigorian, A.; Tian, W.; Demetriou, M. T cell receptor signaling co-regulates multiple Golgi genes to enhance N-glycan branching. *J. Biol. Chem.* **2009**, *284*, 32454–32461. [CrossRef] [PubMed]

43. Hernandez, J.D.; Klein, J.; Van Dyken, S.J.; Marth, J.D.; Baum, L.G. T-cell activation results in microheterogeneous changes in glycosylation of CD45. *Int. Immunol.* **2007**, *19*, 847–856. [CrossRef] [PubMed]

44. Comelli, E.M.; Sutton-Smith, M.; Yan, Q.; Amado, M.; Panico, M.; Gilmartin, T.; Whisenant, T.; Lanigan, C.M.; Head, S.R.; Goldberg, D.; et al. Activation of murine CD4+ and CD8+ T lymphocytes leads to dramatic remodeling of N-linked glycans. *J. Immunol.* **2006**, *177*, 2431–2440. [CrossRef] [PubMed]

45. Piantoni, S.; Regola, F.; Zanola, A.; Andreoli, L.; Dall'Ara, F.; Tincani, A.; Airo', P. Effector T-cells are expanded in systemic lupus erythematosus patients with highdisease activity and damage indexes. *Lupus* **2018**, *27*, 143–149. [CrossRef] [PubMed]

46. Fritsch, R.D.; Shen, X.; Illei, G.G.; Yarboro, C.H.; Prussin, C.; Hathcock, K.S.; Hodes, R.J.; Lipsky, P.E. Abnormal differentiation of memory T cells in systemic lupus erythematosus. *Arthritis Rheum.* **2006**, *54*, 2184–2197. [CrossRef] [PubMed]

47. Stowell, S.R.; Arthur, C.M.; Mehta, P.; Slanina, K.A.; Blixt, O.; Leffler, H.; Smith, D.F.; Cummings, R.D. Galectin-1, -2, and -3 exhibit differential recognition of sialylatedglycans and blood group antigens. *J. Biol. Chem.* **2008**, *283*, 10109–10123. [CrossRef] [PubMed]

48. Dotan, N.; Altstock, R.T.; Schwarz, M.; Dukler, A. Anti-glycan antibodies as biomarkers for diagnosis and prognosis. *Lupus* **2006**, *15*, 442–450. [CrossRef]

49. Harding, C.V.; Kihlberg, J.; Elofsson, M.; Magnusson, G.; Unanue, E.R. Glycopeptides bind MHC molecules and elicit specific T cell responses. *J. Immunol.* **1993**, *151*, 2419.

50. Jensen, T.; Hansen, P.; Galli-Stampino, L.; Mouritsen, S.; Frische, K.; Meinjohanns, E.; Meldal, M.; Werdelin, O. Glycopeptide specific T cell hybridomas raised against an αGalNAc O-glycosylated self peptide are discriminating between highly related carbohydrate groups. *Immunol. Lett.* **1997**, *56*, 449. [CrossRef]

51. Green, R.S.; Stone, E.L.; Tenno, M.; Lehtonen, E.; Farquhar, M.G.; Marth, J.D. Mammalian N-Glycan branching protects against innate immune self-recognition and inflammation in autoimmune disease pathogenesis. *Immunity* **2007**, *27*, 308–320. [CrossRef]

52. Moremen, K.W. Golgi alpha-mannosidase II deficiency in vertebrate systems: Implications for asparagine-linked oligosaccharide processing in mammals. *Biochim. Biophys. Acta* **2002**, *1573*, 225–235. [CrossRef]

53. Hochberg, M.C. Updating the American college of rheumatology revised criteria for the classification of systemic lupus erythematosus. *Arthritis Rheum.* **1997**, *40*, 1725. [CrossRef] [PubMed]

54. Petri, M.; Orbai, A.-M.; Alarcon, G.S.; Gordon, C.; Merrill, J.T.; Fortin, P.R.; Bruce, I.N.; Isenberg, D.; Wallace, D.J.; Nived, O.; et al. Derivation and validation of the Systemic Lupus International Collaborating Clinics classification criteria for systemic lupus erythematosus. *Arthritis Rheum.* **2012**, *64*, 2677–2686. [CrossRef] [PubMed]

55. Rabinovich, G.A.; Ramhorst, R.E.; Rubinstein, N.; Corigliano, A.; Daroqui, M.C.; Kier-Joffé, E.B.; Fainboim, L. Induction of allogenic T-cell hyporesponsiveness by galectin-1-mediated apoptotic and non-apoptotic mechanisms. *Cell Death Differ.* **2002**, *9*, 661–670. [CrossRef] [PubMed]

56. Bustin, S.A.; Benes, V.; Garson, J.A.; Hellemans, J.; Huggett, J.; Kubista, M.; Mueller, R.; Nolan, T.; Pfaffl, M.W.; Shipley, G.L.; et al. The MIQE guidelines: Minimum information for publication of quantitative real-time PCR experiments. *Clin. Chem.* **2009**, *55*, 611–622. [CrossRef] [PubMed]

57. Ma, H.; Zhou, H.; Song, X.; Shi, S.; Zhang, J.; Jia, L. Modification of sialylation is associated with multidrug resistance in human acute myeloid leukemia. *Oncogene* **2015**, *34*, 726–740. [CrossRef] [PubMed]

58. Tringali, C.; Lupo, B.; Cirillo, F.; Papini, N.; Anastasia, L.; Lamorte, G.; Colombi, P.; Bresciani, R.; Monti, E.; Tettamanti, G.; et al. Silencing of membrane-associated sialidase *NEU3* diminishes apoptosis resistance and triggers megakaryocytic differentiation of chronic myeloid leukemic cells K562 through the increase of ganglioside GM3. *Cell Death Differ.* **2009**, *16*, 164–174. [CrossRef]

59. Zhou, H.; Ma, H.; Wei, W.; Ji, D.; Song, X.; Sun, J.; Zhang, J.; Jia, L. B4GALT family mediates the multidrug resistance of human leukemia cells by regulating the hedgehog pathway and the expression of p-glycoprotein and multidrug resistance-associated protein 1. *Cell Death Dis.* **2013**, *4*, e654. [CrossRef]

MDPI

St. Alban-Anlage 66

4052 Basel

Switzerland

Tel. +41 61 683 77 34

Fax +41 61 302 89 18

www.mdpi.com

International Journal of Molecular Sciences Editorial Office

E-mail: ijms@mdpi.com

www.mdpi.com/journal/ijms

9 783039 434664